Other McGraw-Hill Telecommunication Books of Interest

Phase-Locked Loops

Design, Simulation, and Applications

Roland E. Best
Best Engineering
Oberwil, Switzerland

Fourth Edition

McGraw-Hill

New York   San Francisco   Washington, D.C.   Auckland   Bogotá
Caracas   Lisbon   London   Madrid   Mexico City   Milan
Montreal   New Delhi   San Juan   Singapore
Sydney   Tokyo   Toronto

McGraw-Hill

*A Division of The **McGraw·Hill** Companies*

P/N 134904-9
PART OF
ISBN 0-07-134903-0

*The sponsoring editor for this book was Stephen S. Chapman and the
production supervisor was Sherri Souffrance. It was set in Century
Schoolbook by Pro-Image Corporation.*

Printed and bound by R. R. Donnelley & Sons Company.

This book was printed on acid-free paper containing a
minimum of 50% recycled de-inked fiber.

Contents

Preface to the Fourth Edition

When this book first appeared in 1984, the area of phase-locked loop (PLL) applications was dominated by the linear PLL (LPLL) and by the "classical" digital PLL (DPLL). In recent years, two different types of PLL have gained increased attention: the all-digital PLL (ADPLL) and the software PLL (SPLL). The ADPLL, which is entirely built from logical circuits, avoids many of the drawbacks of the DPLL, which is still a semianalog circuit and suffers from the problems related to the component's variations, drift, and aging. This is why the ADPLL now replaces the classical DPLL in many applications, especially in the field of digital communication. Whereas applications of software PLLs were considered rather exotic 12 years ago, the availability of fast digital signal processors (DSP) now widens the field for SPLL applications. For these reasons, the author decided to emphasize equally all four categories of PLLs: the LPLL, the classical DPLL, the ADPLL, and the SPLL.

Another dramatic change in recent years is how the engineer solves problems: As the calculator replaced the slide rule a long time ago, the workhorse of the engineer is now the personal computer. This book includes a CD-ROM with a program simulating all the types of PLL's that are discussed in the following text. It runs under all versions of Microsoft Windows and is installed automatically using a setup program (InstallShield). Using the simulation program, readers can look at the performance of a PLL circuit as if they were checking the waveforms of a breadboard circuit with the scope. The author does not claim, however, that building breadboards has become unnecessary, but the simulation can be used to check whether a design will yield the planned performance. Whereas the program distributed with the second edition supported only second-order LPLLs and DPLLs, the new application also simulates first-order systems. This may be of only academic interest, but at least this allows the operator to omit the loop filter if he or she really wants to.

Chapter 1 provides a short introduction into the principles of PLL operation and classifies the PLL types known today. The four categories of PLL as mentioned above are discussed in Chaps. 2 to 5. Chapter 6 discusses the role of the PLL in the field of communications. It starts with an introduction into the most often used modulation techniques in digital communications and describes a number of special circuits (or systems) that are required for carrier and symbol synchronization (e.g. Costas Loop, Early-Late-Gate etc.). Chapter 7 lists the integrated circuits used to build PLL systems, which became available in the first quarter of 1999. Chapter 8 reviews the techniques used to measure PLL parameters. For the reader interested in mathematics, Appendix A analyzes the nonlinear pull-in process of different PLL types in more detail. Appendix B provides an introduction to the Laplace transform, which is widely used throughout the book. Appendix C presents the fundamentals of digital filters; digital filters have become increasingly important for the design of ADPLLs.

Roland Best

Phase-Locked Loops

Introduction to PLLs

1.1 Operating Principles of the PLL

The phase-locked loop (PLL) helps keep parts of our world orderly. If we turn on a television set, a PLL will keep heads at the top of the screen and feet at the bottom. In color television, another PLL makes sure that green remains green and red remains red (even if the politicians claim that the reverse is true).

A PLL is a circuit which causes a particular system to track with another one. More precisely, a PLL is a circuit synchronizing an output signal (generated by an oscillator) with a reference or input signal in frequency as well as in phase. In the synchronized—often called *locked*—state the phase error between the oscillator's output signal and the reference signal is zero, or remains constant.

If a phase error builds up, a control mechanism acts on the oscillator in such a way that the phase error is again reduced to a minimum. In such a control system the phase of the output signal is actually locked to the phase of the reference signal. This is why it is referred to as a *phase-locked loop*.

The operating principle of the PLL is explained by the example of the linear PLL (LPLL). As will be pointed out in Sec. 1.2, there exist other types of PLLs, e.g., digital PLLs (DPLL), all-digital PLLs (ADPLL), and software PLLs (SPLL). Its block diagram is shown in Fig. 1.1a. The PLL consists of three basic functional blocks:

1. A voltage-controlled oscillator (VCO)

2. A phase detector (PD)

3. A loop filter (LF)

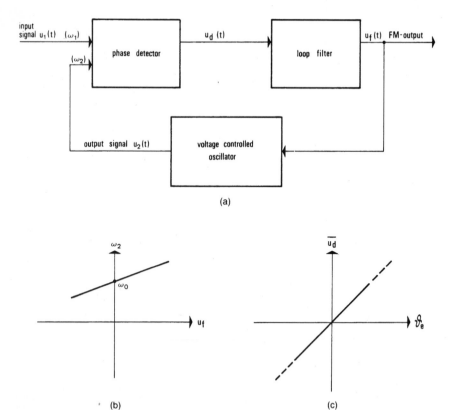

Figure 1.1 (*a*) Block diagram of the PLL. (*b*) Transfer function of the VCO. (u_f = control voltage; ω_2 = angular frequency of the output signal.) (*c*) Transfer function of the PD. ($\overline{u_d}$ = average value of the phase-detector output signal; θ_e = phase error.)

In some PLL circuits a current-controlled oscillator (CCO) is used instead of the VCO. In this case the output signal of the phase detector is a controlled current source rather than a voltage source. However, the operating principle remains the same.

The signals of interest within the PLL circuit are defined as follows:

- The reference (or input) signal $u_1(t)$
- The angular frequency ω_1 of the reference signal
- The output signal $u_2(t)$ of the VCO
- The angular frequency ω_2 of the output signal
- The output signal $u_d(t)$ of the phase detector
- The output signal $u_f(t)$ of the loop filter
- The phase error θ_e, defined as the phase difference between signals $u_1(t)$ and $u_2(t)$

Let us now look at the operation of the three functional blocks in Fig. 1.1a. The VCO oscillates at an angular frequency ω_2, which is determined by the output signal u_f of the loop filter. The angular frequency ω_2 is given by

$$\omega_2(t) = \omega_0 + K_0 u_f(t) \qquad (1.1)$$

where ω_0 is the center (angular) frequency of the VCO and K_0 is the VCO gain in $s^{-1} V^{-1}$.

Equation (1.1) is plotted graphically in Fig. 1.1b. In many textbooks the physical unit rad $s^{-1} V^{-1}$ is used for the VCO gain, since the unit rad s^{-1} is often used for angular frequencies. We shall drop the unit radians in this text. (Note, however, that any phase variables used in this book will have to be measured in *radians* and not in *degrees!*) Therefore, in the equations a phase shift of 180° must always be specified as a value of π.

The PD—also referred to as *phase comparator*—compares the phase of the output signal with the phase of the reference signal and develops an output signal $u_d(t)$ which is approximately proportional to the phase error θ_e, at least within a limited range of the latter

$$u_d(t) = K_d \theta_e \qquad (1.2)$$

Here K_d represents the gain of the PD. The physical unit of K_d is volts. Some textbooks use the unit V rad^{-1} for the reasons discussed above. Figure 1.1c is a graphical representation of Eq. (1.2).

The output signal $u_d(t)$ of the PD consists of a dc component and a superimposed ac component. The latter is undesired; hence it is canceled by the loop filter. In most cases a first-order, low-pass filter is used. (Refer to Chap. 2.)

Let us now see how the three building blocks work together. First we assume that the angular frequency of the input signal $u_1(t)$ is equal to the center frequency ω_0. The VCO then operates at its center frequency ω_0. As we see, the phase error θ_e is zero. If θ_e is zero, the output signal u_d of the PD must also be zero. Consequently the output signal of the loop filter u_f will also be zero. This is the condition that permits the VCO to operate at its center frequency.

If the phase error θ_e were not zero initially, the PD would develop a nonzero output signal u_d. After some delay the loop filter would also produce a finite signal u_f. This would cause the VCO to change its operating frequency in such a way that the phase error finally vanishes.

Assume now that the frequency of the input signal is changed suddenly at time t_0 by the amount $\Delta\omega$. As shown in Fig. 1.2, the phase of the input signal then starts leading the phase of the output signal. A phase error is built up and increases with time. The PD develops a

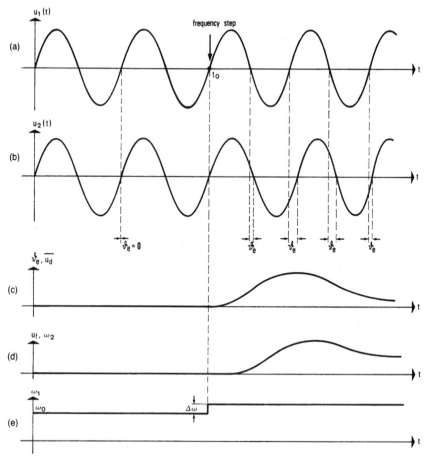

Figure 1.2 Transient response of a PLL onto a step variation of the reference frequency. (a) Reference signal $u_1(t)$. (b) Output signal $u_2(t)$ of the VCO. (c) Signals $\overline{u_d}(t)$ and θ_e (t) as a function of time. (d) Angular frequency ω_2 of the VCO as a function of time. (e) Angular frequency ω_1 of the reference signal $u_1(t)$.

signal $u_d(t)$, which also increases with time. With a delay given by the loop filter, $u_f(t)$ will also rise. This causes the VCO to increase its frequency. The phase error becomes smaller now, and after some settling time the VCO will oscillate at a frequency that is exactly the frequency of the input signal. Depending on the type of loop filter used, the final phase error will have been reduced to zero or to a finite value.

The VCO now operates at a frequency which is greater than its center frequency ω_0 by an amount $\Delta\omega$. This will force the signal $u_f(t)$ to settle at a final value of $u_f = \Delta\omega/K_0$. If the center frequency of the input signal is frequency-modulated by an arbitrary low-frequency signal, then the output signal of the loop filter is the *demodulated signal*.

The PLL can consequently be used as an (FM) detector. As we shall see later, it can be further applied as an AM or PM detector.

One of the most intriguing capabilities of the PLL is its ability to suppress noise superimposed on its input signal. Let us suppose that the input signal of the PLL is buried in noise. The PD tries to measure the phase error between input and output signals. The noise at the input causes the zero crossings of the input signal $u_1(t)$ to be advanced or delayed in a stochastic manner. This causes the PD output signal $u_d(t)$ to jitter around an average value. If the corner frequency of the loop filter is low enough, almost no noise will be noticeable in the signal $u_f(t)$, and the VCO will operate in such a way that the phase of the signal $u_2(t)$ is equal to the average phase of the input signal $u_1(t)$. Therefore, we can state that the PLL is able to detect a signal that is buried in noise. These simplified considerations have shown that the PLL is nothing but a servo system which controls the phase of the output signal $u_2(t)$.

As shown in Fig. 1.2 the PLL was always able to track the phase of the output signal to the phase of the reference signal; this system was locked at all times. This is not necessarily the case, however, because a larger frequency step applied to the input signal could cause the system to "unlock." The control mechanism inherent in the PLL will then try to become locked again, but will the system indeed lock again? We shall deal with this problem in the following chapters. Basically two kinds of problems have to be considered:

- The PLL is initially locked. Under what conditions will the PLL remain locked?
- The PLL is initially unlocked. Under what conditions will the PLL become locked?

If we try to answer these questions, we notice that different PLLs behave quite differently in this regard. We find that there are some fundamentally different types of PLLs. Therefore, we first identify these various types.

1.2 Classification of PLL Types

The very first phase-locked loops (PLL) were implemented as early as 1932 by de Bellescize[22]; this French engineer is considered inventor of the "coherent communication." The PLL found broader industrial applications only when it became available as an integrated circuit. The first PLL ICs appeared around 1965 and were purely analog devices. An analog multiplier (four-quadrant multiplier) was used as the phase detector, the loop filter was built from a passive or active RC filter, and the well-known voltage-controlled oscillator (VCO) was used to generate the output signal of the PLL. This type of PLL is referred to as the "linear PLL" (LPLL) today. In the following years the PLL drifted slowly but steadily into digital territory. The very first digital PLL (DPLL), which appeared around 1970, was in effect a hybrid device: only the phase detector was built from a digital circuit, e.g., from an EXOR gate or a JK-flipflop, but the remaining blocks were still analog. A few years later, the "all-digital" PLL (ADPLL) was invented. The ADPLL is exclusively built from digital function blocks, hence doesn't contain any passive components like resistors and capacitors. In analogy to filters, PLLs can also be implemented "by software." In this case, the function of the PLL is no longer performed by a piece of specialized hardware, but rather by a computer program. This last type of PLL is referred to as SPLL.

Unfortunately, LPLLs, DPLLs, and ADPLLs behave differently, so there is no common theory which covers all of these types. Consequently we must treat the various types of PLLs in separate chapters (Chaps. 2 to 5). Because the software PLL is usually implemented by a microcontroller, microcomputer, or digital signal processor (DSP), it is generally considered to be an "all-digital" vehicle. Ironically, the SPLL can be programmed to perform like an LPLL, a DPLL, or an ADPLL, so the SPLL is, like the software-implemented digital filter, the most universal type of PLL. One would expect at first glance that the SPLL has replaced its hardware counterparts in many cases, but this is not the case. As we will see later, the computer algorithm which performs the PLL function must be executed at least once in every period of the input signal of the PLL. This severely limits the range of frequencies which can be covered by the SPLL.

Let us conclude that we have to deal with four different types of PLLs:

- The LPLL (linear PLL)
- The DPLL ("classical" digital PLL)
- The ADPLL (all-digital PLL)
- The SPLL (software PLL)

2

The Linear PLL (LPLL)

2.1 Building Blocks of the LPLL

Different PLL types are built from different building blocks. To realize an LPLL we have to find appropriate circuits implementing the functions defined in Fig. 1.1a. In linear PLLs, the four-quadrant multiplier is used as a phase detector (Fig. 2.1). In most cases the input signal $u_1(t)$ is a sine wave with angular frequency ω_1, whereas the output signal $u_2(t)$ is a symmetrical square wave with angular frequenty ω_2. In the locked state the two frequencies are equal. As we will see in Sec. 2.3 the output signal $u_d(t)$ of the phase detector then consists of a number of terms; the first of these is a "dc" component and is roughly proportional to the phase error θ_e; the remaining terms are "ac" components having frequencies of 2 ω_1, 4 ω_1.... Because these higher fre-

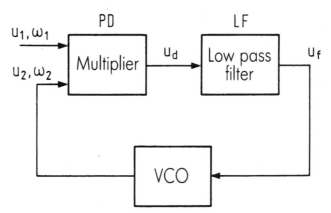

Figure 2.1 Block diagram of the linear PLL (LPLL). The symbols defined here are used throughout this chapter.

quencies are unwanted signals, they are filtered out by the loop filter. Because the loop filter must pass the lower frequencies and suppress the higher, it must be a low-pass filter. In most LPLL designs a first-order low-pass filter is used. Figure 2.2 lists the versions which are most frequently encountered. Figure 2.2a is a passive lag-lead filter having one pole and one zero. Its transfer function $F(s)$ is given by

$$F(s) = \frac{1 + s\tau_2}{1 + s(\tau_1 + \tau_2)} \tag{2.1}$$

where $\tau_1 = R_1C$ and $\tau_2 = R_2C$. [A note on terminology: a *lag-lead* (also called *lead-lag*) filter combines a phase-leading with a phase-lagging network. The phase-leading action comes from the numerator (i.e. from the zero) in the transfer function in Eq. 2.1, whereas the denominator (i.e. the pole) produces the phase lag. All filters that will be

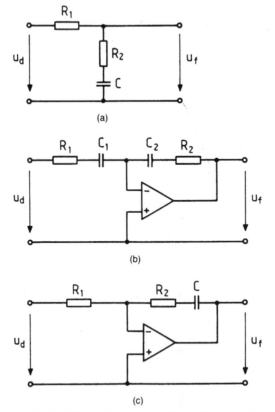

(a)

(b)

(c)

Figure 2.2 Schematic diagram of loop filters used in linear PLLs. (a) Passive lag filter. (b) Active lag filter. (c) Active PI filter.

used as loop filters are lag-lead filters; for simplicity we will refer to them as *lag filters*.] The amplitude response of this filter is shown in Fig. 2.3a. As we will see in Sec. 2.3, the zero of this filter is crucial because it has a strong influence on the damping factor ζ of the LPLL system. Figure 2.2b shows an active lag-lead filter. Its transfer function is very similar to the passive but has an additional gain term K_a, which can be chosen greater than 1. Its transfer function $F(s)$ is given by

$$F(s) = K_a \frac{1 + s\tau_2}{1 + s\tau_1} \tag{2.2}$$

where $\tau_1 = R_1 C_1$, $\tau_2 = R_2 C_2$, and $K_a = -C_1/C_2$. The amplitude response

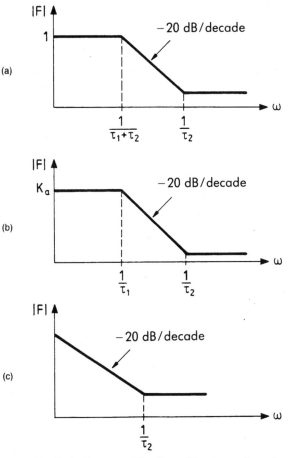

Figure 2.3 Bode diagram of the three filter types shown in Fig. 2.2. (a) Passive lag filter. (b) Active lag filter. (c) Active PI filter.

of the active lag filter is given in Fig. 2.3b. Finally, Fig. 2.2c shows another active low-pass filter, which is commonly referred to as a "PI" filter. It is a *lag-lead* filter as well. The term "PI" is taken from control theory, where it stands for "proportional + integral" action. The transfer function of the PI filter is given by

$$F(s) = \frac{1 + s\tau_2}{s\tau_1} \tag{2.3}$$

where again $\tau_1 = R_1C$ and $\tau_2 = R_2C$. The PI filter has a pole at $s = 0$ and therefore behaves like an integrator. It has—at least theoretically—infinite gain at zero frequency. Its amplitude response is depicted in Fig. 2.3c.

Higher-order low-pass filters could be used instead of simple one-pole filters;[10] as we will see later, this is done in some applications. Because each additional filter pole introduces phase shift, it is much more difficult to maintain stability in higher-order systems.

It is almost needless to say that the VCO used in Fig. 2.1 is the third purely analog part of the LPLL system. Before discussing the dynamic behavior of the LPLL, let us have a look at the nature of "phase signals," because this is a major source of trouble to newcomers in the field of the PLL.

2.2 A Note on Phase Signals

As we have seen in Sec. 2.1, the PLL is nothing more than a control system for phase signals. Dynamic analysis of control systems is normally performed by means of its transfer function $H(s)$. $H(s)$ relates the input and output signals of the system; in conventional electrical networks the input and the output are represented by voltage signals $u_1(t)$ and $u_2(t)$, respectively, so $H(s)$ is given by

$$H(s) = \frac{U_2(s)}{U_1(s)} \tag{2.4}$$

where $U_1(s)$ and $U_2(s)$ are the Laplace transforms of $u_1(t)$ and $u_2(t)$, respectively, and s is the Laplace operator. In the case of the PLL, the input and output signals are *phases*, however, which is less familiar to many electronic engineers.

To see what phase signals really are, we assume for the moment that both input and output signals of the LPLL (Fig. 2.1) are sine waves:

$$u_1(t) = U_{10} \sin [\omega_1 t + \theta_1(t)] \qquad u_2(t) = U_{20} \sin [\omega_2 t + \theta_2(t)]$$

The information carried by these signals is neither the amplitude (U_{10}

or U_{20}, respectively) nor the frequency (ω_1 or ω_2, respectively) but the *phases* $\theta_1(t)$ and $\theta_2(t)$. Let us consider now some simple phase signals; Fig. 2.4 lists a number of phase signals $\theta_1(t)$ which are frequently used to excite a PLL. Figure 2.4a shows the simplest case: the phase $\theta_1(t)$ performs a step change at time $t = 0$, hence is given by

$$\theta_1(t) = \Delta\Phi\ u(t)$$

where $u(t)$ is the unit step function. This case is an example for *phase modulation*.

Let us consider next an example of frequency modulation (Fig. 2.4b). Assume the angular frequency of the reference signal is ω_0 for $t < 0$. At $t = 0$ the angular frequency is abruptly changed by the increment $\Delta\omega$. For $t \geq 0$ the reference signal is consequently given by

$$u_1(t) = U_{10} \sin (\omega_0 t + \Delta\omega\ t) = U_{10} \sin (\omega_0 t + \theta_1)$$

In this case the phase θ_1 can be written as

$$\theta_1(t) = \Delta\omega\ t$$

Consequently the phase signal θ_1 is a ramp signal.

As a last example, consider a reference signal whose angular frequency is ω_0 for $t < 0$ and increases linearly with time for $t \geq 0$ (Fig. 2.4c). For $t \geq 0$ its angular frequency is therefore

$$\omega_1(t) = \omega_0 + \Delta\dot{\omega}\ t$$

where $\Delta\dot{\omega}$ denotes the rate of change of angular frequency. Remember that the angular frequency of a signal is defined as the first derivative of its phase with respect to time:

$$\omega_1 = \frac{d\theta_1}{dt}$$

Hence the phase of a signal at time t is the integral of its angular frequency over the time interval $0 \leq \tau \leq t$, where τ denotes elapsed time. The reference signal can be written as

$$u_1(t) = U_{10} \sin \int_0^t (\omega_0 + \Delta\dot{\omega}\ \tau)\ d\tau = U_{10} \sin \left(\omega_0 t + \Delta\dot{\omega}\ \frac{t^2}{2} \right)$$

Consequently the corresponding phase signal $\theta_1(t)$ is given by

$$\theta_1(t) = \Delta\dot{\omega}\ \frac{t^2}{2}$$

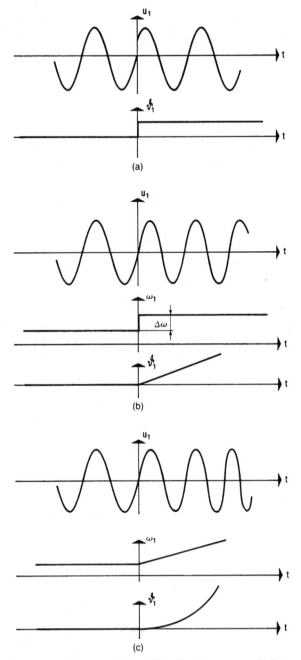

Figure 2.4 Some typical exciting functions as applied to the reference input of a PLL. (a) Phase error applied at $t = 0$; $\theta_1 = \Delta\Phi\, u(t)$. ($b$) Frequency step $\Delta\omega$ applied at $t = 0$; $\theta_1 = \Delta\omega\, t$. ($c$) Frequency ramp starting at $t = 0$; $\theta_1 = \Delta\dot\omega\, t^2/2$.

2.3 LPLL Performance in the Locked State

If we assume that the LPLL has locked and stays locked for the near future, we can develop a linear mathematical model for the system. As will be shown in this section, the mathematical model is used to calculate a phase-transfer function $H(s)$ which relates the phase θ_1 of the input signal to the phase θ_2 of the output signal:

$$H(s) = \frac{\Theta_2(s)}{\Theta_1(s)} \tag{2.5}$$

where $\Theta_1(s)$ and $\Theta_2(s)$ are the Laplace transforms of the phase signals $\theta_1(t)$ and $\theta_2(t)$, respectively. (Note that we are using lowercase symbols for time functions and uppercase symbols for their Laplace transforms throughout the text; this also applies to Greek letters.) $H(s)$ is called a *phase transfer function*. To get an expression for $H(s)$ we must know the transfer functions of the individual building blocks in Fig. 2.1. Let us start with the phase detector.

As we stated in Sec. 2.1, the input signal of an LPLL is usually a sine wave:

$$u_1(t) = U_{10} \sin(\omega_1 t + \theta_1) \tag{2.6a}$$

whereas the output signal is usually a square wave and can therefore be written as a Walsh function[21]

$$u_2(t) = U_{20} w(\omega_2 t + \theta_2) \tag{2.6b}$$

These signals are shown in Fig. 2.5. The dashed curve in Fig. 2.5a is a sine wave having a phase of $\theta_1 = 0$; the solid line has a nonzero phase θ_1. For simplicity we assume here that the phase is constant over time. The dashed curve in Fig. 2.5b shows a symmetrical Walsh function $w(\omega_2 t)$ having a phase $\theta_2 = 0$; the solid line has a nonzero phase θ_2. The output signal of the four-quadrant multiplier is obtained by multiplying the signals u_1 and u_2. To simplify the analysis the Walsh function is replaced by its Fourier series. For $u_2(t)$ we then get

$$u_2(t) = U_{20} \left[\frac{4}{\pi} \cos(\omega_2 t + \theta_2) + \frac{4}{3\pi} \cos(3\omega_2 t + \theta_2) \cdots \right] \tag{2.6c}$$

The first term in the square brackets is the fundamental component; the remaining terms are odd harmonics. For the output signal $u_d(t)$ therefore we get

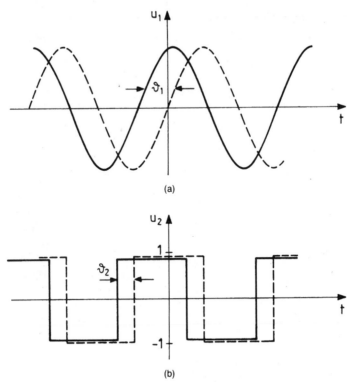

Figure 2.5 Input signals of the multiplier phase detector. (*a*) Signal $u_1(t)$ is a sine wave. Dashed line: phase $\theta_1 = 0$; solid line: phase $\theta_1 > 0$. (*b*) Signal $u_2(t)$ is a symmetrical square wave (Walsh function). Dashed line: $\theta_2 = 0$; solid line: $\theta_2 > 0$.

$$u_d(t) = u_1(t)u_2(t) = U_{10}U_{20} \sin (\omega_1 t + \theta_1)$$

$$\times \left[\frac{4}{\pi} \cos (\omega_2 t + \theta_2) + \frac{4}{\pi} \cos (3\omega_2 t + \theta_2) + \cdots \right] \quad (2.7)$$

When the LPLL is locked, the frequencies ω_1 and ω_2 are identical, and $u_d(t)$ becomes

$$u_d(t) = U_{10}U_{20} \left[\frac{2}{\pi} \sin \theta_e + \cdots \right] \quad (2.8a)$$

where $\theta_e = \theta_1 - \theta_2$ is the phase error. The first term of this series is the wanted "dc" term, whereas the higher-frequency terms will be eliminated by the loop filter. Setting $K_d = 2U_{10}U_{20}/\pi$ and neglecting higher-frequency terms we get

$$u_d(t) \approx K_d \sin \theta_e \quad (2.8b)$$

where K_d is called *detector gain*. When the phase error is small, the sine function can be replaced by its argument, and we have

$$u_d(t) \approx K_d \theta_e \qquad (2.9)$$

This equation represents the linearized model of the phase detector. The dimension of K_d is rad/V. As shown above, K_d is proportional to both amplitudes U_{10} and U_{20}. Normally, U_{20} is constant, so K_d becomes a linear function of the input signal level U_{10}. This is plotted in Fig. 2.6. Because the multiplier saturates when its output signal comes close to the power supply rails, this function flattens out at large signal levels, and K_d approaches asymptotically a limiting value. To conclude the analysis of the four-quadrant multiplier we state that—in the locked state of the LPLL—the phase detector represents a zero-order block having a gain of K_d.

The transfer functions of the three loop filters have been given in Eqs. (2.1) to (2.3). What we still need is the transfer function of the VCO. As stated in Eq. (1.1), the angular frequency of the VCO is given by

$$\omega_2(t) = \omega_0 + \Delta\omega_2(t) = \omega_0 + K_0 u_f(t)$$

where K_0 is called *VCO gain* (dimension: rad s^{-1} V^{-1}). The model of the VCO should yield the output phase θ_2, however, and not the output frequency ω_2. By definition, the phase θ_2 is given by the integral over the frequency variation $\Delta\omega_2$

$$\theta_2(t) = \int \Delta\omega_2 \, dt = K_0 \int u_f \, dt$$

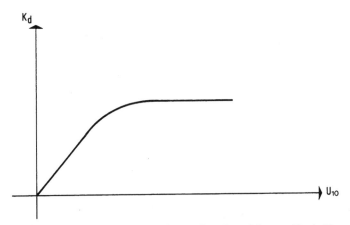

Figure 2.6 Phase-detector gain K_d as a function of the amplitude U_{10} of the reference signal.

In the Laplace transform, integration over time corresponds to division by s, so the Laplace transform of the output phase θ_2 is given by

$$\Theta_2(s) = \frac{K_0}{s} \, U_f(s) \tag{2.10}$$

The transfer function of the VCO is therefore given by

$$\frac{\Theta_2(s)}{U_f(s)} = \frac{K_0}{s} \tag{2.11}$$

For phase signals the VCO simply represents an integrator. We are now in the position to draw the simplified linear model of the LPLL (Fig. 2.7). This model enables us to analyze the tracking performance of the LPLL, i.e., the ability of the system to maintain phase tracking when excited by phase steps, frequency steps, or other excitation signals. From the model in Fig. 2.7 the phase transfer function $H(s)$ is computed. We get

$$H(s) = \frac{\Theta_2(s)}{\Theta_1(s)} = \frac{K_0 K_d F(s)}{s + K_0 K_d F(s)} \tag{2.12}$$

In addition to the phase-transfer function, an *error-transfer function* $H_e(s)$ has been defined. $H_e(s)$ is defined by

$$H_e(s) = \frac{\Theta_e(s)}{\Theta_1(s)} = \frac{s}{s + K_0 K_d F(s)} \tag{2.13}$$

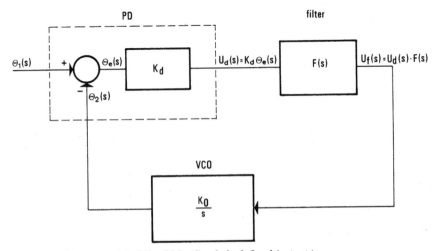

Figure 2.7 Linear model of the PLL. (Symbols defined in text.)

$H_e(s)$ relates phase error θ_e to the input phase θ_1. Between $H_e(s)$ and $H(s)$ we have the simple relation

$$H_e(s) = 1 - H(s) \qquad (2.14a)$$

To analyze the phase-transfer function we have to insert the loop filter transfer function [Eqs. (2.1) to (2.3)] into Eq. (2.12). For the three different loop filters (Fig. 2.2) we get

- For the passive lag filter

$$H(s) = \dfrac{K_0 K_d \dfrac{1 + s\tau_2}{\tau_1 + \tau_2}}{s^2 + s\,\dfrac{1 + K_0 K_d \tau_2}{\tau_1 + \tau_2} + \dfrac{K_0 K_d}{\tau_1 + \tau_2}} \qquad (2.14b)$$

- For the active lag filter

$$H(s) = \dfrac{K_0 K_d K_a \dfrac{1 + s\tau_2}{\tau_1}}{s^2 + s\,\dfrac{1 + K_0 K_d K_a \tau_2}{\tau_1} + \dfrac{K_0 K_d K_a}{\tau_1}} \qquad (2.14c)$$

- For the active PI filter

$$H(s) = \dfrac{K_0 K_d \dfrac{1 + s\tau_2}{\tau_1}}{s^2 + s\,\dfrac{K_0 K_d \tau_2}{\tau_1} + \dfrac{K_0 K_d}{\tau_1 + \tau_2}} \qquad (2.14d)$$

In circuit and control theory it is common practice to write the denominator of the transfer function in the so-called *normalized form*,[3]

$$\text{Denominator} = s^2 + 2\zeta\omega_n s + \omega_n^2$$

where ω_n is the *natural frequency* and ζ is the *damping factor*. The denominator of Eqs. (2.14) will take this form if the following substitutions are made:

- For the passive lag filter

$$\omega_n = \sqrt{\dfrac{K_0 K_d}{\tau_1 + \tau_2}} \qquad \zeta = \dfrac{\omega_n}{2}\left(\tau_2 + \dfrac{1}{K_0 K_d}\right) \qquad (2.15a)$$

■ For the active lag filter

$$\omega_n = \sqrt{\frac{K_0 K_d K_a}{\tau_1}} \qquad \zeta = \frac{\omega_n}{2}\left(\tau_2 + \frac{1}{K_0 K_d K_a}\right) \qquad (2.15b)$$

■ For the active PI filter

$$\omega_n = \sqrt{\frac{K_0 K_d}{\tau_1}} \qquad \zeta = \frac{\omega_n \tau_2}{2} \qquad (2.15c)$$

(The natural frequency ω_n must never be confused with the center frequency ω_0 of the PLL.) Inserting these substitutions into Eqs. (2.14), we get the following phase-transfer functions:

■ For the passive lag filter

$$H(s) = \frac{s\omega_n\left(2\zeta - \dfrac{\omega_n}{K_0 K_d}\right) + \omega_n^2}{s^2 + 2s\zeta\omega_n + \omega_n^2} \qquad (2.16a)$$

■ For the active lag filter

$$H(s) = \frac{s\omega_n\left(2\zeta - \dfrac{\omega_n}{K_0 K_d K_a}\right) + \omega_n^2}{s^2 + 2s\zeta\omega_n + \omega_n^2} \qquad (2.16b)$$

■ For the active PI filter

$$H(s) = \frac{2s\zeta\omega_n + \omega_n^2}{s^2 + 2s\zeta\omega_n + \omega_n^2} \qquad (2.16c)$$

Aside from the parameters ω_n and ζ, only the parameters K_d, K_0, and K_a appear in Eqs. (2.16). The term $K_d K_0$ in Eqs. (2.16a) and (2.16c) is called loop gain and has the dimension of angular frequency (s^{-1}). In Eq. (2.16b) the term $K_d K_0 K_a$ is called loop gain. If the condition

$$K_d K_0 \gg \omega_n \qquad \text{or} \qquad K_d K_0 K_a \gg \omega_n$$

is true, this LPLL system is said to be a *high-gain loop*. If the reverse is true, the system is called a *low-gain loop*. Most practical LPLLs are high-gain loops. For high-gain loops, Eqs. (2.16a) to (2.16c) become approximately identical and read

$$H(s) \approx \frac{2\zeta\omega_n s + \omega_n^2}{s^2 + 2s\zeta\omega_n + \omega_n^2} \qquad (2.17)$$

for all filters shown in Fig. 2.2. Similarly, assuming a high-gain loop, we get for the error-transfer function $H_e(s)$ for all three filter types the approximate expression

$$H_e(s) \approx \frac{s^2}{s^2 + 2s\zeta\omega_n + \omega_n^2} \qquad (2.18)$$

To investigate the transient response of a control system, it is customary to plot a Bode diagram of its transfer function. The Bode diagram of the phase-transfer function is obtained by putting $s = j\omega$ in Eq. (2.17) and by plotting the magnitude (absolute value) $|H(j\omega)|$ as a function of angular frequency ω (Fig. 2.8). Both scales are usually logarithmic. The frequency scale is further normalized to the natural frequency ω_n. Thus the graph is valid for every second-order PLL system. We can see from Fig. 2.8 that the second-order PLL is actually a low-pass filter for input phase signals $\theta_1(t)$ whose frequency spectrum is flat between zero and approximately the natural frequency ω_n. This means that the second-order PLL is able to track for phase and frequency modulations of the reference signal as long as the modulation

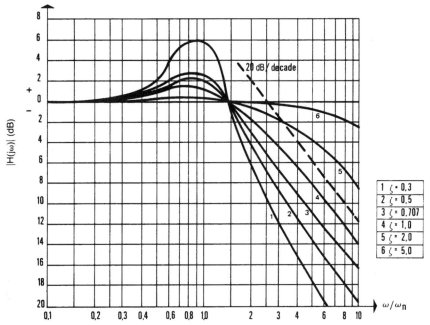

Figure 2.8 Bode diagram of the phase-transfer function $H(j\omega)$. (*Adapted from Gardner*[1] *with permission.*)

frequencies remain within an angular frequency band roughly between zero and ω_n.

The damping factor ζ has an important influence on the dynamic performance of the LPLL. For $\zeta = 1$ the system is critically damped. If ζ is made smaller than unity, the transient response becomes oscillatory; the smaller the damping factor, the larger becomes the overshoot. In most practical systems, an optimally flat frequency-transfer function is the goal. The transfer function is optimally flat for $\zeta = 1/\sqrt{2} \approx 0.7$, which corresponds to a second-order Butterworth lowpass filter. If ζ is made considerably larger than unity, the transfer function flattens out, and the dynamic response becomes sluggish.

A Bode plot of $H_e(s)$ is shown in Fig. 2.9. The value of 0.707 has been chosen for ζ. The diagram shows that for modulation frequencies smaller than the natural frequency ω_n the phase error remains relatively small. For larger frequencies, however, the phase error θ_e becomes as large as the reference phase θ_1, which means that the PLL is no longer able to maintain phase tracking.

As in amplifiers, the bandwidth of a PLL is often specified by the 3 db corner frequency ω_{3db}. This is the radian frequency where the gain has dropped by 3 db referred to the gain at dc. ω_{3db} is given by

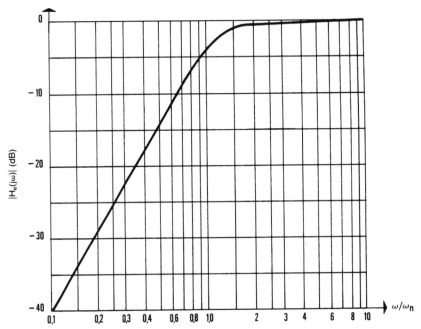

Figure 2.9   Bode diagram of the error-transfer function $H_e(j\omega)$.

$$\omega_{3db} = \omega_n[1 + 2\zeta^2 + \sqrt{(1 + 2\zeta^2)^2 + 1}]^{1/2}$$

For a damping factor $\zeta = 0.7$, ω_{3db} becomes $\omega_{3db} = 2.06\ \omega_n$ which is about twice the natural frequency.

Knowing that a second-order PLL in the locked state behaves very much like a servo or follow-up control system, we can plot a simple model for the locked PLL (Fig. 2.10). The model consists of a reference potentiometer G, a servo amplifier, and a follow-up potentiometer F whose shaft is driven by an electric motor. In this model the reference phase θ_1 is represented by the shaft position of the reference potentiometer G. The phase of the output signal of the VCO $\theta_2(t)$ is represented by the shaft position of the follow-up potentiometer. If the shaft position of the reference potentiometer is varied slowly, the servo system will be able to maintain tracking of the follow-up potentiometer. If $\theta_1(t)$ is changed too abruptly, the servo system will lose tracking and large phase errors θ_e will result.

So far we have seen that a linear model is best suited to explain the tracking performance of the PLL *if it is assumed that the PLL is initially locked.* If the PLL is initially unlocked, however, the phase error θ_e can take on arbitrarily large values, and the linear model is no longer valid. When we try to calculate the *acquisition process* of the PLL itself, we must use a model which also accounts for the nonlinear effect of the phase detector. This is done in Sec. 2.5.

Knowing the phase transfer function $H(s)$ and the error transfer function $H_e(s)$ of the LPLL, we can calculate its response on the most important excitation signals. We therefore analyze the LPLLs answer to

■ A phase step

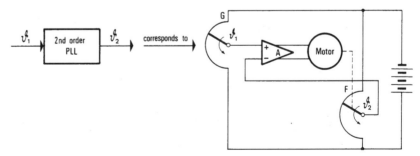

Figure 2.10 Simple electromechanical analogy of the linearized second-order PLL. In this servo system, the angles θ_1 and θ_2 correspond to the phases θ_1 and θ_2, respectively, of the LPLL system.

■ A frequency

■ A frequency ramp

applied to its reference input.

Phase step applied to the reference input.   A reference signal performing a phase step at time $t = 0$ has been shown in Fig. 2.4a. In this case the phase signal $\theta_1(t)$ is a step function,

$$\theta_1(t) = u(t)\,\Delta\Phi$$

where $u(t)$ is the unit step function and $\Delta\Phi$ is the size of the phase step. For the Laplace transform $\Theta_1(s)$ we get therefore

$$\Theta_1(s) = \Delta\Phi/s$$

The phase error θ_e is obtained from

$$\Theta_e(s) = H_e(s)\Theta_1(s) = H_e(s)\,\Delta\Phi/s \qquad (2.19)$$

Inserting Eq. (2.18) into Eq. (2.19) yields

$$\Theta_e(s) = \frac{\Delta\Phi}{s}\,\frac{s^2}{s^2 + 2s\zeta\omega_n + \omega_n^2} \qquad (2.20)$$

Applying the inverse Laplace transform to Eq. (2.20) we get the phase error functions $\theta_e(t)$ shown in Fig. 2.11. The phase error has been normalized to the phase step $\Delta\Phi$. Phase error functions have been plotted for different damping factors. The initial phase error $\theta_e(0)$ equals $\Delta\Phi$ for any value of ζ. For $t \to \infty$ the phase error $\theta_e(\infty)$ approaches zero. This can also be shown by the final value theorem of the Laplace transform,[3] which reads

$$\theta_e(\infty) = \lim_{s \to 0} s\Theta_e(s) = 0$$

Frequency step applied to the reference input.   If a frequency step is applied to the input of the LPLL, the angular frequency of the reference signal becomes

$$\omega_1(t) = \omega_0 + \Delta\omega\,u(t)$$

where $\Delta\omega$ is the magnitude of the frequency step. Because the phase $\theta_1(t)$ is the integral over the frequency variation $\Delta\omega$, we have

$$\theta_1(t) = \Delta\omega\,t$$

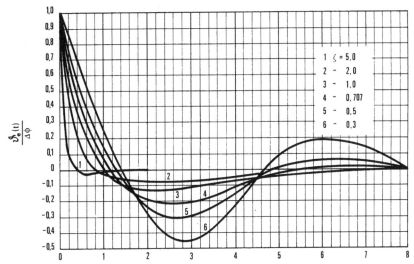

Figure 2.11 Transient response of a linear second-order PLL to a phase step $\Delta\Phi$ applied at $t = 0$. The PLL is assumed to be a high-gain loop. (*Adapted from Gardner*[1] *with permission.*)

i.e., the phase is a ramp function now. For the Laplace transform $\Theta_1(s)$ we get therefore

$$\Theta_1(s) = \Delta\omega/s^2$$

Performing the same calculation as above, we get for the phase error

$$\Theta_e(s) = \frac{\Delta\omega}{s^2}\frac{s^2}{s^2 + 2s\zeta\omega_n + \omega_n^2} \qquad (2.21)$$

Applying the inverse Laplace transform to Eq. (2.21) we get the phase error curves shown in Fig. 2.12. If we apply the final value theorem again into Eq. (2.21), it turns out that the phase error approaches 0 when $t \to \infty$. This is only true, however, for high-gain loops. For low-gain loops, the numerator of Eq. (2.21) would have an additional first-order term in s; hence the phase error $\theta_e(\infty)$ becomes nonzero.

Frequency ramp applied to the reference input. If a frequency ramp is applied to the LPLLs input, the angular frequency ω_1 is given by

$$\omega_1(t) = \omega_0 + \Delta\dot{\omega}\,t$$

where $\Delta\dot{\omega}$ is the rate of change of the reference frequency. Because the phase $\theta_1(t)$ is the integral over the frequency variation, we get

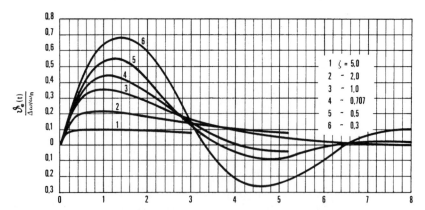

Figure 2.12 Transient response of a linear second-order PLL to a frequency step $\Delta\omega$ applied to its reference input at $t = 0$. (*Adapted from Gardner[1] with permission.*)

$$\theta_1(t) = \Delta\dot\omega \frac{t^2}{2}$$

The Laplace transform $\Theta_1(s)$ now becomes

$$\Theta_1(s) = \frac{\Delta\dot\omega}{s^3}$$

Assuming a high-gain loop again and applying the final value theorem of the Laplace transform, the final phase error $\theta_e(\infty)$ is

$$\theta_e(\infty) = \lim sH_e(s)\Theta_1(s) = \frac{\Delta\dot\omega}{\omega_n^2}$$

Now we remember that our linear model is valid for small phase errors only. Because for large phase error the output signal of the phase detector is not proportional to phase error, but to the sine of phase error, the last equation can be shown to read actually[1]

$$\sin\theta_e(\infty) = \frac{\Delta\dot\omega}{\omega_n^2}$$

Because the sine function cannot exceed unity, the maximum rate of change of the reference frequency that does not cause lock-out is

$$\Delta\dot\omega_{\max} = \omega_n^2 \qquad (2.22a)$$

This result has two consequences:

1. If the reference frequency is swept at a rate larger than ω_n^2, the system will unlock.

2. If the system is initially unlocked, it cannot become locked if the reference frequency is simultaneously swept at a rate larger than ω_n^2.

Practical experience with LPLLs has shown that Eq. (2.22a) presents a theoretical limit which is normally not practicable. If the reference frequency is swept in the presence of noise, the rate at which an initially unlocked LPLL can become locked is markedly less than ω_n^2.[1] A more practical design limit for $\Delta\omega_{\mathrm{max}}$ is considered to be

$$\Delta\dot{\omega}_{\mathrm{max}} = \frac{\omega_n^2}{2} \qquad (2.22b)$$

2.4 The Order of the LPLL System

2.4.1 Number of poles

The LPLLs considered in Sec. 2.3 were second-order PLLs. The loop filter had one pole, and the VCO had a pole at $s = 0$, so the whole system had two poles. Generally the order of a PLL is always by 1 higher than the order of the loop filter. If the loop filter is omitted, i.e., if the output of the phase detector directly controls the VCO, we obtain a first-order PLL. We will discuss the first-order LPLL in Sec. 2.4.2. In LPLL applications, first-order systems are rarely used, because they offer little noise suppression (more about noise in later sections). Because higher-order loop filters offer better noise cancellation, loop filters of order 2 and more are used in critical applications. As mentioned, it is much more difficult to obtain a stable higher-order system than a lower-order one. We will encounter higher-order PLLs in Chap. 3 when frequency synthesizer applications of DPLLs are discussed.

2.4.2 A special case: The first-order PLL

The first-order LPLL is an extremely simple system. When we omit the loop filter, $F(s)$ in Eq. (2.12) becomes 1, and for the phase transfer function we obtain

$$H(s) = \frac{1}{1 + s/K_0K_d}$$

This is the transfer function of a first-order low-pass filter having a 3-dB angular corner frequency $\omega_{\mathrm{3dB}} = K_0K_d$. As we will see in Sec.

2.6.1, the product $K_0 K_d$ is identical with the *hold range* of the PLL. Because the hold range is generally much larger than the natural frequency ω_n of second-order PLLs, the first-order PLL has large bandwidth and hence tracks phase and frequency variations of the input signal very rapidly. Due to its high bandwidth, however, the first-order PLL does not suppress noise superimposed to the input signal. Because this is an undesirable property in most LPLL applications, the first-order LPLL is rarely utilized here. First-order PLLs do really exist in applications where noise is not a primary concern. In Chap. 4 we will deal with a first-order all-digital PLL system which is frequently used in FSK modems.

2.5 LPLL Performance in the Unlocked State

As we have seen in Sec. 2.3, the linear model of the LPLL is valid only when the PLL is in the locked state. When the LPLL is out of lock, its model becomes much more complicated and is nonlinear, of course. We will not lose too much time considering nonlinear LPLL models but will show the most important phenomena by means of a simple analogy. For the engineer it is not of major concern to know exactly what the LPLL does when it is in the unlocked state. The interesting questions are rather

- Under what conditions will the LPLL get locked?
- How much time does the lock-in process need?
- Under what conditions will the LPLL lose lock?

Let us consider again the LPLL shown in Fig. 2.1. When the passive lag filter (Fig. 2.2a) is used for the loop filter, the behavior of the LPLL is described by the nonlinear differential equation[4]

$$\ddot{\theta}_e + \dot{\theta}_e \, \frac{1 + K_0 K_d \tau_2 \cos \theta_e}{\tau_1 + \tau_2} + \frac{K_0 K_d}{\tau_1 + \tau_2} \sin \theta_e = \ddot{\theta}_1 + \dot{\theta}_1 \, \frac{1}{\tau_1 + \tau_2} \quad (2.23)$$

This equation can be simplified. First the substitutions of Eqs. (2.15a) are made for τ_1 and τ_2. Next, in most practical cases the inequality

$$\frac{1}{\tau_2} \ll K_0 K_d$$

holds. This leads to the simplified differential equation

$$\ddot{\theta}_e + 2\zeta\omega_n\dot{\theta}_e\cos\theta_e + \omega_n^2\sin\theta_e = \ddot{\theta}_1 + \dot{\theta}_1\frac{\omega_n^2}{K_0K_d} \qquad (2.24)$$

The nonlinearities in this equation stem from the trigonometric terms $\sin\theta_e$ and $\cos\theta_e$.

As already stated there is no exact solution for this problem. We find, however, that Eq. (2.24) is almost identical to the differential equation of a somewhat special mathematical pendulum, as shown in Fig. 2.13. A beam having a mass M is rigidly fixed to the shaft of a cylinder which can rotate freely around its axis. A thin rope is attached at point P to the surface of the cylinder and is then wound several times around the latter. The outer end of the rope hangs down freely and is attached to a weighing platform. If there is no weight on the platform, the pendulum is assumed to be in a vertical position with $\Phi_e = 0$. If some weight G is placed on the platform, the pendulum will be deflected from its quiescent position and will eventually settle

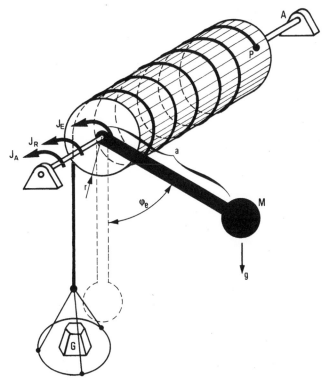

Figure 2.13 Mechanical analogy illustrating the linear and the nonlinear performance of the LPLL. (Symbols defined in text.)

at a final deflection angle Φ_e. The dynamic response of the pendulum can be calculated by Newton's third law,

$$T\ddot{\varphi}_e = \sum_i J_i \tag{2.25}$$

where T is the moment of inertia of the pendulum plus cylinder, φ_e is the angle of deflection, and J_i is a driving torque. Three different torques can be identified in the mechanical system of Fig. 2.13:

1. The torque J_E generated by gravitation of the mass M; $J_E = -Mag$ sin φ_e, where a is the length of the beam and g is acceleration due to gravity.

2. A friction torque J_R, which is assumed to be proportional to the angular velocity $\dot{\varphi}_e$ (viscous friction); $J_R = -\rho\dot{\varphi}_e$, where ρ is the coefficient of friction.

3. The torque J_A generated by the gravity of the weight G; $J_A = rG$, where r is the radius of the cylinder.

Introducing these individual torques into Eq. (2.25) yields

$$\ddot{\varphi}_e + \frac{\rho}{T}\dot{\varphi}_e + \frac{Mag}{T}\sin\varphi_e = \frac{r}{T}G(t) \tag{2.26}$$

As in the case of the PLL, we can write this equation in a normalized form. If we introduce the substitutions

$$\omega_n' = \left(\frac{Mag}{T}\right)^{1/2} \tag{2.27}$$

$$\zeta' = \frac{\rho}{2\sqrt{Mag\ T}}$$

Equation (2.26) is converted into

$$\ddot{\varphi}_e + 2\zeta'\omega_n'\dot{\varphi}_e + \omega_n'^2\sin\varphi_e = \frac{r}{T}G(t) \sim G(t) \tag{2.28}$$

This nonlinear differential equation for the deflection angle φ_e looks very much like the nonlinear differential equation of the PLL according to Eq. (2.24). There is a slight difference in the second term, however. In the case of the PLL the second term contains the factor cos θ_e, whereas for the pendulum the coefficient of the second term ($\dot{\varphi}_e$) is the constant $2\zeta'\ \omega_n'$. Strictly speaking, the pendulum of Fig. 2.13 would only be an accurate analogy of the PLL if the friction varied with the cosine of the deflection angle. This would be true if the damping factor ζ' were not a constant but would vary with cos φ_e. As a consequence, the momentary friction torque would be positive for

$$-\frac{\pi}{2} < \varphi_e < +\frac{\pi}{2}$$

that is, when the position of the pendulum is in the lower half of a circle around the cylinder shaft. On the other hand, the momentary friction torque would be negative for

$$\frac{\pi}{2} < \varphi_e < \frac{3\pi}{2}$$

that is, when the position of the pendulum is in the upper half of this circle. A negative friction is hard to imagine, of course, but let us assume for the moment that $\zeta' \sim \cos \varphi_e$ is valid. Imagine further that the weight G is large enough to make the pendulum tip over and continue to rotate forever around its axis (provided the rope is long enough). Because of the nonconstant torque generated by the mass M of the pendulum this oscillation will be nonharmonic. During the time when the pendulum swings through the lower half of the circle ($-\pi/2 \leq \varphi_e \leq \pi/2$), its average angular velocity is greater than its velocity during the time when it swings through the upper half ($\pi/2 \leq \varphi_e \leq 3\pi/2$). The *positive* friction torque averaged over the lower semicircle is therefore greater in magnitude than the *negative* friction torque averaged over the upper semicircle. This means that the friction torque averaged over one full revolution stays positive; hence it is acceptable to state that the coefficient of friction ρ varies with the cosine of φ_e. The mathematical pendulum is therefore a reasonable approximation for the PLL.

Comparing the PLL with this mathematical pendulum, we find the following analogies:

1. The phase error θ_e of the PLL corresponds to the angle of deflection φ_e of the pendulum.

2. The natural frequency ω_n of the PLL corresponds to the natural (or resonant) frequency ω'_n of the pendulum.

3. The damping factor ζ of the PLL corresponds to the damping factor ζ' of the pendulum, which results from viscous friction.

4. The weight G on the platform corresponds to a reference phase disturbance according to the relation

$$\ddot{\theta}_1 + \dot{\theta}_1 \frac{\omega_n^2}{K_0 K_d} \sim G(t) \tag{2.29}$$

Let us now see what is the physical meaning of the term $\ddot{\theta}_1 + \dot{\theta}_1(\omega_n^2/K_0 K_d)$ in Eq. (2.29). We assume first that the frequency of the reference signal has an arbitrary value:

$$\omega_1 = \omega_0 + \Delta\omega(t)$$

where $\Delta\omega(t)$ can be considered the frequency offset of the reference signal. The reference signal $u_1(t)$ can therefore be written in the form

$$u_1(t) = U_{10} \sin \left\{ \int_0^t [\omega_0 + \Delta\omega(t)] \, dt \right\}$$

$$= U_{10} \sin [\omega_0 t + \theta_1(t)]$$

Consequently the phase $\theta_1(t)$ is given by

$$\theta_1(t) = \int_0^t \Delta\omega(t) \, dt$$

From this it becomes immediately evident that the first derivative $\dot{\theta}_1(t)$ represents the momentary frequency offset:

$$\dot{\theta}_1(t) = \Delta\omega(t)$$

whereas the second derivative $\ddot{\theta}_1(t)$ signifies the rate of change of the frequency offset:

$$\ddot{\theta}_1(t) = \frac{d}{dt} \Delta\omega(t) = \Delta\dot{\omega}$$

The weight G placed on the platform is thus equivalent to a weighted sum of the frequency offset $\Delta\omega(t)$ and its rate of change $\Delta\dot{\omega}$:

$$G(t) \sim \Delta\dot{\omega} + \frac{\omega_n^2}{K_0 K_d} \Delta\omega \tag{2.30}$$

This simple correspondence paves the way toward understanding the quite complex dynamic performance of a PLL in the locked and unlocked states. To see what happens to a PLL when phase and/or frequency steps of arbitrary size are applied to its reference input, we have to place the corresponding weight $G(t)$ given by Eq. (2.30) on the platform and observe the response of the pendulum. The notation $G(t)$

should emphasize that G must not necessarily be a constant, but can also be a function of time, as would be the case when an impulse is applied.

Let us first consider the trivial case of no weight on the platform. The pendulum is then at rest in a vertical position, $\varphi_e = 0$. This corresponds to the PLL operating at its center frequency ω_0 with zero frequency offset ($\Delta\omega = 0$) and zero phase error ($\theta_e = 0$).

What happens if the frequency of the reference signal is changed *slowly*? The rate of change of the reference frequency is assumed to be so low that the derivative term $\Delta\dot{\omega}$ in Eq. (2.30) is negligible. A slow variation of the reference frequency corresponds to a slow increase of weight G, achieved by very carefully pouring a fine powder onto the platform. The analogy is given in this case by

$$G \sim \frac{\omega_n^2}{K_0 K_d} \Delta\omega$$

The pendulum now starts to deflect, indicating that a finite phase error is established within the PLL. For small offsets of the reference frequency the phase error θ_e will be proportional to $\Delta\omega$. If the frequency offset reaches a critical value, called the *hold range*, the deflection of the pendulum is just 90°. This is the *static limit of stability*.

With the slightest disturbance the pendulum would now tip over and rotate around its axis forever. This corresponds to the case where the PLL is no longer able to maintain phase tracking and consequently unlocks. One full revolution of the pendulum equals a phase error of 2π. Because the pendulum is now rotating permanently, the phase error increases toward infinity.

Another interesting case is given by a step change of the reference frequency at the input of the PLL. When a frequency step of the size $\Delta\omega$ is applied at $t = 0$, the angular frequency of the reference signal is

$$\omega_1(t) = \omega_0 + \Delta\omega\, u(t)$$

where $u(t)$ is the unit-step function. The first derivative $\Delta\dot{\omega}$ therefore shows a delta function at $t = 0$; this is written

$$\Delta\dot{\omega}(t) = \Delta\omega\, \delta(t)$$

and is plotted in Fig. 2.14. What will now be the weight $G(t)$ required to simulate this condition? As also shown in Fig. 2.14, the weight function should be a superposition of a step function and a delta (impulse) function. In practice this can be simulated by dropping an appropriate weight from some height onto the platform. The impulse is generated

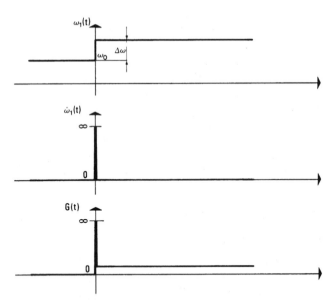

Figure 2.14 Weight function $G(t)$ required to simulate a frequency step applied to the reference input of the PLL. (ω_1 = angular frequency of the reference signal; $\Delta\omega$ = frequency step applied at $t = 0$.)

when the weight hits the platform. To get a narrow and steep impulse, the stroke should be *elastic*. If this is done, the pendulum will show a transient response, mostly in the form of damped oscillation. If a relatively small weight is *dropped* onto the platform, the final deflection of the pendulum will be the same as if the weight had been placed *smoothly* onto the platform. If the pendulum is not heavily overdamped, however, its *peak deflection* $\hat{\varphi}_e$ will be considerably greater than its final deflection. If we increase the weight to be dropped onto the platform, we will observe a situation where the peak deflection exceeds 90°, but not 180°, and the final deflection is less than 90°. We thus conclude that a linear PLL can operate stably when the phase error *momentarily* exceeds the value of 90°. If the weight dropped onto the platform is increased ever further, the peak deflection will exceed 180°. The pendulum now tips over and performs a number of revolutions around its axis, but it will probably come to rest again after some time. The *weight* which *caused the system to unlock* (at least temporarily) is observed to be considerably *smaller* than the weight that represented the *hold range*. We therefore have to define another critical-frequency offset, i.e., the offset that causes the LPLL to unlock when it is applied as a step. This frequency step is called the *pull-out range*. Keep in mind that the pull-out range of an LPLL is markedly smaller than its hold range. The pull-out range may be considered the *dynamic limit of stability*. The LPLL always stays locked as long as the frequency steps applied to the system do not exceed the pull-out range.

There is a third way to cause an LPLL to unlock from an initially stable operating point. If the frequency of the reference signal is increased linearly with time, we have

$$\Delta\omega = \Delta\dot{\omega}\,t$$

where $\Delta\dot{\omega}$ is the rate of change of the angular frequency. In the mechanical analogy this corresponds to a weight G that also builds up linearly with time. This can be realized by feeding a material onto the platform at an appropriate feed rate. It becomes evident that too rapid a feed rate acts on the pendulum like the impulse which was generated in the last example by dropping a weight onto the platform. As has been shown in Sec. 2.3, the rate of change of the reference frequency must always be smaller than ω_n^2 to keep the system locked:

$$\Delta\dot{\omega} < \omega_n^2$$

These examples have demonstrated that three conditions are necessary if a PLL system is to maintain phase tracking:

1. The angular frequency of the reference signal must be within the hold range.
2. The maximum frequency step applied to the reference input of a PLL must be smaller than the pull-out range.
3. The rate of change of the reference frequency $\Delta\dot{\omega}$ must be lower than ω_n^2.

Whenever a PLL has lost tracking because one of these conditions has not been fulfilled, the question arises whether it will return to stable operation when all the necessary conditions are met again. The answer is clearly *no*. When the reference frequency exceeded the hold range, the pendulum in our analogy tipped over and started rotating around its axis. If the weight on the platform were reduced to a value slightly less than the critical limit that caused instability, the pendulum would nevertheless continue to rotate because there is enough kinetic energy stored in the mass to maintain the oscillation. If there were no friction at all, the pendulum would continue to rotate even if the weight G were reduced to zero. Fortunately friction exists, so the pendulum will decelerate if the weight is decreased to an appropriate level. For a PLL, this means that buildup of the phase error will decelerate if the reference-frequency offset is decreased below another critical value, the *pull-in frequency*. If the slope of the average phase error becomes smaller, the frequency ω_2 of the VCO more and more approaches the frequency of the reference signal, and the system will finally lock. The pull-in frequency $\Delta\omega_P$ is markedly smaller than the hold range, as can be expected from the mechanical analogy. The pull-in process is a relatively slow one, as will be demonstrated in Sec. 2.6.

In most practical applications it is desired that the locked state be obtained within a short time period. Suppose again that the weight put on the platform of the mechanical model is large enough to cause sustained oscillation of the pendulum. It is easily shown that the pendulum can be brought to rest *within one single revolution* if the weight is suddenly decreased below another critical value. This implies that a PLL can become locked within one single-beat note between reference frequency and output frequency, provided the frequency offset $\Delta\omega$ is reduced below a critical value called the *lock range*. This latter process is called the *lock-in process*. The lock-in process is much faster than the pull-in process, but the lock range is smaller than the pull-in range.

The mechanical model has shown that there are four key parameters specifying the frequency range in which the LPLL can be operated. Figure 2.15 is a graphical representation of these parameters. The four key parameters can be summarized as follows:

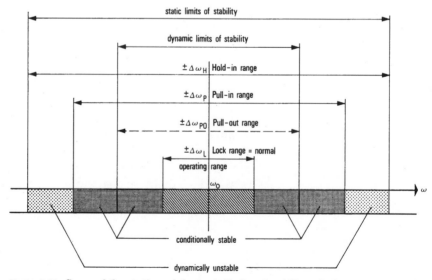

Figure 2.15 Scope of the static and dynamic limits of stability of a linear second-order PLL.

1. *The hold range* $\Delta\omega_H$. This is the frequency range in which an LPLL can statically maintain phase tracking. A PLL is conditionally stable only within this range.

2. *The pull-out range* $\Delta\omega_{PO}$. This is the dynamic limit for stable operation of a PLL. If tracking is lost within this range, an LPLL normally will lock again, but this process can be slow if it is a *pull-in process*.

3. *The pull-in range* $\Delta\omega_P$. This is the range within which an LPLL will always become locked, but the process can be rather slow.

4. *The lock range* $\Delta\omega_L$. This is the frequency range within which a PLL locks within one single-beat note between reference frequency and output frequency. *Normally the operating-frequency range of an LPLL is restricted to the lock range.*

The discussion of the mechanical analogy did not provide numerical solutions for these four key parameters. We will derive such expressions in Sec. 2.6. The quantitative relationships among these four parameters are plotted in Fig. 2.15 for most practical cases. We can state in advance that the *hold range* $\Delta\omega_H$ is greater than the three remaining parameters. Furthermore we know that the pull-in range $\Delta\omega_P$ must be greater than the lock range $\Delta\omega_L$. The pull-in range $\Delta\omega_P$ is greater than the pull-out range $\Delta\omega_{PO}$ in most practical designs, so we get the simple inequality

$$\Delta\omega_L < \Delta\omega_{PO} < \Delta\omega_P < \Delta\omega_H$$

Keep in mind that the results gained by means of the mechanical analogy are valid for the *linear* PLL only. In the case of the *digital* PLL the situation is very much different, though simpler, as will be demonstrated in Chap. 3.

In many texts the term *capture range* can also be found. In most cases capture range is an alternative expression for lock range; sometimes it is also used to mean pull-in range. The differentiation between lock-in and pull-in ranges is not clearly established in some books, but we will maintain it throughout this text.[1]

2.6   Key Parameters of the LPLL

In Sec. 2.5 it has been demonstrated that the dynamic performance of the LPLL is governed by a set of key parameters,

- The lock range $\Delta\omega_L$
- The pull-out range $\Delta\omega_{PO}$

■ The pull-in range $\Delta\omega_P$

■ The hold range $\Delta\omega_H$

To design an actual LPLL system, we need equations which tell us how these key parameters depend on the parameters of the circuit, i.e., the time constants τ_1 and τ_2 of the loop filter and the gain factors K_d, K_0, and K_a. The last of these is used only for the active lag filter (Fig. 2.2b).

Because life is always difficult, the formulas used to compute the key parameters depend on the type of loop filter used. In Secs. 2.6.1 to 2.6.4 we give approximate design equations for the four mentioned parameters.

2.6.1 The hold range

First of all, let us state that the hold range is a parameter of more academic interest. The hold range is the frequency range in which a PLL is able to maintain lock *statically*. As we have seen in Sec. 2.5, the LPLL locks out forever when the frequency of the input signal exceeds the hold range, so most practitioners don't even worry about the actual value of this parameter.

The magnitude of the hold range is obtained by calculating that frequency offset at the reference input which causes a phase error θ_e of $\pi/2$. In this case we have

$$\omega_1 = \omega_0 + \Delta\omega_H \tag{2.31}$$

where $\Delta\omega_H$ is the hold range. For the phase signal $\theta_1(t)$ we get

$$\theta_1(t) = \Delta\omega_H \cdot t \tag{2.32}$$

The Laplace transform of the phase signal therefore becomes

$$\Theta_1(s) = \frac{\Delta\omega}{s^2} \tag{2.33}$$

The phase error can now be calculated according to Eq. (2.13):

$$\Theta_e(s) = \Theta_1(s)H_e(s) = \frac{\Delta\omega}{s^2} \frac{s}{s + K_0 K_d F(s)} \tag{2.34}$$

Using the final-value theorem of the Laplace transform, we calculate the final phase error in the time domain:

$$\lim_{t \to \infty} \theta_e(t) = \lim_{s \to 0} s\Theta_e(s) = \frac{\Delta\omega}{K_0 K_d F(0)} \qquad (2.35)$$

Remember that the PLL network was linearized when the Laplace transform was introduced. Consequently Eq. (2.35) is valid for small values of θ_e only. For greater values of phase error we would have to write

$$\lim_{t \to \infty} \sin \theta_e(t) = \frac{\Delta\omega_H}{K_0 K_d F(0)} \qquad (2.36)$$

At the limit of the hold range, $\theta_e = \pi/2$ and $\sin \theta_e$ is exactly 1. Therefore we obtain for the hold range the expression

$$\Delta\omega_H = K_0 K_d F(0) \qquad (2.37)$$

The dc gain $F(0)$ of the loop filter depends on the filter type. For the passive lag filter [Eq. (2.1)] the dc gain $F(0) = 1$. For the active lag filter, the dc gain is $K(0) = K_a$. For the active PI filter, $F(0)$ is—at least theoretically—infinite. Thus we get for the hold range

- For the passive lag filter

$$\Delta\omega_H = K_0 K_d \qquad (2.38a)$$

- For the active lag filter

$$\Delta\omega_H = K_0 K_d K_a \qquad (2.38b)$$

- For the active PI filter

$$\Delta\omega_H \to \infty \qquad (2.38c)$$

When the active PI filter is used, the actual hold range is limited by the frequency range covered by the VCO.

2.6.2 The lock range

The magnitude of the lock range can be obtained by a simple consideration. We assume that the LPLL is initially not locked and that the reference frequency is $\omega_1 = \omega_0 + \Delta\omega$. The reference signal of the LPLL is then given by

$$u_1(t) = U_{10} \sin (\omega_0 t + \Delta\omega t)$$

and the output signal by

$$u_2(t) = U_{20}w(\omega_0 t)$$

where $w(\omega_0 t)$ is the Walsh function shown by the dashed curve in Fig. 2.5b. Using Eq. (2.6) the phase detector will deliver an output signal given by

$$u_d(t) = K_d \sin (\Delta\omega t) + \text{higher-frequency terms}$$

The higher-frequency terms can be discarded because they will be almost entirely filtered out by the loop filter. At the output of the loop filter there appears a signal $u_f(t)$ given by

$$u_f(t) \approx K_d|F(\Delta\omega)| \sin (\Delta\omega t) \qquad (2.39)$$

This is an ac signal which causes a frequency modulation of the VCO. The peak frequency deviation is equal to $K_d K_0 |F(\Delta\omega)|$.

In Fig. 2.16, the frequency ω_2 of the VCO is plotted against time for two cases. In Fig. 2.16a the peak frequency deviation is less than the

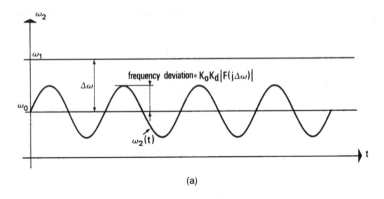

(a)

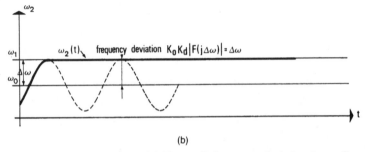

(b)

Figure 2.16 Lock-in process. (a) The peak frequency deviation is smaller than the offset $\Delta\omega$; therefore a fast lock-in process cannot take place. (b) The peak frequency deviation is exactly as large as the actual frequency offset; thus the PLL will become locked after a very short time.

offset $\Delta\omega$ between the reference frequency ω_1 and the frequency of the VCO ω_2. Hence a lock-in process will not take place, at least not instantaneously. Figure 2.16b shows a special case, however, where the peak frequency deviation is just as large as the frequency offset $\Delta\omega$. The frequency ω_2 of the VCO output signal develops as shown by the solid line. When the frequency deviation is at its largest, ω_2 exactly meets the value of the reference frequency ω_1. Consequently the PLL locks within *one single-beat note* between the reference and output frequencies. This corresponds exactly to the lock-in process described in Sec. 2.5 by means of the mechanical analogy.

The condition for locking is therefore

$$K_0 K_d F(\Delta\omega) \geq \Delta\omega \tag{2.40}$$

The lock range $\Delta\omega_L$ itself is consequently given by

$$\Delta\omega_L = K_0 K_d |F(\Delta\omega_L)| \tag{2.41}$$

This is a nonlinear equation for $\Delta\omega_L$. Its solution becomes very simple, however, if an approximation is introduced for $|F(\Delta\omega_L)|$. It follows from practical considerations that the lock range is always much greater than the corner frequencies $1/\tau_1$ and $1/\tau_2$ of the loop filter. For the filter gain $F(\Delta\omega_L)$ we can therefore make the following approximations:

$$F(\Delta\omega_L) \approx \tau_2/(\tau_1 + \tau_2) \qquad \text{for the passive lag filter}$$

$$F(\Delta\omega_L) \approx K_a \tau_2/\tau_1 \qquad \text{for the active lag filter}$$

$$F(\Delta\omega_L) \approx \tau_2/\tau_1 \qquad \text{for the active PI filter}$$

Moreover, τ_2 is normally much smaller than τ_1, so we can use the simplified relationship

$$F(\Delta\omega_L) \approx \tau_2/\tau_1 \qquad \text{for the passive lag filter}$$

$$F(\Delta\omega_L) \approx K_a \tau_2/\tau_1 \qquad \text{for the active lag filter} \tag{2.42}$$

$$F(\Delta\omega_L) \approx \tau_2/\tau_1 \qquad \text{for the active PI filter}$$

Making use of the substitutions Eqs. (2.15) and assuming high-gain loops, we get

$$\Delta\omega_L \approx 2\zeta\omega_n \tag{2.43}$$

for all types of loop filters in Fig. 2.2.

Knowing the approximate size of the lock range we are certainly interested in having some indication of the lock-in time. When the

LPLL locks in quickly, the signals u_d and u_f perform (for $\zeta < 1$) a damped oscillation, whose angular frequency is approximately ω_n. As Fig. 2.12 shows, the transients die out in about one cycle of this oscillation; hence it is reasonable to state that the lock-in time is

$$T_L \approx \frac{2\pi}{\omega_n} \qquad (2.44)$$

which is valid for any type of loop filter.

T_L is also referred to as *settling time*.

2.6.3 The pull-in range

We are now going to determine the size of the pull-in range, $\Delta\omega_p$. We assume again that the LPLL is not locked initially, that the frequency of the reference signal is $\omega_1 = \omega_0 + \Delta\omega$, and that the VCO initially operates at the center frequency ω_0. Consequently the output signal u_d of the phase detector is a sine wave having the frequency $\Delta\omega$, hence is an "ac" signal. We now assume that the frequency offset $\Delta\omega$ is so large that a lock-in process will not take place. Let us assume furthermore that a passive lag filter is used for the loop filter. The u_d signal will therefore be attenuated by the loop filter. Its output signal u_f is an ac signal as well and will frequency-modulate the VCO, as is shown in Fig. 2.16a. In the positive half-cycle of u_f the output frequency ω_2 increases, whereas it decreases in the negative half-cycles. Because the peak output frequency ω_2 never meets the input frequency ω_1, we would think at a first glance that the LPLL never could get locked.

Fortunately, we have overlooked the fact that the difference $\Delta\omega$ between reference frequency ω_1 and output frequency $\omega_2(t)$ is not a constant; it is also varied by the frequency modulation of the VCO output signal. If the frequency $\omega_2(t)$ is modulated in the positive direction, the difference $\Delta\omega$ becomes smaller and reaches some minimum value $\Delta\omega_{\min}$; if $\omega_2(t)$ is modulated in the negative direction, however, $\Delta\omega$ becomes greater and reaches some maximum value $\Delta\omega_{\max}$. Because $\Delta\omega(t)$ is not a constant, the VCO frequency is modulated nonharmonically, that is, the duration of the half-period in which $\omega_2(t)$ is modulated in the positive direction becomes *longer* than that of the half-period in which $\omega_2(t)$ is modulated in the negative sense. This is shown graphically in Fig. 2.17. As a consequence, the average frequency $\overline{\omega_2}$ of the VCO is now higher than it was without any modulation; i.e., the VCO frequency is pulled in the direction of the reference signal.

The asymmetry of the waveform $\omega_2(t)$ is greatly dependent on the value of the average offset $\Delta\omega$; the asymmetry becomes more marked

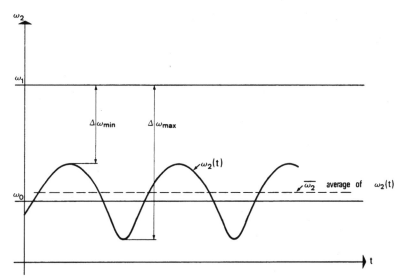

Figure 2.17 In the unlocked state of the PLL the frequency modulation of the VCO output signal is nonharmonic. This causes the average value $\overline{\omega_2}$ of the VCO output frequency to be pulled in the direction of the reference frequency.

as $\Delta\omega$ is decreased. If the average value of $\omega_2(t)$ is pulled somewhat in the direction of ω_1 (which is assumed to be greater than $\overline{\omega_2}$), the asymmetry of the $\omega_2(t)$ waveform becomes stronger. This in turn causes $\overline{\omega_2}$ to be pulled even more in the positive direction. This process is regenerative under certain conditions, so that the output frequency ω_2 finally reaches the reference frequency ω_1. This phenomenon is called the *pull-in process* (Fig. 2.18). Mathematical analysis shows that a pull-in process occurs whenever the initial frequency offset $\Delta\omega$ is smaller than a critical value, the *pull-in range* $\Delta\omega_p$. If, on the other hand, the initial frequency offset $\Delta\omega$ is larger than $\Delta\omega_p$, a pull-in process does not take place because the pulling effect is not then regenerative.

The mathematical treatment of the pull-in process is quite cumbersome and is treated in more detail in Appendix A; here we only give the final results. It is very important to note that the pull-in range depends on the type of loop filter. The first edition of this book gave a formula for the pull-in range of the LPLL using a passive loop filter. Here are the formulas for the pull-in range:

- For the passive lag filter
  Low-gain loops

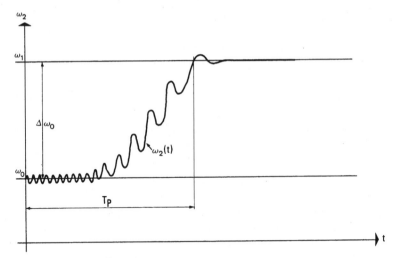

Figure 2.18 The pull-in process.

$$\Delta\omega_p \approx \frac{4}{\pi}\sqrt{2\,\zeta\omega_n K_0 K_d - \omega_n^2} \qquad (2.45a)$$

High-gain loops

$$\Delta\omega_p \approx \frac{4\sqrt{2}}{\pi}\sqrt{\zeta\omega_n K_0 K_d}$$

■ For the active lag filter
Low-gain loops

$$\Delta\omega_p \approx \frac{4}{\pi}\sqrt{2\,\zeta\omega_n K_0 K_d - \frac{\omega_n^2}{K_a}} \qquad (2.45b)$$

High-gain loops

$$\Delta\omega_p \approx \frac{4\sqrt{2}}{\pi}\sqrt{\zeta\omega_n K_0 K_d}$$

■ For the active PI filter (all loops)

$$\Delta\omega_p \rightarrow \infty \qquad (2.45c)$$

These formulas are also shown in Table 2.1. Obviously, the LPLL pulls in under any conditions when the loop filter is an active PI filter. As we know, the dc gain of this filter is (theoretically) infinite; hence the slightest nonharmonicity of the u_d signal is sufficient to pull in the loop.

The pull-in process depicted in Fig. 2.18 can be easily explained by the mechanical analogy of Fig. 2.13. Initially the frequency offset $\Delta\omega_0$ is fairly large, and in the analogy the pendulum rotates at $\Delta\omega_0/2\pi$ revolutions per second. The angular velocity decelerates slowly, however, and the pendulum comes to rest after some time. The "pumping" of the instantaneous frequency $\omega_2(t)$ is very characteristic of this process and is easily explained by the nonharmonic rotation of the pendulum caused by the gravity of its mass M. In fact its angular velocity is greater at the lower "dead point" than at the higher one.

The duration of the pull-in process can also be computed from the mathematical analysis of the acquisition process (see Appendix A). The result also slightly depends on the type of loop filter used. The values given by the following formulas agree quite well with measurements on actual LPLL circuits and with computer simulations. They are valid only, however, if the initial frequency offset $\Delta\omega$ (difference between reference frequency and initial frequency of the VCO) is distinctly smaller than the pull-in range, typically less than 0.8 times the pull-in range. When $\Delta\omega$ approaches the pull-in range $\Delta\omega_P$, the pull-in time T_P approaches infinity; thus the formula cannot be used. The analysis gives the results

■ For the passive lag filter

$$T_p = \frac{\pi^2}{16} \frac{\Delta\omega_0^2}{\zeta\omega_n^3} \tag{2.46a}$$

■ For the active lag filter

$$T_p = \frac{\pi^2}{16} \frac{\Delta\omega_0^2 K_a}{\zeta\omega_n^3} \tag{2.46b}$$

■ For the active PI filter

$$T_p = \frac{\pi^2}{16} \frac{\Delta\omega_0^2}{\zeta\omega_n^3} \tag{2.46c}$$

In this formula $\Delta\omega_0$ is the initial frequency offset $\omega_1 - \omega_2$ for $t = 0$. The quadratic and cubic terms in Eq. (2.46) show that the pull-in process is highly nonlinear. The pull-in time T_p is normally much longer than the lock-in time T_L. This is demonstrated easily by a numerical example.

Numerical Example A second-order PLL having a passive lag loop filter is assumed to operate at a center frequency f_0 of 100 kHz. Its natural frequency $f_n = \omega_n/2\pi$ is 3 Hz; this is a very narrow-band system. The damping factor is chosen to be $\zeta = 0.7$. The loop gain $K_0 K_d$ is assumed to be $2\pi \cdot 1000$ s^{-1}. We shall now calculate the lock-in time T_L and the pull-in time T_P for an initial frequency offset Δf of 30 Hz.

According to Eqs. (2.44) and (2.46a) we get

$$T_L = \frac{1}{f_n} = 0.33 \text{ s}$$

$$T_P = 4.67 \text{ s}$$

T_P is much larger than T_L.

2.6.4 The pull-out range

The pull-out range is by definition that frequency step which causes a lock-out if applied to the reference input of the PLL. In the mechanical analogy of Fig. 2.13 the pull-out frequency corresponds to that weight that causes the pendulum to tip over if the weight is suddenly dropped onto the platform.

An exact calculation of the pull-out range is not possible for the linear PLL. However, simulations on an analog computer[6] have led to an approximation:

$$\Delta\omega_{PO} = 1.8\omega_n(\zeta + 1) \tag{2.47}$$

In most practical cases the pull-out range is between the lock range and the pull-in range

$$\Delta\omega_L < \Delta\omega_{PO} < \Delta\omega_P \tag{2.48}$$

If, in an FM system the LPLL is pulled out by too large a frequency step, we can expect that LPLL to come back to stable operation—by means of a relatively slow pull-in process—provided the frequency offset $\Delta\omega = \omega_1 - \omega_0$ is smaller than $\Delta\omega_P$. If the corresponding pull-in time is considered to be too long, the peak frequency deviation $\Delta\omega$ must be confined to the lock range $\Delta\omega_L$.

2.6.5 The steady-state error of the LPLL

In control systems, the *steady-state error* is defined as the deviation of the controlled variable from the setpoint after the transient response has died out. For the PLL, the steady-state error is simply the phase error $\theta_e(\infty)$. To see how the LPLL settles on various excitation signals applied to its input, we will calculate the steady-state error for a phase step, a frequency step, and a frequency ramp. When the input phase $\theta_1(t)$ is given, the phase error $\theta_e(t)$ is computed from the error transfer function $H_e(s)$ as defined in Eq. (2.13). It turns out that the steady-state error depends largely on the number of "integrators" which are present in the control system, i.e., on the number of poles at $s = 0$ of the open-loop transfer function. Using Eq. (2.13) and the final value theorem of the Laplace transform [see Appendix B, Eq. (B.19)], $\theta_e(\infty)$ is given by

$$\theta_e(\infty) = \lim_{s \to 0} s \; \Theta_1(s) \; \frac{s}{s + K_0 K_d F(s)}$$

For the transfer function $F(s)$ of the loop filter, we choose a generalized expression

$$F(s) = \frac{P(s)}{Q(s)s^N}$$

where $P(s)$ and $Q(s)$ can be any polynomials in s, and N is the number of poles at $s = 0$. For many loop filters, $N = 0$. For the active PI loop filter as defined in Eq. (2.3), $N = 1$. It is possible to build loop filters having more than one pole at $s = 0$, but the larger the number of integrators, the more difficult it becomes to get a stable system. In most cases, $P(s)$ is a first-order polynomial, and $Q(s)$ is a polynomial of order 0 or 1. When inserting $F(s)$ into the equation for the steady-state error, we get

$$\theta_e(\infty) = \lim_{s \to 0} \frac{s^2 s^N Q(s) \Theta_1(s)}{s \cdot s^N Q(s) + K_0 K_d P(s)}$$

Let us calculate now the steady-state error for different excitations at the reference input of the LPLL.

Case study 1: Phase step applied to the reference input. If a phase step of size $\Delta\Phi$ is applied to the reference input, we have (cf. Sec. 2.3.)

$$\Theta_1(s) = \frac{\Delta\Phi}{s}$$

Hence the steady-state error becomes

$$\theta_e(\infty) = \lim_{s \to 0} \frac{s^2 s^N Q(s) \, \Delta\Phi}{s(s \cdot s^N Q(s) + K_0 K_d P(s))}$$

This quantity becomes 0 for any N, i.e., even for $N = 0$. Hence the phase error settles to zero for any type of loop filter.

Case study 2: Frequency step applied to the reference input. In case of a frequency step, the phase $\theta_1(t)$ becomes a ramp function as shown in Sec. 2.3, and for its Laplace transform we get

$$\Theta_1(s) = \frac{\Delta\omega}{s^2}$$

where $\Delta\omega$ is the size of the (angular) frequency step. The steady-state error is given by

$$\theta_e(\infty) = \lim_{s \to 0} \frac{s^2 s^N Q(s)\, \Delta\omega}{s^2(s \cdot s^N Q(s) + K_0 K_d P(s))}$$

This only becomes 0 if N is greater or equal to 1, i.e., if the loop filter has at least one pole at $s = 0$. In other words, the open-loop transfer function of the LPLL must have at least 2 poles at $s = 0$ if the steady-state error caused by a frequency step must become zero.

Case study 3: Frequency ramp applied to the reference input. As shown in Sec. 2.3, for a frequency ramp the Laplace transform of the phase signal $\theta_1(t)$ becomes

$$\Theta_1(s) = \frac{\Delta\dot{\omega}}{s^3}$$

where $\Delta\dot{\omega}$ is the rate of change of angular frequency. For the steady-state error we get

$$\theta_e(\infty) = \lim_{s \to 0} \frac{s^2 s^N Q(s)\, \Delta\dot{\omega}}{s^3(s \cdot s^N Q(s) + K_0 K_d P(s))}$$

Here the steady-state error only approaches zero if the loop filter has at least two poles at $s = 0$. For $N = 2$ and $Q(s) = 1$, the order of the LPLL becomes 3. Because each pole of the transfer function causes a phase shift of nearly 90° at higher frequencies, the overall phase shift can approach 270°, which means that the system can get unstable. To get a stable loop, the poles must be compensated for by zeros, i.e., the loop filter must have at least one zero, which must be correctly placed, of course.[3]

2.7   Noise in Linear PLL Systems

2.7.1   Simplified noise theory

The theory of noise in PLLs is very cumbersome. Exact solutions for noise performance have been derived for first-order PLLs only;[4] for second-order PLLs, computer simulations have been made which provided approximate results.[5] We deal here only with some final results of the noise theory and discuss a few key parameters which have a marked effect on the practical design of a PLL system. The interested reader is pointed toward the specialized noise literature.[1,4]

Noise performance is most easily considered by means of the well-known mechanical analogy shown in Fig. 2.19. The noise theory reveals that a noise signal $u_{n1}(t)$ superimposed on the (noiseless) refer-

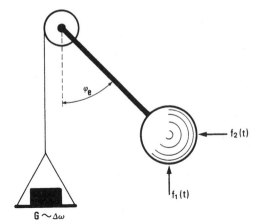

$G \sim \Delta\omega$

Figure 2.19 Mechanical model for a PLL having a noisy reference signal; f_1 and f_2 are stochastic forces (disturbances) as explained in text.

ence signal $u_1(t)$ is characterized in the model by two stochastic perpendicular forces $f_1(t)$ and $f_2(t)$; $f_1(t)$ and $f_2(t)$ are totally uncorrelated. These disturbing forces are likely to have an adverse effect on the acquisition and tracking performances of the PLL.

For a low noise amplitude the pendulum in Fig. 2.19 jitters around its quiescent position only slightly; the system remains stable. If the noise amplitude is made larger, the pendulum will tip over from time to time on a larger noise peak, but nevertheless it stays near its quiescent position most of the time.

The probability that the pendulum will be found near its point of equilibrium is much higher than the probability of its being seen at a completely different location. If the noise amplitude is increased further, however, no stable operating point will be established. The pendulum is now jittering around irregularly, and the probability that it will be found in the sector $\varphi_i \cdots \varphi_e + \Delta\varphi_e$ (where ϕ_e is the deflection angle) at a time t is identical for all angles ϕ_e in the range 0 to 2π. Now, of course, the PLL can no longer become locked.

For the designer of a PLL system the following questions are of interest:

1. How large must the signal-to-noise ratio (SNR) at the reference input be to enable safe acquisition of the PLL?

2. Provided the SNR is large enough to enable the lock-in process, how often on the average then does the PLL temporarily unlock (how many times per second?)

These questions are answered by the noise theory (at least to a useful approximation).

In the following analysis we consider amplitude and power density spectra of information signals and noise signals. Two types of power density spectra have been defined, one-sided and two-sided. If an information signal has one single spectral line at a frequency f_0, e.g., the one-sided power density spectrum shows up a line at f_0, but the two-sided spectrum would have two spectral lines at f_0 and at $-f_0$, respectively. The same applies for noise spectra. It is a matter of taste whether to use one-sided or two-sided power spectra. When using two-sided power spectra, total power of a signal can be calculated by integrating its power density spectrum over the frequency interval $-\infty < f < +\infty$. For one-sided spectra, however, this interval becomes $0 < f < +\infty$, of course. We follow Gardner's terminology in the following and use one-sided spectra only.

The most important parameters describing a noise signal are explained with the aid of Fig. 2.20. The corresponding power spectra are shown in Fig. 2.21. All noise signals considered here are defined as white noise, i.e., the power spectrum is flat; this means that the same noise power ΔP is contained in every frequency interval Δf (dP/df is a constant). (Refer to Fig. 2.21a.) The bandwidth of the noise spectrum is limited to B_i, mostly by a prefilter. Even without a prefilter the noise bandwidth would be limited for other reasons, for the total noise power cannot be infinite. In most practical applications the prefilter is a bandpass or low-pass filter; B_i is the bandwidth of such a filter. The noise signal is assumed to have a total noise power P_n (in watts). The reference (input) signal of the PLL is assumed to be narrow-band, so its spectrum in Fig. 2.21a may be characterized by one single line. Its power is defined by P_s, where P_s is normally given by

$$P_s = \frac{U_1^2 \, (\text{rms})}{R_1}$$

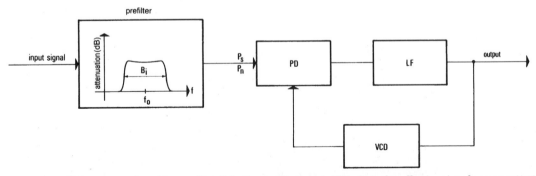

Figure 2.20   Noise suppression. ($f_0 = \omega_0/2\pi$; B_i is the prefilter bandwidth; P_s is the effective signal power output at the PLL input; P_n is the noise power output at the PLL input; the rest of the symbols are used in Fig. 1.1a.)

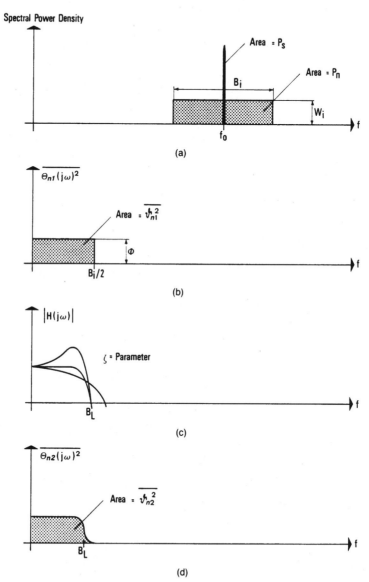

Figure 2.21 Method of calculating the phase jitter θ_{n2}^2 at the output of the PLL. (*a*) Power spectra of the reference signal $u_1(t)$ and the superimposed noise signal $u_n(t)$. (*b*) Spectrum of the phase noise at the input of the PLL. (*c*) Bode plot of the phase-transfer function $H(j\omega)$. (*d*) Spectrum of the phase noise at the output of the PLL.

where R_1 is the input impedance of the LPLL and U_1 (rms) is the rms value of the input signal. The prefilter is assumed to have an ideal transfer function; i.e., its gain is unity inside the passband and zero outside of it.

The spectral density of the noise signal is defined as

$$W_i = \frac{P_n}{B_i} \quad \text{W Hz}^{-1} \tag{2.49}$$

(Refer to Fig. 2.21a.). It is postulated now that the reference signal $u_1(t)$ is a sine wave. If a noise signal is added to $u_1(t)$, the zero crossings of the resulting signal are displaced back or forward depending on the instantaneous polarity of the noise signal. A so-called phase jitter is generated, designated $\theta_{n1}(t)$ (phase noise).

It can be shown[1] that the rms value of the phase jitter $\overline{\theta_{n1}^2}$ (or, more exactly, the square of the rms phase noise) is given by the simple expression

$$\overline{\theta_{n1}^2} = \frac{P_n}{2P_s} \tag{2.50}$$

We now define the SNR at the input of the PLL as

$$(\text{SNR})_i = \frac{P_s}{P_n} \tag{2.51}$$

For the phase jitter at the input of the PLL we get the simple relation

$$\overline{\theta_{n1}^2} = \frac{1}{2(\text{SNR})_i} \quad \text{rad}^2 \tag{2.52}$$

that is, the square of the rms value of the phase jitter is inversely proportional to the SNR at the input of the PLL. According to the theory of noise[1] the phase jitter $\theta_{n1}^2(t)$ can also be represented as a frequency spectrum. This spectrum reaches from zero to $B_i/2$ and is shown in Fig. 2.21b. The spectral density of the phase jitter is denoted as $\Theta_{n1}(j\omega)$. The square of the spectral density $\Theta_{n1}^2(j\omega)$ is then

$$\overline{\Theta_{n1}^2(j\omega)} = \Phi = \frac{\overline{\theta_{n1}^2}}{B_i/2} \quad \text{rad}^2 \text{ Hz}^{-1} \tag{2.53}$$

Because the frequency spectrum of phase noise $\Theta_{n1}(j\omega)$ is known at the input of the PLL, we are able to calculate the frequency spectrum at its output, too. The rms phase jitter at the output is defined as $\sqrt{\theta_{n2}^2(t)}$, its frequency spectrum as $\Theta_{n2}(j\omega)$. Using Eq. (2.12) we can write

$$\sqrt{\Theta_{n2}^2(j\omega)} = |H(j\omega)| \sqrt{\Theta_{n1}^2(j\omega)}$$

$$= |H(j\omega)| \sqrt{\Phi} \tag{2.54}$$

This relationship is demonstrated in Fig. 2.21. Fig. 2.21c shows the Bode diagram $|H(j\omega)|$, and Fig. 2.21d shows the output phase noise spectrum obtained by the multiplication of the curves shown in Fig. 2.21b and c.

   We are effectively looking for the rms value of the phase noise $\theta_{n2}^2(t)$ at the PLLs output. The phase noise is calculated by integrating $\Theta_{n2}(j\omega)$ over the bandwidth of the PLL

$$\overline{\theta_{n2}^2} = \int_0^\infty \overline{\Theta_{n2}^2} \, (j2\pi f) \, df \tag{2.55}$$

where $2\pi f = \omega$. $\overline{\theta_{n2}^2}$ is the area under the curve in Fig. 2.21d. Making use of Eqs. (2.54) and (2.55), we get

$$\overline{\theta_{n2}^2} = \int_0^\infty \Phi |H(j2\pi f)|^2 \, df$$

$$= \frac{\Phi}{2\pi} \int_0^\infty |H(j\omega)|^2 \, d\omega \tag{2.56}$$

The integral $\int_0^\infty |H(j2\pi f)|^2 \, df$ is called the *noise bandwidth* B_L,

$$B_L = \int_0^\infty |H(j2\pi f)|^2 \, df \tag{2.57}$$

The solution of this integral reads[1]

$$B_L = \frac{\omega_n}{2} \left(\zeta + \frac{1}{4\zeta} \right) \tag{2.58}$$

Thus B_L is proportional to the natural frequency ω_n of the PLL; furthermore, it depends on the damping factor ζ. Figure 2.22 is a plot of B_L/ω_n vs. ζ, and B_L has a minimum at $\zeta = 0.5$. In this case we have

$$B_L = B_{L \text{ min}} = \frac{\omega_n}{2} \tag{2.59}$$

   In Sec. 2.3 we showed that the transient response of the LPLL is best at $\zeta \approx 0.7$. Because the function $B_L(\zeta)$ is fairly flat in the neighborhood of $\zeta = 0.5$, the choice of $\zeta = 0.7$ does not noticeably worsen the noise performance. For $\zeta = 0.7$, B_L is $0.53\omega_n$ instead of the minimum value $0.5\omega_n$. For this reason $\zeta = 0.7$ is chosen for most applications. For the output phase jitter $\overline{\theta_{n2}^2(t)}$ we can now write

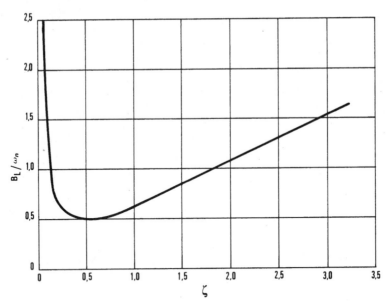

Figure 2.22 Noise bandwidth of a linear second-order PLL as a function of the damping factor ζ. (*Adapted from Gardner*[1] *with permission.*)

$$\overline{\theta_{n2}^2} = \Phi B_L \qquad (2.60)$$

where Φ is already known from Eq. (2.53). Combining Eqs. (2.50), (2.53), and (2.60), we obtain

$$\overline{\theta_{n2}^2} = \frac{P_n}{P_s} \cdot \frac{B_L}{B_i} \qquad (2.61)$$

We saw in Eq. (2.52) that the phase jitter at the input of the LPLL is inversely proportional to the $(SNR)_i$. By analogy we can also define a signal-to-noise ratio at the output, which will be denoted by $(SNR)_L$ (SNR of the loop). We define this analogy in Eq. (2.62),

$$\overline{\theta_{n2}^2} = \frac{1}{2(SNR)_L} \qquad \text{rad}^2 \qquad (2.62)$$

Comparing Eqs. (2.61) and (2.62), we get

$$(SNR)_L = \frac{P_s}{P_n} \cdot \frac{B_i}{2B_L} = (SNR)_i \cdot \frac{B_i}{2B_L} \qquad (2.63)$$

Equation (2.63) says that the PLL *improves* the SNR of the input signal by a factor of $B_i/2B_L$. The narrower the noise bandwidth B_L of the PLL, the greater the improvement.

All this sounds very theoretical. Let us therefore look at some numerical data. In radio and television the SNR is used to specify the quality of information transmission. For a stereo receiver a minimum SNR of 20 dB is considered a fair design goal.* The same holds true for PLLs. Practical experiments performed with second-order PLLs have demonstrated some very useful results.[1]

1. For $(SNR)_L = 1$ (0 dB), a lock-in process will not occur because the output phase noise $\sqrt{\theta_{n2}^2}$ is excessive.

2. At $(SNR)_L = 2$ (3 dB), lock-in is eventually possible.

3. For $(SNR)_L = 4$ (6 dB), stable operation is generally possible.

In quantitative terms according to Eq. (2.62), for $(SNR)_L = 4$ the output phase noise becomes

$$\overline{\theta_{n2}^2} = \frac{1}{2 \cdot 4} = \frac{1}{8} \quad \text{rad}^2$$

Hence the rms value $\sqrt{\overline{\theta_{n2}^2}}$ becomes

$$\sqrt{\overline{\theta_{n2}^2}} = \sqrt{\tfrac{1}{8}} = 0.353 \text{ rad} \ (\cong 20°)$$

Since the rms value of phase jitter is about 20°, the value of 180° (limit of dynamic stability) is rarely exceeded. Consequently the PLL does not unlock frequently. At $(SNR)_L = 1$, however, the effective value of output phase jitter would be as large as 40°, and the dynamic limit of stability of π would be exceeded on every major noise peak, thus making stable operation impossible.

As a rule of thumb,

$$(SNR)_L \leq 4 \ (\cong 6 \text{ dB}) \tag{2.64}$$

is a convenient design goal.

The designer of practical PLL circuits is vitally interested in how often on the average a system will temporarily unlock. The probability of unlocking is decreased with increasing $(SNR)_L$. We now define T_{av} to be the average time interval between two lock-outs. For example, if $T_{av} = 100$ ms, the PLL unlocks on an average 10 times per second.

---

*The SNR of a signal can be specified either numerically or in decibels. The $(SNR)_{dB}$ is calculated from

$$(SNR)_{dB} = 10 \log_{10} \frac{P_s}{P_n} = 10 \log_{10} \frac{U_s^2 \text{ (rms)}}{U_n^2 \text{ (rms)}} = 20 \log_{10} \frac{U_s \text{ (rms)}}{U_n \text{ (rms)}}$$

where U_s (rms) and U_n (rms) are the rms value of signal and noise, respectively.

For second-order PLLs T_{av} has been found experimentally as a function of $(SNR)_L$.[1] The resulting curve is plotted in Fig. 2.23.

To illustrate the theory, let us calculate a numerical example.

Numerical Example   A second-order PLL is assumed to have the following specifications:

$$f_n = 10 \text{ kHz}$$

$$\omega_n = 62.8 \times 10^3 \text{ s}^{-1}$$

$$(SNR)_L = 1.5$$

From Fig. 2.23 we read $\omega_n T_{av} = 200$. Consequently $T_{av} \approx 3$ ms. This means

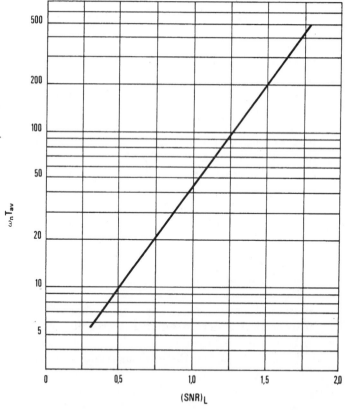

Figure 2.23   Average time T_{av} between two succeeding unlocking events plotted as a function of the $(SNR)_L$ at the output of the PLL. T_{av} is normalized to the natural frequency ω_n of the PLL. Note that this result is valid only for linear second-order PLLs. (*Adapted from Gardner*[1] *with permission.*)

that the PLL unlocks about 300 times per second, which looks quite bad. However, the lock-in time T_L is found from Eq. (2.44) to be $T_L \approx \dfrac{2\pi}{\omega_n} = 100 \ \mu s$; that is, the PLL nevertheless is locked for 97 percent of the total time.

A summary of noise theory. Although the noise theory of the PLL is difficult, for practical design purposes it suffices to remember some rules of thumb:

1. Stable operation of the LPLL is possible if $(\text{SNR})_L$ is approximately 4.

2. $(\text{SNR})_L$ is calculated from

$$(\text{SNR})_L = \frac{P_s}{P_n} \cdot \frac{B_i}{2B_L}$$

where P_s = signal power at the reference input
P_n = noise power at the reference point
B_i = bandwidth of the prefilter, if any; if no prefilter is used, B_i is the bandwidth of the signal source (e.g., antenna, repeater)
B_L = noise bandwidth of the PLL

3. The noise bandwidth B_L is a function of ω_n and ζ. For $\zeta = 0.7$, $B_L = 0.53\omega_n$.

4. The average time interval T_{av} between two unlocking events gets longer as $(\text{SNR})_L$ increases.

2.7.2 The lock-in process in the presence of noise

In an LPLL application, if noise is superimposed on the reference signal, conflicting demands could arise if the designer tries to specify the key parameters of the system, such as lock range $\Delta\omega_L$, noise bandwidth B_L, and so on. If it is used as a communication receiver, the PLL is required to lock onto an input signal which is known to be within the frequency range $\omega_0 - \Delta\omega$ to $\omega_0 + \Delta\omega$. In order to lock onto this signal, the lock range $\Delta\omega_L$ should be at least as large as $\Delta\omega$:

$$\Delta\omega_L \geq \Delta\omega$$

When a suitable value has been chosen for the lock range $\Delta\omega_L$, the designer will calculate the natural frequency ω_n using Eq. (2.43). Assuming a damping factor of $\zeta = 0.7$ we get

$$\Delta\omega_L = 1.4\omega_n$$

With ω_n known, the noise bandwidth B_L is known, too. According to Fig. 2.22, B_L becomes

$$B_L = 0.53\omega_n$$

The designer will now have to check whether or not this noise bandwidth yields a satisfactory $(\text{SNR})_L$. From the given data it will be possible for the designer to specify the signal power P_s, the noise power P_n, and the prefilter bandwidth B_i. (Refer to Sec. 2.7.1.) From these data $(\text{SNR})_L$ can then be calculated by using Eq. (2.63):

$$(\text{SNR})_L = \frac{P_s}{P_n} \cdot \frac{B_i}{2B_L}$$

If it turns out that $(\text{SNR})_L$ is at least approximately 4, we can be satisfied. If the reverse is true, however, we are in the dilemma of reducing the noise bandwidth B_L accordingly. But this also reduces ω_n, and consequently the lock range $\Delta\omega_L$, and thus the signal may possibly no longer be captured by the PLL. The contradiction seems insurmountable, but the problem can be circumvented by the special pull-in techniques to be discussed in the next section.

2.7.3 Pull-in techniques for noisy signals

In this section we summarize the most popular pull-in techniques. A special pull-in procedure becomes mandatory if the noise bandwidth of a PLL system must be made so narrow that the signal may possibly not be captured at all.

The sweep technique.   If this procedure is chosen, the noise bandwidth B_L is made so small that the SNR of the loop $(\text{SNR})_L$ is sufficiently large to provide stable operation. As a consequence the lock range $\Delta\omega_L$ might become smaller than the frequency interval $\Delta\omega$ within which the input signal is expected to be. To solve the locking problem, the center frequency of the VCO is swept by means of a sweeping signal u_{sweep} (see Fig. 2.24) over the frequency range of interest. Of course the sweep rate $\Delta\dot{\omega}$ must be held within the limits specified by Eq. (2.22b); otherwise the PLL could not become locked. (In Sec. 2.3 we considered the case where the center frequency ω_0 was constant and the reference frequency ω_1 was swept; in the case considered *here* the reference frequency is assumed to be constant, whereas the center frequency is swept. For the LPLL both situations are equivalent, because the frequency offset $\Delta\omega = \omega_1 - \omega_0$ is the only parameter of importance.)

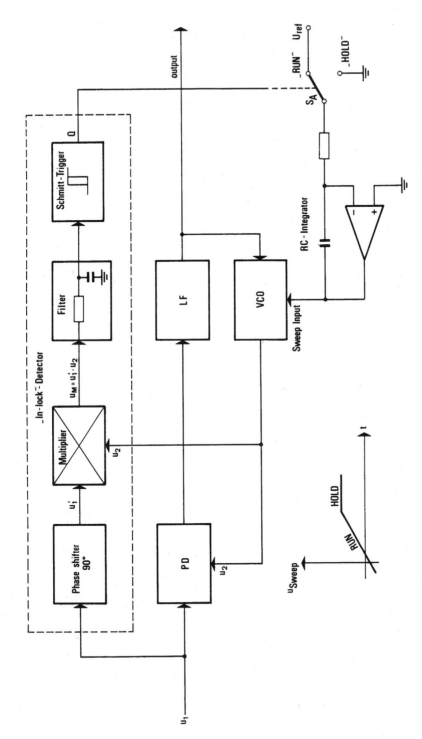

Figure 2.24 Simplified block diagram of a PLL using the sweep technique for the acquisition of signals buried in noise. (S_A = analog switch.)

As shown in the block diagram of Fig. 2.24, the center frequency of the VCO can be tuned by the signal applied to its sweep input. A linear sawtooth signal generated by a simple RC integrator is used as the sweep signal. Assume the LPLL has not yet locked. The integrator is in its RUN mode, and hence the sweep signal builds up in the positive direction. As soon as the frequency of the VCO approaches the frequency of the input signal, the PLL suddenly locks. The sweep signal should now be frozen at its present value (otherwise the VCO frequency would run away). This is realized by throwing the analog switch in Fig. 2.24 to the HOLD position. To control the analog switch we need a signal which tells us whether or not the PLL is in the locked state. Such a control signal is generated by the *in-lock detector* shown in Fig. 2.24. The in-lock detector is a cascade connection of a 90° phase shifter, an analog multiplier, a low-pass filter, and a Schmitt trigger. If the PLL is locked, there is a phase offset of approximately 90° between input signal u_1 and VCO output signal u_2. The phase shifter then outputs a signal u_1', which is nearly in phase with the VCO output signal u_2. The average value of the output signal of the multiplier $u_M = u_1' \cdot u_2$ is positive. If the PLL is not in the locked state, however, the signals u_1' and u_2 are uncorrelated, and the average value of u_M is zero. Thus the output signal of the multiplier is a clear indication of lock. To eliminate ac components and to inhibit false triggering, the u_M signal is conditioned by a low-pass filter. The filtered signal is applied to the input of the Schmitt trigger.

If the PLL has locked, the integrator is kept in the HOLD mode, as mentioned. Of course, an analog integrator would drift away after some time. To avoid this effect, an antidrift circuit has to be added; this is not shown, however, in Fig. 2.24.

The switched-filter technique. This locking method is depicted in Fig. 2.25. This configuration uses a loop filter whose bandwidth can be switched by a binary signal. The control signal for the switched filter is also derived from an in-lock detector, as shown in the previous example. In the unlocked state of the PLL, the output signal Q of the in-lock detector is zero. In this state the bandwidth of the loop filter is so large that the lock range exceeds the frequency range within which the input signal is expected. The noise bandwidth is then too large to enable stable operation of the loop. There is nevertheless a high probability that the PLL will lock spontaneously at some time. To avoid repeated unlocking of the loop, the filter bandwidth has to be reduced instantaneously to a value where the noise bandwidth B_L is small enough to provide stable operation. This is done by switching the loop filter to its low-bandwidth position by means of the Q signal.

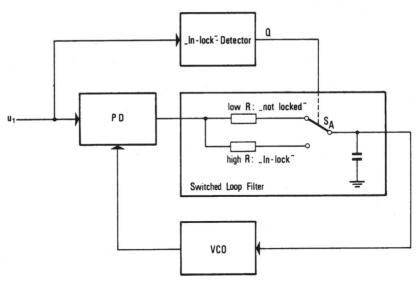

Figure 2.25 Simplified block diagram of a PLL using a switched loop filter for the acquisition of noisy signals. (S_A = analog switch.)

2.8   LPLL Design

When analyzing the dynamic performance of the LPLL we defined a number of key parameters and developed formulas for them. These formulas are summarized in Table 2.1; they are the basis of every LPLL design. In the following we give a stepwise procedure for designing an LPLL.

When designing an LPLL, an integrated circuit (IC) is used in most applications. Table 6.1 lists the presently available PLL ICs with a short description of each type. All hardware is normally included in these ICs, but some external components have to be added such as resistors and capacitors. Many LPLL ICs contain an additional operational amplifier which could be used to build an active loop filter, for example.

Having selected a suitable IC, the designer must specify the remaining components. An LPLL design usually starts with the specification of the most important key parameters such as center frequency ω_0, lock range $\Delta\omega_L$, damping factor ζ, and the required frequency range of the application. Given these parameters, the remaining parameters are calculated using the formulas listed in Table 2.1. A straightforward design procedure follows. The procedure is also shown graphically in Fig. 2.31.

Step 1. Specify the center (angular) frequency ω_0. In some cases ω_0 remains constant. In other applications the LPLL is required to lock onto different signal channels, so you then must specify a range for

TABLE 2.1 Summary of Parameters and Formulas for Linear PLLs

Parameter category	Symbol	Parameter	Definition	Formulas for second-order PLLs Type of loop filter		
				Passive lag	Active lag	Active PI
General	ω_0	Center frequency of the VCO	Angular frequency of the VCO at $u_f = 0$			
	τ_1, τ_2	Time constants of loop filter				
	ω_n	Natural frequency of the PLL	ω_n is the natural frequency of the PLL system. The PLL responds to an excitation at its input with a transient, normally a damped oscillation with angular frequency ω_n	$\omega_n = \sqrt{\dfrac{K_0 K_d}{\tau_1 + \tau_2}}$	$\omega_n = \sqrt{\dfrac{K_0 K_d K_a}{\tau_1}}$	$\omega_n = \sqrt{\dfrac{K_0 K_d}{\tau_1}}$
	ζ	Damping factor	$1/\omega_n\zeta$ = time constant of the damped oscillation	$\zeta = \dfrac{\omega_n}{2}\left(\tau_2 + \dfrac{1}{K_0 K_d}\right)$	$\zeta = \dfrac{\omega_n}{2}\left(\tau_2 + \dfrac{1}{K_0 K_d K_a}\right)$	$\zeta = \dfrac{\omega_n \tau_2}{2}$
Acquisition	$\Delta\omega_H$	Hold range	Frequency range within which PLL operation can be statically stable	$\Delta\omega_H = K_0 K_d$	$\Delta\omega_H = K_0 K_d K_a$	$\Delta\omega_H = \infty$
	$\Delta\omega_L$	Lock range	If the frequency offset of the reference signal is smaller than the lock range, the PLL locks within one single-beat note between reference and output frequencies	$\Delta\omega_L \approx 2\zeta\omega_n$ (all types)		
	T_L	Lock-in time	Time required for the lock-in process	$T_L \approx \dfrac{2\pi}{\omega_n}$ (all types)		

				Low-gain loops	Low-gain loops	All loops
	$\Delta\omega_P$	Pull-in range	If the frequency offset of the reference signal is larger than the lock range but smaller than the pull-in range, the PLL will slowly lock after a number of beat notes between reference and output frequencies	$\Delta\omega_P = \dfrac{4}{\pi}\sqrt{2\zeta\omega_n K_0 K_d - \omega_n^2}$ High-gain loops $\Delta\omega_P = \dfrac{4\sqrt{2}}{\pi}\sqrt{\zeta\omega_n K_0 K_d}$	$\Delta\omega_P = \dfrac{4}{\pi}\sqrt{2\zeta\omega_n K_0 K_d - \dfrac{\omega_n^2}{K_a}}$ High-gain loops $\Delta\omega_P = \dfrac{4\sqrt{2}}{\pi}\sqrt{\zeta\omega_n K_0 K_d}$	$\Delta\omega_P = \infty$
	T_P	Pull-in time	Time required for a pull-in process	All loops $T_P = \dfrac{\pi^2}{16}\dfrac{\Delta\omega_0^2}{\zeta\omega_n^3}$	All loops $T_P = \dfrac{\pi^2}{16}\dfrac{\Delta\omega_0^2 K_a}{\zeta\omega_n^3}$	All loops $T_P = \dfrac{\pi^2}{16}\dfrac{\Delta\omega_0^2}{\zeta\omega_n^3}$
Tracking	$\Delta\omega_{PO}$	Pull-out range	Dynamic limit of stable operation of the PLL. The system unlocks if a frequency step larger than $\Delta\omega_{PO}$ is applied to the reference input	$\Delta\omega_{PO} \approx 1.8\omega_n(\zeta + 1)$		
	$\Delta\dot\omega$	Rate of change of frequency offset	Maximum allowable rate of change of (angular) reference frequency	$\Delta\dot\omega < \omega_n^2$ (see text)		
Noise	$P_s,\ P_n$	Signal, noise power	Power of input signal and noise signal applied to the input of a PLL			
	B_i	Prefilter bandwidth	Bandwidth of the prefilter (or the input signal source)			
	B_L	Noise bandwidth		$B_L \approx \dfrac{\omega_n}{2}\left(\zeta + \dfrac{1}{4\zeta}\right)$		
	$(\mathrm{SNR})_i$	Signal-to-noise ratio of the input signal		$\mathrm{SNR}_i = \dfrac{P_s}{P_n}$		
	$(\mathrm{SNR})_L$	Signal-to-noise ratio of the loop		$\mathrm{SNR}_L = \mathrm{SNR}_i\,\dfrac{B_i}{2B_L}$		

the center frequency. Let ω_{0min} be the minimum and ω_{0max} be the maximum value of the center frequency.

Step 2. Specify the damping factor ζ. ζ should be chosen to obtain fast response of the LPLL onto phase or frequency steps applied to its reference input, small overshoot, and minimum noise bandwidth B_L. If ζ is chosen very small (e.g., 0.1), large overshoot will result. If it is chosen very large (e.g., 5), the response will become sluggish. Selecting $\zeta = 0.7$ is a good compromise in most applications.

Step 3. Specify the lock range $\Delta\omega_L$ or the noise bandwidth B_L. In step 7 we will have to calculate the natural frequency ω_n of the LPLL. As shown in Eq. (2.43), ω_n depends directly on $\Delta\omega_L$ and ζ which has been chosen in step 2. On the other hand, ω_n can also be calculated from the noise bandwidth B_L and ζ, Eq. (2.58). This signifies that $\Delta\omega_L$ and B_L cannot be specified independently. In cases where noise is not of concern, we may discard noise bandwidth and directly specify the lock range $\Delta\omega_L$. In those cases, however, where noise must be considered, we should first specify B_L such that locking of the LPLL is guaranteed. Depending on whether or not noise is important, we differentiate two cases:

   Case 1. Noise can be neglected. The lock range $\Delta\omega_L$ is immediately specified. It is not so easy to give a general rule for the specification of this parameter. If the LPLL is expected to lock onto a carrier frequency of say 300 kHz and this carrier could drift by ± 3 kHz, for example, the lock range Δf_L should be chosen larger than 3 kHz.

   Case 2. Noise must be considered. In the second case the signal-to-noise ratio $(SNR)_i$ at the input of the LPLL must be known or at least estimated. Furthermore the noise bandwidth B_i at the input of the LPLL should be determined [see Eq. (2.63)]. Now, using Eq. (2.63), the loop noise bandwidth B_L must be chosen such that the signal-to-noise ratio $(SNR)_L$ at the output of the LPLL is at least 4, which enables safe locking in the presence of noise.

Step 4. Specify the frequency range of the LPLL. Let ω_{2min} be the minimum and ω_{2max} be the maximum frequency of the VCO output signal. To be able to lock within the given range of center frequencies, we must guarantee that

$$\omega_{2min} < \omega_{0min} - \Delta\omega_L$$

and

$$\omega_{2max} > \omega_{0max} + \Delta\omega_L$$

It is practical to choose

$$\omega_{2min} = \omega_{0min} - 1.5\,\Delta\omega_L$$

and

$$\omega_{2max} = \omega_{0max} + 1.5\,\Delta\omega_L$$

Step 5.   Given the frequency range the parameters and components of the VCO can be specified now. First the supply voltage(s) of the VCO must be determined (or chosen from the data sheet of the IC used). In most cases the IC will be powered from a unipolar +5-V supply. The control signal u_f is usually limited to a range which is smaller than the supply voltage(s). Let u_{fmin} and u_{fmax} be the minimum and maximum value allowed for u_f, respectively. Now you can draw the transfer characteristics of the VCO (Fig. 2.26). The VCO is required to generate the frequency ω_{2min} when $u_f = u_{fmin}$ and the frequency ω_{2max} when $u_f = u_{fmax}$. U_B is the supply voltage, which is assumed to be unipolar in this example. Now determine the angular frequency at $u_f = U_B/2$. From now on this frequency is considered to be the center frequency ω_0 of the LPLL, irrespective of the fact that the center frequency could be varying. All data of the VCO are known now, and the data sheet of the selected circuit will tell you how to specify its external components. A numerical example will be discussed in Sec. 2.9. From Fig. 2.26 the VCO gain K_0 can be calculated:

$$K_0 = \frac{\omega_{2max} - \omega_{2min}}{u_{fmax} - u_{fmin}}$$

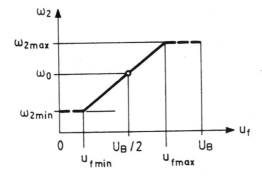

Figure 2.26 Characteristic curve of the VCO. Output frequency ω_2 is plotted vs. loop filter output signal u_f.

Step 6. Determining phase detector gain K_d. As noted in Sec. 2.3, the phase detector gain depends on the level of the input signal (see also Fig. 2.6). The data sheet will tell you how K_d depends on the signal level; in most cases it will show a plot of K_d vs. signal level as shown by Fig. 2.6. Fix the signal level and determine the actual phase detector gain K_d. If the signal level can be varied, choose a large level to get a large value of K_d. If K_d becomes too small, there is danger that the desired lock range $\Delta\omega_L$ cannot be realized.

Step 7. Determine the natural frequency ω_n. Check step 3 again and see whether you specified the lock range $\Delta\omega_L$ directly (case 1) or had to determine noise bandwidth B_L first (case 2). Then proceed as follows.

Case 1. Lock range $\Delta\omega_L$ has been specified in step 3. Using Eq. (2.43), calculate the natural frequency from

$$\omega_n = \frac{\Delta\omega_L}{2\zeta}$$

Case 2. Noise bandwidth has been specified in step 3. Using Eq. (2.58), calculate the natural frequency from

$$\omega_n = \frac{2B_L}{\zeta + 1/4\zeta}$$

Step 8. Selecting the type of loop filter: Different loop filters can be used (see Fig. 2.1). Equations (2.15) are used now to calculate the time constants τ_1 and τ_2 of the loop filter. When the active lag filter is used, an additional parameter K_a has to be specified. The procedure depends on the filter type chosen.

Case 1. Passive lag filter requires use of the second equation of Eqs. (2.15a) to get the time constant τ_2. Use the first equation of Eqs. (2.15a) to get τ_1. Normally τ_1 will become considerably greater than τ_2, typically 5 to 10 times. Should τ_1 become smaller than τ_2, try another type of loop filter.

Case 2. Active lag filter requires use of Eq. (2.15b) to determine τ_1, τ_2, and K_a. This set of equations is overdetermined, because we have to calculate three unknowns from two equations, hence, one of the unknowns can be chosen arbitrarily. It is recommended to start with $K_a = 1$ and to calculate the two time constants from Eqs. (2.15b). Ideally, τ_1 should be about 5 to 10 times larger than τ_2. If this is not the case, select a larger K_a (e.g., $K_a = 2$) and repeat the calculation until τ_1 is markedly larger than τ_2.

Case 3. Active PI filter requires use of Eqs. (2.15c) to calculate the time constants τ_1 and τ_2. Because this filter has a pole at $s = 0$, the relationship between τ_1 and τ_2 is not of concern. Don't worry if τ_1 is not larger than τ_2.

Step 9. Determine the values of R_1, R_2, and C of the loop filter (Fig. 2.1). Use the equations

$$\tau_1 = R_1\,C$$

$$\tau_2 = R_2\,C$$

to get the values of R_1 and R_2. Because these are again three equations for two unknowns, C can be chosen arbitrarily. Select C such that the values of R_1 and R_2 are "reasonable," i.e., that R_1 is in the range of say 3 to 300 kΩ. If R_1 becomes too small, the filter presents a heavy load to the phase detector. If R_1 becomes too large, leakage currents could produce adverse effects (charging or discharging the capacitor and the like).

2.9 LPLL Applications

Most applications of LPLLs are in the field of communications. The very first application was described as early as 1932 by de Bellescize[22] and dealt with "coherent communication." De Bellescize tried to improve the noise performance of AM receivers—the frequency modulation was invented much later. His vacuum tube LPLL circuit locked onto the carrier of an AM signal and generated an auxiliary carrier which was exactly in phase with the received carrier. The auxiliary carrier was then used to synchronously demodulate the AM signal. A synchronous demodulator has by far better noise suppression capabilities than the peak rectifier, which has been exclusively used in these times.

The PLL became popular, however, only in the age of television. In a TV receiver, one PLL is used for horizontal synchronization. Another PLL serves for vertical synchronization, and still another reconstructs the color subcarrier.[2] The LPLL found widespread application in the field of satellite communications. Here the LPLL's capabilities of extracting a signal from noise have proved very successful.

Because it would take too much space to cover all these applications in detail, we concentrate on one typical LPLL application. We consider a multichannel telemetry system, where one single voice-grade communication line is used to transmit a number of signal channels. This example will serve as a case study; i.e., we apply the design procedure presented in Sec. 2.8. Figure 2.27 shows the block diagram of the te-

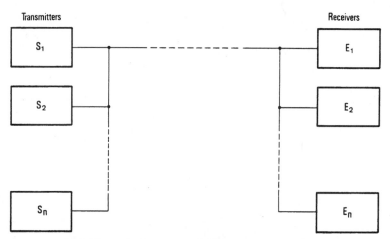

Figure 2.27 Multichannel telemetry with PLLs operating as FM transmitters and FM receivers.

lemetry system. Each transmitter has to transmit a binary signal which has a baud rate of 50 bits/s. This signal is encoded in the so-called NRZ format (non-return to zero); i.e., a logical 1 is represented as a "high" level and a logical 0 as a "low" level. It is known that the bandwidth of an NRZ signal is half the baud rate,[20] which is 25 Hz in our case. Each transmitter generates a different carrier frequency; the carrier signal of each transmitter is frequency-modulated by the corresponding binary signal. The spectrum of the FM-modulated carrier consists of the carrier frequency and a number of sidebands, which are displaced by ± 25 Hz, $\pm 2 \cdot 25$ Hz, ... from the carrier. We assume furthermore that "narrow-band" FM is used[20]; in this case only the first pair of sidebands has appreciable power. Thus every transmitter produces a spectrum consisting of the carrier plus an upper and lower sideband; the bandwidth of each transmitter signal is therefore 50 Hz. The communication line is an ordinary telephone cable; it is capable of transmitting frequencies in the range from 300 to 3000 Hz. Because each receiver of the system in Fig. 2.27 must be able to pick up its corresponding signal, the spectra of the transmitter signals must not overlap. For this reason, the individual carriers are given a channel spacing of 60 Hz (Fig. 2.28). The overall bandwidth of the transmission line is $B_i = 3000 - 300 = 2700$ Hz. Of course, B_i is the noise bandwidth of the LPLL system. The maximum number of telemetry channels is obtained by dividing overall bandwidth by channel spacing, i.e.,

$$\text{max number of channels} = \frac{B_i}{60} = 45$$

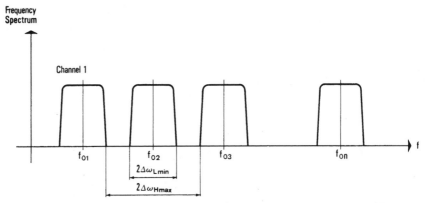

Figure 2.28 Frequency spectrum of the multichannel FM telemetry system of Fig. 2.27.

For each receiver an LPLL is used. We now design one of these receivers, using the step-by-step procedure described in Sec. 2.8. It is assumed that the carrier frequency of this receiver is 1000 Hz. We decide furthermore that the passive lag filter will be used as a loop filter.

Step 1. Determine center (angular) frequency ω_0 of the LPLL: $\omega_0 = 2 \pi\ 1000 = 6280\ \text{s}^{-1}$.

Step 2. Choose damping factor ζ: For optimum LPLL performance, we specify $\zeta = 0.7$.

Step 3. Specify noise bandnwidth B_L of the loop: In this application noise cannot be neglected, so we have to specify noise bandwidth B_L and not lock range $\Delta\omega_L$. The signal-to-noise ratio $(\text{SNR})_i$ at the input of the LPLL is given by

$$(\text{SNR})_i = \frac{P_s}{P_n}$$

Because the other 44 channels of the telemetry system represent noise for our particular channel, we have $P_n = 44\ P_s$; i.e., we get

$$(\text{SNR})_i = \frac{1}{44} \approx 0.023$$

To enable locking onto the carrier, the signal-to-noise ratio of the loop $(\text{SNR})_L$ should be 4. Thus we can calculate the required noise bandwidth B_L from

$$B_L = \frac{(\text{SNR})_i}{(\text{SNR})_L} \frac{B_i}{2} = \frac{0.023 \cdot (3000 - 300)}{4 \cdot 2} = 7.67 \text{ Hz}$$

Step 4. Determine the frequency range of the LPLL. Because the noise bandwidth B_L is very small, the lock range will also become very narrow. We will see in step 7 that the lock range will be about $\Delta\omega_L \approx 20 \text{ s}^{-1}$. Therefore, the VCO is not required to generate a large range of frequencies. The final frequency range will be fixed in step 5 only.

Step 5. Determine the parameters and components of the VCO. The data sheet of the XR-215 shows a schematic diagram of the VCO including its external components (Fig. 2.30). Two resistors and one capacitor must be specified. The data sheet gives two design equations

$$f_0 = \frac{200}{C_0}\left(1 + \frac{0.6}{R_x}\right)$$

$$K_0 = \frac{700}{C_0 R_0}$$

In these equations the resistors must be specified in kΩ and the capacitors in μF. Choosing $C_0 = 0.27 \ \mu$F and $R_x = 1.71$ kΩ gives the required center frequency of $f_0 = 1000$ Hz. From the second equation the VCO gain K_0 must be determined. The data sheet specifies that resistor R_0 should be in the range 1 . . . 10 kΩ. Thus K_0 can be in the range 260 . . . 2600 rad s^{-1} V^{-1} in this example. We choose $R_0 = 10$ kΩ; hence we get $K_0 = 260$ rad s^{-1} V^{-1}. In other words, the VCO will change its frequency by $260/2\pi = 41.4$ Hz for a variation of the phase detector output signal of 1 V. The frequency range of the VCO is therefore large enough to enable locking within the lock range of $\Delta\omega_L = 20 \text{ s}^{-1}$ (see step 7).

Step 6. Determine phase detector gain K_d. The data sheet shows a curve which plots K_d vs. signal level U_1 (rms); see also Fig. 2.6. In our application the level of the input signal is 3 mV (rms). Hence we get $K_d = 0.2$ rad V^{-1}.

Step 7. Calculate natural frequency ω_n. ω_n is calculated from B_L and ζ and becomes $\omega_n = 14.5 \text{ s}^{-1}$. Now the lock range $\Delta\omega_L$ can also be determined from $\Delta\omega_L = 2\zeta\omega_n = 20.3 \text{ s}^{-1}$.

Step 8. Determine the time constants τ_1 and τ_2. Using the formulas given in step 8 of Sec. 2.8 we get $\tau_1 = 170$ ms and $\tau_2 = 80$ ms.

Step 9. Determine R_1, R_2, and C. The data sheet of the XR-215 tells us that resistor R_1 is already integrated on the chip ($R_1 = 6$ kΩ); refer also to the schematic of the LPLL receiver in Fig. 2.29. Furthermore

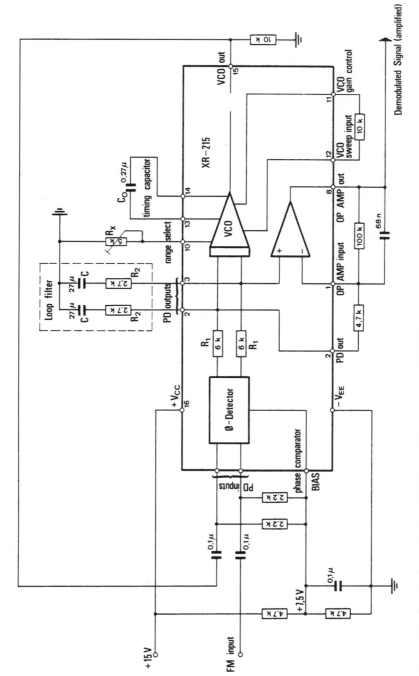

Figure 2.29 Practical circuit of an FSK demodulator using a PLL IC.

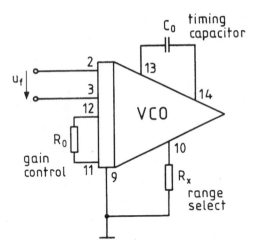

Figure 2.30 Simplified block diagram of the VCO used in the integrated circuit XR-215. See text for details.

we see that the output of the phase detector is differential, so we have to implement the passive lag filter twice. We first calculate C and get

$$C = \frac{\tau_1}{R_1} = \frac{0.170}{6000} = 28.3 \ \mu\text{F} \rightarrow 27 \ \mu\text{F}$$

R_2 then becomes

$$R_2 = \frac{\tau_2}{C} = \frac{0.08}{28.3} \cdot 10^{-6} = 2.82 \ \text{k}\Omega \rightarrow 2.7 \ \text{k}\Omega$$

The design is completed now. As can be seen from Fig. 2.29, a trimmer has been chosen for R_x to enable exact tuning of the center frequency.

2.10   Simulating the LPLL on the PC

After much work with the difficult theory of the LPLL, it is time to relax. Let us do some simulations with the enclosed disk and see whether the simulated LPLL does what theory says. As mentioned in the preface the book includes a CD-ROM with a Windows application that simulates various types of PLL's.

Program Installation   Insert the CD-ROM. If you have a plug-and-play CD program installation starts automatically. Follow the instructions on screen. If your CD drive is not plug-and-play, start D:\SETUP.EXE from Windows Explorer or from My Computer.

Simulating an LPLL on the PC.   You are ready now to start a simulation. After program start-up, a main window is displayed which has a menu bar with four topics: *CONFIG, SIMULATION, OPTIONS,* and *HELP.*

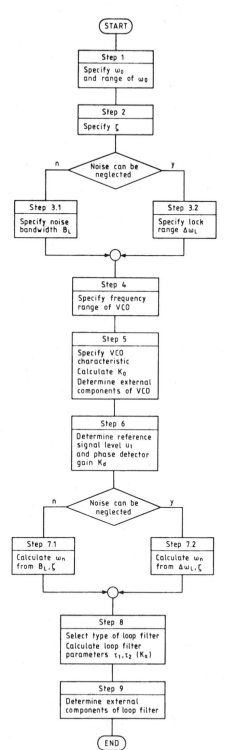

Figure 2.31 Flowchart.

CONFIG is used to set up the parameters of the PLL. With *SIMU-LATION* we will perform PLL simulations. *OPTIONS* enables you to choose individual colors, fonts, line widths, etc., to customize the appearance of the graphics. Finally, *HELP* provides a choice of help texts on the types of PLLs which can be simulated.

On start-up, a logo will show up, defining the most important variables used in the simulation. Click on the *Continue* button at bottom right to delete the logo. To see what is in the *HELP* menu, click on the *Help* topic. This brings up one single menu item named *General Info*. When *General Info* is clicked, a standard Windows help file is displayed (WinHelp). This is very much the same help structure as used in many popular programs, such as the WinWord or WordPerfect text editors, Corel Draw, Power Point, Harvard Graphics, Excel, Access, and many others. The first page of the help file shows the contents. When dragging the mouse over the titles of the content page, the mouse arrow changes to a hand. Clicking at a title immediately displays the corresponding page. There are two browse buttons ($<<$ and $>>$) on top of the help window. These buttons allow you to browse forward and backward through the pages of the help file. If the *Search* button is clicked, you can search for keywords. Instead of calling up help by clicking at *Help* and then at *General Info,* you can use an accelerator key (CTL G) to display the help file directly. There are accelerator keys for all other menu items. The accelerators are listed on the right side of the pop-up menus.

To start a PLL simulation, the PLL system must be configured first. Note that the program sets initial default values for all parameters. It is recommended that you use these defaults in the first simulation. Whenever you specify a different value for a parameter, this overwrites the default. When the program is terminated, the actual parameters are saved on a configuration file named PARAMS.DAT. When the program is started next, the parameters from the configuration file are loaded. Occasionally it may become desirable to delete the PARAMS.DAT file and restart the program from scratch. This procedure is recommended if somebody entered so many "stupid" parameter values that the simulation only provides "garbage," or if unsuitable colors, fonts, or pen widths have been entered.

To set up a particular configuration, select the *CONFIG* menu. Clicking at *Config* displays a pop-up menu having two items, *Params* and *StartLogo*. The latter can be used to redisplay the start logo. When *Params* is clicked, a dialog box (Fig. 2.32) for *PLL Type Selection* appears. It shows three radio buttons labelled *LPLL, DPLL* and *ADPLL,* respectively. To simulate an LPLL, press the *LPLL* button and then the *OK* button (note that the LPLL is already selected by default). Pressing *Cancel* aborts the setup dialog. When you exited the

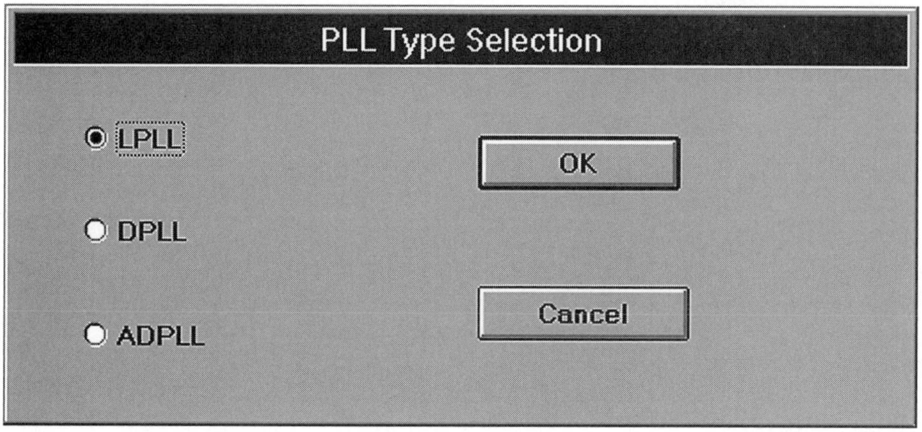

Figure 2.32 Dialog box for the specification of the PLL type.

dialog box with *OK*, a dialog box entitled *LPLL Circuit Selection* appears, Fig. 2.33. In this dialog you have to select the desired type of loop filter (passive lag, active lag, or PI) and the desired type of oscillator (VCO or VCO + Scaler). Normally you would select VCO. When using option *VCO + Scaler*, a divide-by-N counter is inserted at the output of the VCO, which scales down the frequency of the VCO output

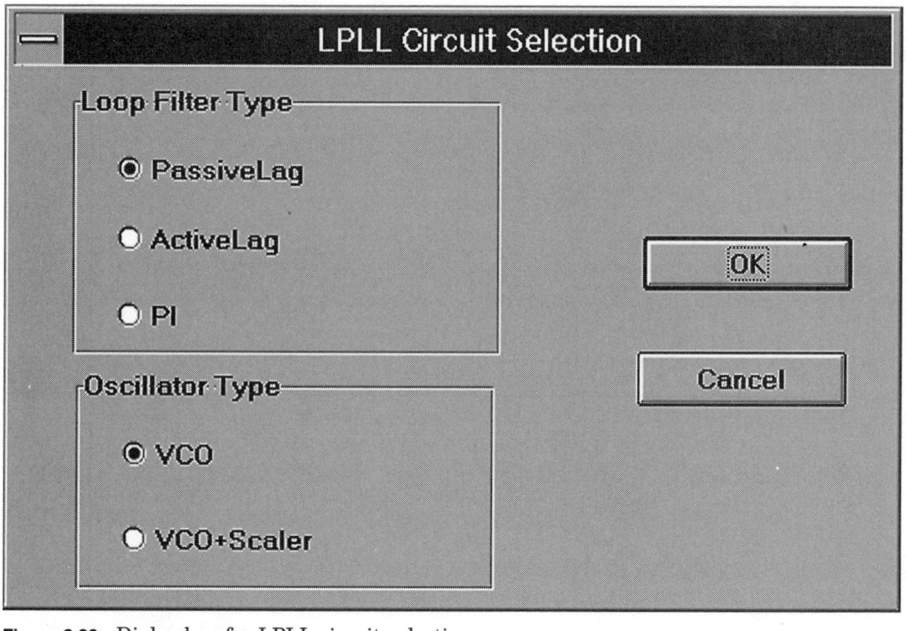

Figure 2.33 Dialog box for LPLL circuit selection.

signal by a factor N. This feature will be used mainly in frequency-synthesizer applications, i.e., in combination with DPLLs, and is discussed in Chap. 3. Use the default settings (passive loop filter and VCO without scaler) and press the *OK* button. (The *Cancel* button aborts the configuration dialog.) The next dialog box *PLL Parameter Selection* is presented, Fig. 2.34. On the left are four group boxes labeled *PLL Supply Voltages, Phase Detector, Loop Filter,* and *Oscillator,* respectively. In the box *PLL Supply Voltages,* the rail voltages have to be specified. For most devices used today, the positive supply voltage is 5 V, and the negative is 0 V (ground). Some PLLs, however, have symmetrical supply voltages, e.g., 7.5 V and -7.5 V. In the *Phase Detector* box, phase detector gain K_d (in V/rad) and the saturation voltages of the phase detector have to be entered. For a given type of LPLL, K_d is usually specified on its data sheet. Note that for LPLLs the phase detector gain depends on the input signal level. Most data

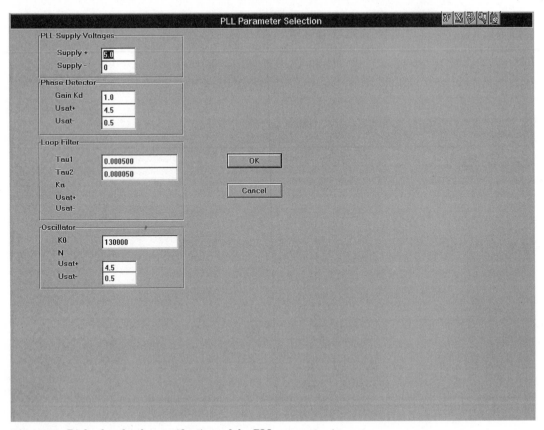

Figure 2.34 Dialog box for the specifications of the PLL parameters.

sheets specify therefore phase detector gain versus input level, mostly in the form of a curve. Moreover, in most phase detector circuits, the output signal u_d does not reach up to the rail voltages but saturates perhaps at 4.5 V for positive signal swings and at 0.5 V for "negative" swings. In the group box *Loop Filter* the parameters of the loop filter must be entered. When a passive loop filter has been chosen (as shown in Fig. 2.34), only the time constants τ_1 and τ_2 have to be entered. In the example shown, $\tau_1 = 500$ μs and $\tau_2 = 50$ μs. Note that all entries have to be done in international units, i.e., times in seconds, voltages in V, frequencies in Hz, etc. The lower three edit windows are blanked here. If you had selected an active lag filter, you would have to specify in addition its DC gain K_a and its saturation voltages Usat+ and Usat−. Finally, in the group box *Oscillator* the VCO gain K_0 and the saturation voltages must be specified. Use the default parameter values and press the *OK* button.

Having entered (or checked) the parameters, exit the dialog box with the *OK* button. (Again, the *Cancel* button would abort the configuration.) On exiting the dialog, a message box (Fig. 2.35) displays the computed key parameters of the LPLL such as natural (angular) frequency ω_n and damping factor ζ. Having chosen the default parameters, the message box indicates $\omega_n = 15{,}374$ s^{-1} and $\zeta = 0.443$. The message box also gives some hints on how to make these parameters larger or smaller. If you accept your choice, press the *Yes* button in the message box. This stops the configuration. When the *No* button is pressed, the parameter selection is resumed.

You are ready now to perform the simulation. (Actually, you could have started the simulation immediately, because the default parameter values had already been loaded.) To start the simulation, click on the *SIMULATION* menu. This brings up a pop-up menu consisting of one item only (*Run*). Pressing *Run* displays a dialog box as shown in Fig. 2.36. When the program runs the first time, default values are loaded for all parameters. It is recommended again to use these defaults in the very first simulation. The frame window at the left is empty at start. The edit windows and buttons at right are used to control the simulation. Topmost are two radio buttons *f-Step* and *Ph-Step*. If *f-Step* is checked (clicked with the left mouse button), the lock-in process for a frequency step applied to the input signal is simulated. If *Ph-Step* is checked, however, a phase step is simulated. Under the radio buttons are five edit windows labeled *f0* (center frequency of the PLL in Hz), *df* (size of frequency step applied to input, in Hz), *dphi* (size of phase step applied to input, in degrees), *nSamp* (number of data samples calculated in one cycle of the input signal), and *T* (the duration of the simulation in seconds). The first three parameters are self-explanatory. Parameter *nSamp* will be discussed shortly; the du-

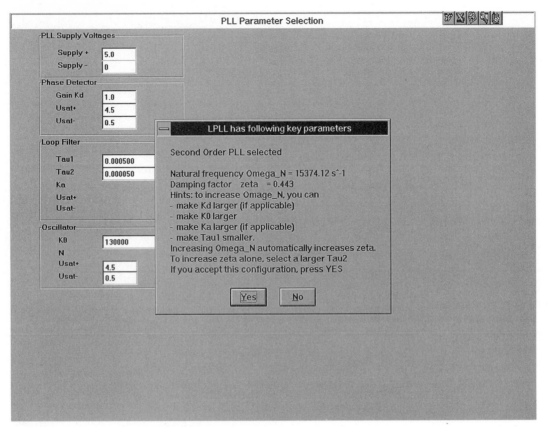

Figure 2.35 Having completed the parameter specification (cf. Fig. 2.34), a message box displays the most important key parameters of the PLL.

ration T of the simulation is precomputed as soon as the operator has specified the parameters of the PLL. It is calculated to show the "complete" transient of the lock-in process. Roughly, the duration T corresponds to two oscillations of the (hopefully) damped oscillation when the PLL pulls in. In cases where the transient's duration is longer, T must be made larger.

Let us discuss the meaning of $nSamp$ now. In the coarse of a simulation the program calculates the signals $u_d(t)$ and $u_f(t)$. These signals are continuous functions of time, but a digital computer can only calculate discrete-time signals. Hence a suitable sampling frequency f_s must be chosen. Because we are going to sample the signals $u_d(t)$ and $u_f(t)$, the sampling frequency must be at least twice the highest frequency component existing in these signals; this is a consequence of the famous sampling theorem, also called the "Nyquist theorem" or the "Shannon theorem." If the frequency of the reference signal is f_1 and the frequency of the output signal f_2, then the spectrum of $u_d(t)$

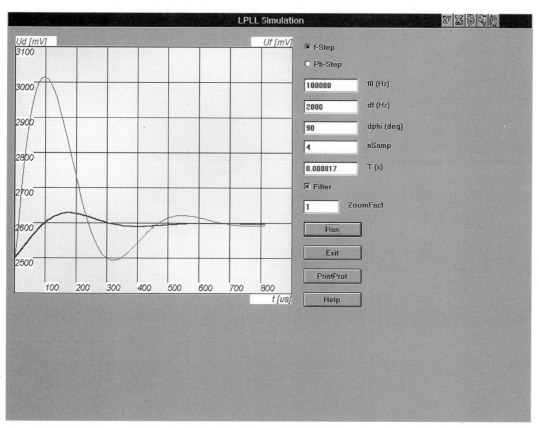

Figure 2.36 Dialog box for LPLL simulation. The frame window displays the simulated lock process.

shows up lines at $f_1 - f_2$ and $f_1 + f_2$. In addition, the spectrum contains higher-order harmonics, which are discarded here for the moment. When the LPLL operates near the center frequency f_0, the first line is nearly at f_0, whereas the other is nearly twice the center frequency ($2 f_0$). To avoid "aliasing," the sampling frequency f_s should be at least $4 f_0$. The parameter *nSamp* is defined as the ratio of sampling frequency to center frequency, i.e., $nSamp = f_s/f_0$. Default for this parameter is 4. Choosing the default causes the program to compute four values for $u_d(t)$ and $u_f(t)$ per cycle of the reference signal. Thus the high-frequency component in the u_d signal becomes clearly visible. Because the amplitude of the $f_1 + f_2$ component is usually quite large, this can "smear" the display of u_d all over the screen. To avoid that, the program offers a filter option. The filter is enabled when the checkbox *Filter* is checked. If the filter is used, the average of four u_d and u_f values is computed in each reference cycle, and only the averaged values are displayed instead of four data pairs. This greatly improves the appearance of the signals, but the user should be aware that the averaged signals look much "nicer" than they would be in reality.

When the user wants to see how the signals look really, she or he can work without the filter, simply by unchecking the *Filter* checkbox. Sampling the $f_1 + f_2$ component with four times the center frequency displays four points of u_d within one reference cycle, which looks a little bit crude. To enhance the resolution, the ratio of sampling to reference can be increased up to 64. When a ratio greater than 4 is selected, no filtering is performed, because sampling at higher rates only makes sense when the waveform of u_d and u_f should be displayed with greater resolution.

Below the checkbox there is an edit window labeled *ZoomFact* (zoom factor). By default, *ZoomFact* is set at 1. A *ZoomFact* greater than 1 is used to scroll the time axis. If the duration is $T = 0.001$ (1 ms) and the *ZoomFact* is 10, e.g., the time axis is expanded by a factor of 10. At the start of the simulation, the time axis ranges from 0 to 100 μs. Whenever *ZoomFact* is >1, a slider is displayed under the frame window. Moving the slider scrolls the time axis right or left, so the user can look at waveform details on an expanded time scale. There are different ways to move the slider. You can either click on the arrows to the left or to the right of the slider. This moves the slider by about 1 percent of full scale. Holding down the left mouse button on one of the arrows continually moves the slider. The slider moves faster if you click in the area between the thumb and one of the arrows. Alternatively, you could position the cursor on the thumb and drag the thumb while holding down the left mouse button.

In our first simulations we will use a *ZoomFact* of 1. The slider is hidden in this case. Let us first simulate the LPLL's response onto a frequency step applied to its input. When using the default parameters, their values will be as given below:

Supply voltages:	
Positive supply	5 V
Negative supply	0 V
Phase detector:	
Phase detector gain	1 V/rad
Positive saturation level	4.5 V
Negative saturation level	0.5 V
Loop filter (passive lag):	
τ_1	500 μs
τ_2	50 μs
Oscillator (without scaler):	
VCO gain K_0	130,000 s^{-1}V^{-1}
Positive saturation level	4.5 V
Negative saturation level	0.5 V

Choosing $K_0 = 130{,}000$ signifies that the frequency created by the VCO changes by about 20 kHz if u_f changes by 1 V.

In the *Simulation* dialog, the default parameters are as follows:

f-Step radio button	Checked
f_0	100,000 Hz
df	2000 Hz
dphi	Any value (not used)
nSamp	4
Filter checkbox	Checked
T	0.000817
ZoomFact	1

Hitting the *Run* button performs the simulation and displays the waveforms for u_d and u_f in the frame window, Fig. 2.36. On the computer screen, the curves are plotted with different colors. The y-axis labels u_d and u_f are displayed with the same colors, so the user easily sees which curve is u_d and which is u_f. Because the figures are printed in black and white in this book, a larger line width has been used for u_f. The simulation shows the transient of a typical second-order linear control system, as can be expected from theory. Note that there are three other buttons on the right of the screen, *PrintProt, Help,* and *Exit*. When the *Exit* button is pressed, the simulation is stopped. Using *PrintProt,* you can make a hard copy of this simulation on the line printer. Clicking at the *PrintProt* starts a common Windows printer dialog. If you have more than one printer available, press the *Select* button in the printer dialog to set up the desired printer. In this dialog a number of printer features can be specified in addition, such as page format (Portrait or Landscape), printer resolution (in dots/inch) and more. When making a hard copy, the printer will paint the frame window plus a number of text lines that list time and date of the simulation and the most important parameters of the simulation. Any printer for which a Windows driver is available can be used. Best results are obtained with color printers, of course.

The appearance of the screen can be customized for your personnel flavors. To set up different colors, line widths, fonts, etc., exit the simulation and click at the *OPTIONS* menu. Two menu items are displayed now, *Color + Pen* and *Set Default*. Click at *Color + Pen* to alter the setup. A dialog box is displayed. On the left top, a box is shown which lists the items that can be changed:

AxisTickLabels

Curve1

Curve2

FrameBackGround

Grid

When *AxisTickLabels* is selected (highlighted), you can change the font and the font color of the tick labels. Standard Windows dialogs are used throughout. For the curves (*Curve1* and *Curve2*), you can change color and pen width. *FrameBackGround* enables you to choose another background color for the frame window. With *Grid,* the color of the grid can be customized. When using Windows color dialogs, you can either select one of the predetermined colors in the palette or mix up a *custom color*. For details, refer to *Windows User's Guide*. The *Set Default* item of the *OPTIONS* menu restores initial values for all colors, pens, fonts, etc.

The *Simulation* dialog box has its own *Help* facility. Hitting the *Help* button provides extensive information on all the controls and parameters that are used in the *Simulation* dialog.

Let us continue the simulation of LPLLs now. When configuring the LPLL (menu *CONFIG*), the key parameters dialog box told us that our LPLL would have natural frequency ω_n = 15,374 s^{-1} and damping factor ζ = 0.443. Using the formulas in Table 2-1 we can compute the remaining key parameters of the LPLL:

Lock range:
  $\Delta\omega_L$ = 13,621 s^{-1}
  Δf_L = 2169 Hz
Pull-out range:
  $\Delta\omega_{PO}$ = 39,932 s^{-1}
  Δf_{PO} = 6358 Hz
Pull-in range:
  The ratio $\omega_n/(K_0\,K_d)$ is 0.12, hence the loop can be considered
  a high-gain loop. The formula for high-gain loops yields
    $\Delta\omega_P$ = 53,597 s^{-1}
    Δf_P = 8534 Hz
Hold range:
  $\Delta\omega_H$ = 130,000 s^{-1}
  Δf_H = 20,700 Hz

In the next simulations we will verify these predictions. The first key parameter calculated above is the lock range $\Delta\omega_L$. Although this is the most straightforward key parameter of a PLL, it is extremely difficult to measure it! Our program assumes that the PLL has been locked at the start of the simulation. To measure the lock range we must force the PLL to get unlocked first and to get locked again. This will be demonstrated later in this session. A parameter which can easily be measured is the pull-out range, however. To get the pull-out range, we apply frequency steps at the reference input and make the steps successively larger, until the PLL gets unlocked. Let us try that first. Applying a frequency step of $\Delta f_1 = 2$ kHz gave the result of Fig. 2.36. As stated, this looks very much like the response of a linear second-order system, but concluding that our LPLL is a linear system would be premature. We repeat the simulation with df set to 4 kHz now and get a picture as shown in Fig. 2.37. If the LPLL were a linear system, the amplitudes of u_d and u_f would have been doubled. This did not occur, however, but the u_d became flat-topped. From Sec. 2.3 we know that the output signal of the phase detector is not proportional to the phase error θ_e, but to $\sin(\theta_e)$. Obviously the phase error is approaching $\pi/2$ in this experiment. This nonlinearity becomes even more evident when we increase the frequency step to 5000 Hz, Fig. 2.38. The u_d waveform shows a dip in the first positive half-wave. The phase error exceeded $\pi/2$ in this simulation. This corresponds to the case where the pendulum (Fig. 2.13) in our mechanical analogy made a deflection larger than 90° but without tipping over. When we increase the frequency step to 5700 Hz, the picture of Fig. 2.39 is obtained. Now the dip has become deeper. In the mechanical analogy this corresponds to the case where the pendulum almost reached its culmination point without tipping over. Increasing the frequency step to 5800 Hz finally pulls out the LPLL, Fig. 2.40. In the analogy the pendulum swung over the culmination point, tipped over, and came down on the left side of the cylinder. The pull-out range of this simulation must therefore be slightly less than $\Delta f_{PO} = 5800$ Hz. The predicted value was 6358 Hz, thus the estimation error is less than 10 percent, which is fairly good for a simulation. Because the frequency step applied to the LPLL is smaller than the pull-in range, the LPLL becomes locked again after short time. Increasing the frequency step to 7000 Hz shows the picture of Fig. 2.41. The LPLL is pulled out again, but now the pull-in process takes much more time than in the previous simulation. The pull-in process becomes even slower when we increase the frequency step to 8000 Hz, Fig. 2.42. Finally, with a frequency step of 9000 Hz, the LPLL is no longer able to pull in (Fig. 2.43). Some further simulation showed that the LPLL still pulls in for $df = 8400$ Hz but not for $df = 8500$ Hz, so the actual pull-in range is near 8500 Hz. This comes very close to the predicted value of 8534 Hz.

When applying such large frequency steps to the reference input, we happened to bring the LPLL out of lock. We remember that the lock range has been defined to be the frequency offset which causes

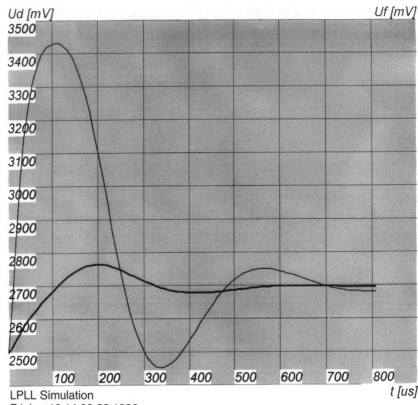

Figure 2.37 Response of the LPLL to a frequency step of 4 kHz applied to its reference input.

the PLL to get locked within one cycle of the u_d signal (or within one revolution of the pendulum in the mechanical analogy). Hence we can try to estimate the lock range from a simulation where the PLL first locked out and became locked again. This happened to occur when we applied a frequency step of df = 8000 Hz, Fig. 2.42. During a time interval of about 2.5 ms the signal u_f slowly "pumped" up. When it reached about 2.8 V, the PLL became locked within one oscillation of the u_d signal; u_f settled at about 2.9 V at the end of the process. Now we simply compute the frequency difference which corresponds to this variation of the signal u_f. A difference of 0.1 V corresponds to a fre-

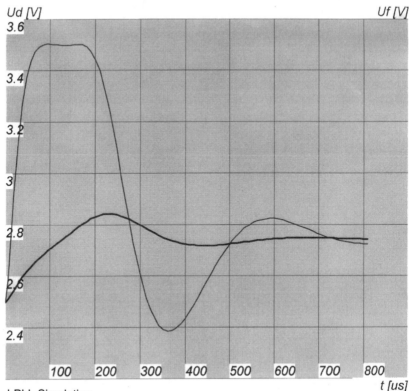

Figure 2.38 Response of the LPLL to a frequency step of 5 kHz applied to its reference input.

quency change of 2 kHz, thus the lock range Δf_L is about 2 kHz, which agrees well with the predicted value of 2118 Hz. We must admit, however, that this estimation is very crude, because we cannot see clearly where the pull-in process stops and the lock-in process starts.

In the last simulations we calculated one pair of u_d, u_f values in each cycle of the input signal. Now we will see what happens when more values are calculated. We perform a simulation having the parameters shown in Fig. 2.36 but uncheck the *Filter* checkbox and choose a *ZoomFact* of 8. For *nSamp* we enter 32, i.e., 32 samples are computed in each cycle of the input signal. The result of the simulation

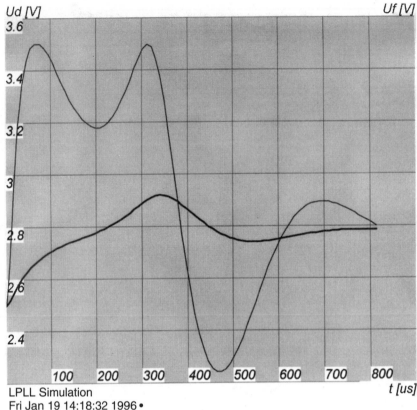

Figure 2.39 Response of the LPLL to a frequency step of 5.7 kHz applied to its reference input.

is shown in Fig. 2.44. We clearly recognize that the u_d signal has a frequency component around 200 kHz now. The same holds true for the u_f signal, but here the amplitude is attenuated by the loop filter. By dragging the thumb of the slider, we could scroll the curves on the time axis.

Suggestions for other case studies. Familiar with the simulation program, now try to simulate other kinds of LPLLs. Here are some suggestions:

Case study 1: Influence of loop filter type. Use the active lag instead of the passive, specify a dc gain K_a greater than 1, and see how the key parameters such as pull-out range and pull-in range are influ-

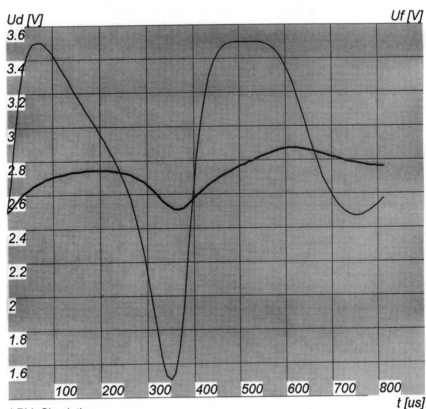

LPLL Simulation
Fri Jan 19 14:19:22 1996•
PD = Multiplier
LF = passive lag
OSC = VCO
Center frequency = 100000 Hz
Response on frequency step 5800 Hz

Figure 2.40 Response of the LPLL to a frequency step of 5.8 kHz applied to its reference input. The PLL pulls out first but locks in again.

enced. Use the active PI filter then. What is the major difference compared with other filter types? What happens to the pull-in range?

Case study 2: Phase step. Apply a phase step to the reference input. Vary the size of the phase step.

Case study 3: VCO with divide-by-N counter. Simulate an LPLL whose output frequency is N times larger than the reference frequency. Check how the damping factor ζ depends on the scaling factor N.

Case study 4: Pull-in processes. Using different types of loop filters, measure the pull-in time of the LPLL under various conditions (different values for ζ, K_0, K_d, K_a, different size of frequency step Δf_1

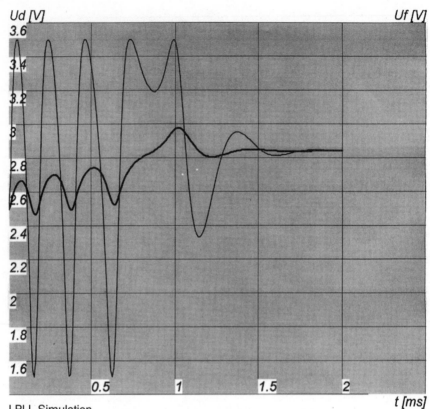

Figure 2.41 Response of the LPLL to a frequency step of 7 kHz applied to its reference input. The PLL pulls out first but pulls in again. The pull-in process shown here is markedly slower than the lock process in Fig. 2.40.

applied to the reference input). Compare the results of the simulation with the values predicted by theory; use the formulas in Table 2-1 to compute pull-in time T_P. Where can you find the largest deviations between theory and practice?

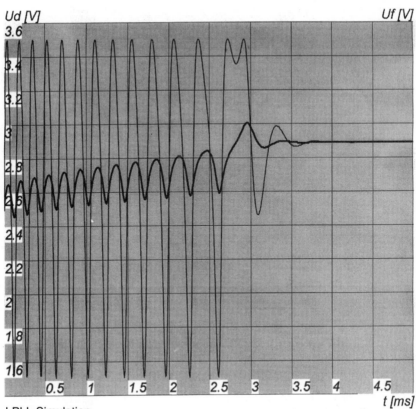

LPLL Simulation
Fri Jan 19 14:21:12 1996 •
PD = Multiplier
LF = passive lag
OSC = VCO
Center frequency = 100000 Hz
Response on frequency step 8000 Hz

Figure 2.42 Same as Fig. 2.40, but frequency step is 8 kHz, and the pull-in process takes even more time.

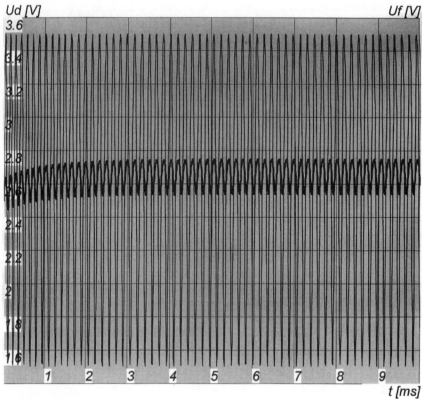

LPLL Simulation
Fri Jan 19 14:22:00 1996 •
PD = Multiplier
LF = passive lag
OSC = VCO
Center frequency = 100000 Hz
Response on frequency step 9000 Hz

Figure 2.43 When the frequency step is increased to 9 kHz, pull-in is no longer possible. The system locks out definitely.

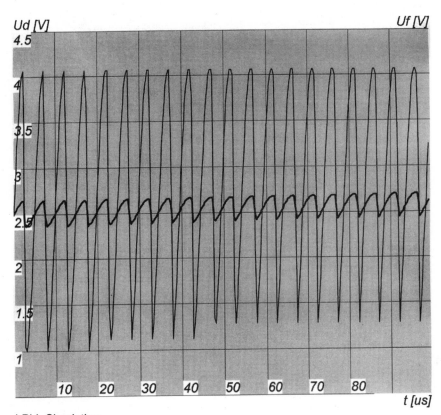

LPLL Simulation
Fri Jan 19 14:35:44 1996 •
PD = Multiplier
LF = passive lag
OSC = VCO
Center frequency = 100000 Hz
Response on frequency step 2000 Hz

Figure 2.44 Zoomed display of waveforms. The response of the LPLL onto a frequency step of 2 kHz is simulated. 32 samples for u_d and u_f are computed in each cycle of the input signal. One clearly recognizes a 200 kHz oscillation on the signals. The display is zoomed (*ZoomFact* = 8).

3

The Classical Digital PLL (DPLL)

As mentioned in Sec. 1.2, the classical DPLL is actually a hybrid system built from analog and digital function blocks. The only part of the DPLL that is really digital is the phase detector. In many aspects the DPLL performs similarly to the LPLL, so some parts of the LPLL theory can be adopted; in some particular aspects, however, DPLL behavior is completely different. Ironically, it will show up that the classical DPLL has more in common with the LPLL than with the all-digital PLL (ADPLL). Consequently we will not treat the ADPLL as a descendant object of the DPLL, but rather discuss it in a separate chapter (Chap. 4).

3.1 Building Blocks of the DPLL

The block diagram of the DPLL is shown in Fig. 3.1. Like the LPLL, it consists of the three known function blocks *phase detector*, *loop filter*, and *voltage-controlled oscillator*. In many DPLL applications (e.g.,

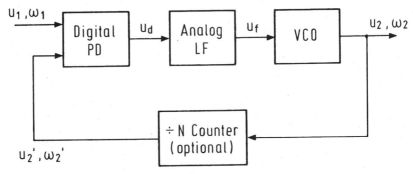

Figure 3.1 Block diagram of the DPLL.

PLL frequency synthesizers) a *divide-by-N counter* is inserted between VCO and phase detector. When such a counter is used, the VCO generates a frequency which is N times the reference frequency. The loop filters used in DPLLs are the same as those already discussed in Chap. 2; this also holds true for the VCO. For the phase detector, a number of different logical circuits can be used; the three most important of these are shown in Fig. 3.2:

- The EXOR gate (Fig. 3.2*a*)
- The (edge-triggered) JK-flipflop (Fig. 3.2*b*)
- The "phase-frequency detector" (PFD) (Fig. 3.2*c*)

The PFD turns out to be the workhorse of the DPLL because it offers a virtually unlimited pull-in range, which guarantees PLL acquisition under even the worst operating conditions. The operation of the EXOR phase detector is most similar to that of the linear multiplier and is discussed first. The signals in DPLLs are always binary signals, i.e., square waves. We assume for the moment that both signals u_1 and u_2' are symmetrical square waves. Figure 3.3 depicts the waveforms of the EXOR phase detector for different phase errors θ_e. At zero phase error the signals u_1 and u_2' are out of phase by exactly 90°, as shown in Fig. 3.3*a*. Then the output signal u_d is a square wave whose frequency is twice the reference frequency; the duty cycle of the u_d signal is exactly 50 percent. Because the high-frequency component of this signal will be filtered out by the loop filter, we consider only the average value of u_d, as shown by the dashed line in Fig. 3.3*a*. The average value $\overline{u_d}$ is the arithmetic mean of the two logical levels; if the EXOR is powered from an asymmetrical 5-V power supply, $\overline{u_d}$ will be approximately 2.5 V. This voltage level is considered the quiescent point of the EXOR and will be denoted as $\overline{u_d} = 0$ from now on. When the output signal u_2' lags the reference signal u_1, the phase error θ_e becomes positive by definition; this case is shown in Fig. 3.3*b*. Now the duty cycle of u_d becomes larger than 50 percent, i.e., the average value of u_d is considered positive, as shown by the dashed line in the u_d waveform. Clearly, the mean of u_d reaches its maximum value for a phase error of $\theta_e = 90°$ and its minimum value for $\theta_e = -90°$. If we plot the mean of u_d vs. phase error θ_e, we get the characteristic shown in Fig. 3.4*a*. Whereas the output signal of the four-quadrant multiplier varied with the sine of phase error, the average output $\overline{u_d}$ of the EXOR is a triangular function of phase error. Within a phase error range of $-\pi/2 < \theta_e < \pi/2$, u_d is exactly proportional to θ_e and can be written as

$$\overline{u_d} = K_d \theta_e \qquad (3.1)$$

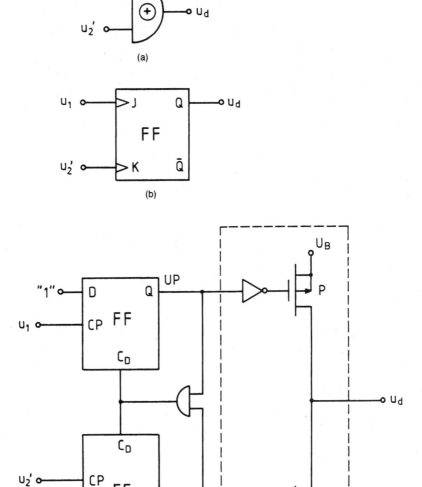

Figure 3.2 The most often used phase detectors in DPLLs. (a) EXOR gate. (b) JK-flipflop. (c) Phase-frequency detector (PFD).

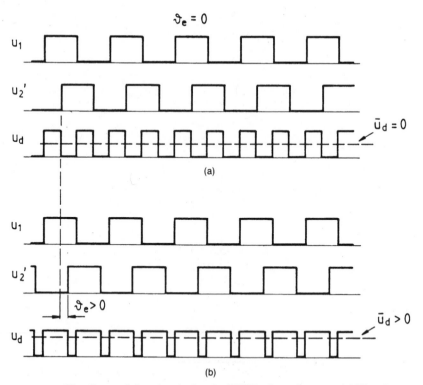

Figure 3.3 Waveforms of the signals for the EXOR phase detector. (*a*) Waveforms at zero phase error ($\theta_e = 0$). (*b*) Waveforms at positive phase error ($\theta_e > 0$).

In case of the EXOR phase detector, the phase detector gain K_d is constant. When the supply voltages of the EXOR are U_B and 0, respectively, and when we assume that the logic levels are U_B and 0, K_d is given by

$$K_d = \frac{U_B}{\pi}$$

When the output signal of the EXOR does not reach the supply rails but rather saturates at some higher level U_{sat+} (in the high state) and some lower level U_{sat-} (in the low state), K_d must be calculated from

$$K_d = \frac{U_{sat+} - U_{sat-}}{\pi}$$

Like the four-quadrant multiplier, the EXOR phase detector can maintain phase tracking when the phase error is confined to the range

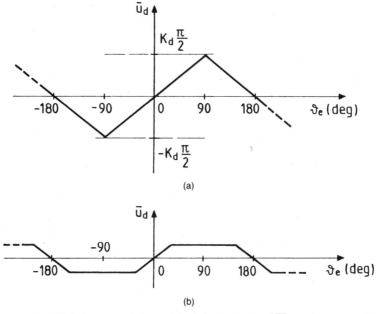

Figure 3.4 Plot of averaged phase detector output signal $\overline{u_d}$ vs. phase error θ_e. (*a*) Normal case: waveforms u_1 and u_2' in Fig. 3.3 are symmetrical square waves. (*b*) Waveforms u_1 and u_2' are asymmetrical. The characteristic of the phase detector is clipped.

$$\frac{-\pi}{2} < \theta_e < \frac{\pi}{2}$$

The performance of the EXOR phase detector becomes severely impaired if the signals u_1 and u_2' become asymmetrical. If this happens, the output signal $\overline{u_d}$ gets clipped at some intermediate level, as shown by Fig. 3.4*b*. This reduces the loop gain of the DPLL and results in smaller lock range, pull-out range, etc.

Waveform symmetry is unimportant, however, when the JK-flipflop is used as phase detector (Fig. 3.2*b*). This JK-flipflop differs from conventional JK-flipflops, because it is edge-triggered. A positive edge appearing at the J input triggers the flipflop into its "high" state ($Q = 1$), a positive edge at the K input into its "low" state ($Q = 0$). Figure 3.5*a* shows the waveforms of the JK-flipflop phase detector for the case $\theta_e = 0$. With no phase error, u_1 and u_2' have opposite phase. The output signal u_d then represents a symmetrical square wave whose frequency is identical with the reference frequency (and not twice the reference frequency). This condition is considered as $\overline{u_d}$ being zero. If the phase error becomes positive (Fig. 3.5*b*), the duty cycle of the u_d signal becomes greater than 50 percent, i.e., $\overline{u_d}$ becomes positive. Clearly, $\overline{u_d}$

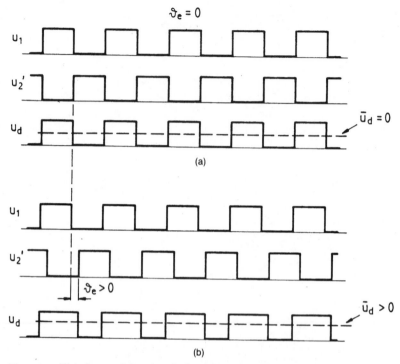

Figure 3.5 Waveforms of the signals for the JK-flipflop phase detector. (*a*) Waveforms at zero phase error. (*b*) Waveforms at positive phase error.

becomes maximum when the phase error reaches 180° and minimum when the phase error is −180°. If the mean value of u_d is plotted vs. phase error θ_e, the sawtooth characteristic of Fig. 3.6 is obtained. Within a phase error range of $-\pi < \theta_e < \pi$ the average signal u_d is proportional to θ_e and is given by

$$\overline{u_d} = K_d\theta_e$$

Obviously the JK-flipflop phase detector is able to maintain phase tracking for phase errors within the range

$$-\pi < \theta_e < \pi$$

By an analogous consideration, the phase detector gain of the JK-flipflop phase detector is given by

$$K_d = \frac{U_B}{2\pi}$$

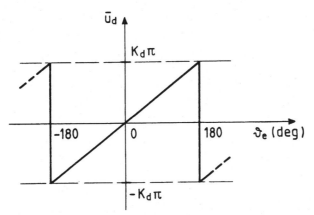

Figure 3.6 Plot of averaged phase detector output signal $\overline{u_d}$ vs. phase error θ_e. In contrast to the EXOR, $\overline{u_d}$ does not depend on the duty cycle of the signals.

when the logic levels are U_B or 0, respectively. If, however, these levels are limited by saturation, phase detector gain must be computed from

$$K_d = \frac{U_{\text{sat}+} - U_{\text{sat}-}}{2\pi}$$

In contrast to the EXOR gate, the symmetry of the u_1 and u_2' signals is irrelevant, because the state of the JK-flipflop is altered only by the positive transitions of these signals. In all other aspects, the EXOR and the JK-flipflop behave very much the same. Both of these circuits show the shortcomings observed with the four-quadrant multiplier: If the loop filter used within the DPLL does not contain an integrating term, the pull-in range stays severely limited for the reasons explained in Sec. 2.6.3.

The PFD differs greatly from the phase detector types discussed hitherto. As its name implies, its output signal depends not only on phase error θ_e but also on frequency error $\Delta\omega = \omega_1 - \omega_2'$, when the DPLL has not yet acquired lock. Figure 3.2c shows the schematic diagram of the PFD. It is built from two D-flipflops, whose outputs are denoted "UP" and "DN" (down), respectively. The PFD can be in one of four states:

- UP = 0, DN = 0
- UP = 1, DN = 0
- UP = 0, DN = 1
- UP = 1, DN = 1

The fourth state is inhibited, however, by an additional AND gate. Whenever both flipflops are in the 1 state, a logic "high" level appears at their C_D ("clear direct") inputs which resets both flipflops. Consequently the device acts as a tristable device ("triflop"). We assign the symbols -1, 0, and 1 to these three states:

- DN = 1, UP = 0 $\longrightarrow$ state = -1
- UP = 0, DN = 0 $\longrightarrow$ state = 0
- UP = 1, DN = 0 $\longrightarrow$ state = $+1$

The actual state of the PFD is determined by the positive-going transients of the signals u_1 and u_2', as explained by the state diagram in Fig. 3.7. (In this example we assumed that the PFD acts on the positive edges of these signals exclusively; we could have reversed the definition by saying that the PFD acts on the negative transitions only.) As Fig. 3.7 shows, a positive transition of u_1 forces the PFD to go into its next higher state, unless it is already in the $+1$ state. In analogy, a positive edge of u_2' forces the PFD into its next lower state, unless it is already in the -1 state. The output signal u_d is a logical function of the PFD state. When the PFD is in the $+1$ state, u_d must be positive; when it is in the -1 state, u_d must be negative, and when it is in the 0 state, u_d must be zero. Theoretically, u_d is a *ternary signal*. Most logical circuits used today generate *binary signals,* however, but the third state ($u_d = 0$) can be substituted by a "high-impedance" state. The circuitry within the dashed box of Fig. 3.2c shows how the u_d signal is generated. When the UP signal is high, the P-channel MOS transistor conducts, so u_d equals the positive supply voltage U_B. When

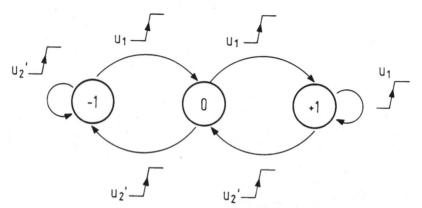

Figure 3.7 State diagram for the phase-frequency detector (PFD). This drawing shows the events causing the PFD to change its current state.

the DN signal is high, the N-channel MOS transistor conducts, so u_d is on ground potential. If neither signal is high, both MOS transistors are off, and the output signal floats, i.e., is in the high-impedance state. Consequently, the output signal u_d represents a tristate signal.

To see how the PFD works in a real DPLL system, we consider the waveforms in Fig. 3.8. Figure 3.8a shows the (rather theoretical) case where the phase error is zero. It is assumed that the PFD has been in the 0 state initially. The signals u_1 and u_2' are "exactly" in phase here; both positive edges of u_1 and u_2' occur "at the same time"; hence their effects will cancel. The PFD then will stay in the 0 state forever. Figure 3.8b shows the case where u_1 leads u_2'. The PFD now toggles between the states 0 and +1. If u_1 lags u_2' as shown in Fig. 3.8c, the PFD toggles between states −1 and 0. It is easily seen from the waveforms in Fig. 3.8b and c that u_d becomes largest when the phase error is positive and approaches 360° (Fig. 3.8b) and smallest when the phase error is negative and approaches −360° (Fig. 3.8c). If we plot the average u_d signal vs. phase error θ_e, we get a sawtooth function as shown in Fig. 3.9. Figure 3.9 also shows the average detector output signal for phase errors greater than 2π or smaller than -2π. When the phase error θ_e exceeds 2π, the PFD behaves as if the phase error recycled at zero; hence the characteristic curve of the PFD becomes periodic with period 2π. An analogous consideration can be made for phase errors smaller than -2π. When the phase error is restricted to the range $-2\pi < \theta_e < 2\pi$, the average of u_d becomes

$$\overline{u_d} = K_d \theta_e$$

In analogy to the JK-flipflop, phase detector gain is computed by

$$K_d = \frac{U_B}{4\pi}$$

when the logic levels are U_B or 0, respectively. If, however, these levels are limited by saturation, phase detector gain must be computed from

$$K_d = \frac{U_{\text{sat}+} - U_{\text{sat}-}}{4\pi}$$

A comparison of the PFD characteristic (Fig. 3.9) with the characteristic of the JK-flipflop (Fig. 3.6) does not yet reveal exciting properties. To recognize the bonus offered by the PFD, we must assume that the DPLL is unlocked initially. Furthermore we make the assumption that the reference frequency ω_1 is higher that the output frequency ω_2'. The u_1 signal then generates more positive transitions per unit of time than the signal u_2'. Looking at Fig. 3.7, we see that the PFD can toggle

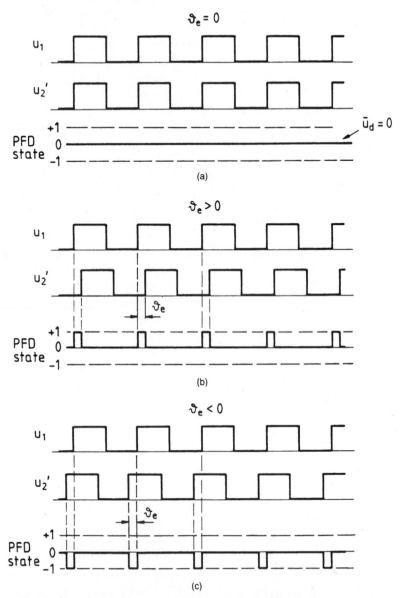

Figure 3.8 Waveforms of the signals for the PFD. (*a*) Waveforms for zero phase error. The output signal of the PFD is permanently in the 0 state (high-impedance). (*b*) Waveforms for positive phase error. The PFD output signal is pulsed to the +1 state. (*c*) Waveforms for negative phase error. The PFD output signal is pulsed to the −1 state.

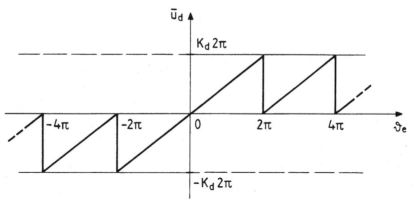

Figure 3.9 Plot of the averaged PFD output signal $\overline{u_d}$ vs. phase error θ_e. $\overline{u_d}$ does not depend on the duty cycle of u_1 and u_2'.

only between the states 0 and $+1$ under this condition but will never go into the -1 state. If ω_1 is much higher than ω_2' furthermore, the PFD will be in the $+1$ state most of the time. When ω_1 is smaller than ω_2' however, the PFD will toggle between the states -1 and 0. When ω_1 is much lower than ω_2', the PFD will be in the -1 state most of the time. We conclude therefore that the average output signal u_d of the PFD varies monotonically with the *frequency error* $\Delta\omega = \omega_1 - \omega_2'$, when the DPLL is out of lock. This leads to the term *phase-frequency detector*. It is possible to calculate the duty cycle of the u_d signal as a function of the frequency ratio ω_1/ω_2;[23] the result of this analysis is shown in Fig. 3.10. For the case $\omega_1 > \omega_2$ the duty cycle δ is defined as the average fraction of time the PFD is in the $+1$ state; for $\omega_1 < \omega_2$, δ is by definition *minus* the average fraction of time the PFD is in the -1 state. As expected, δ approaches -1 when $\omega_1 \ll \omega_2$ and $+1$ when $\omega_1 \gg \omega_2$. Furthermore, δ is nearly 0.5 when ω_1 is greater than ω_2 but both frequencies are close together, and δ is nearly -0.5 when ω_1 is lower than ω_2 but both frequencies are close together. This property will greatly simplify the determination of the pull-in range (Sec. 3.2.3). It must be emphasized that no such characteristic (Fig. 3.10) can be defined for the EXOR and for the JK-flipflop. Because the output signal $\overline{u_d}$ of the PFD depends on phase error in the locked state of the DPLL and on frequency error in the unlocked state, a DPLL which uses the PFD will lock under any condition, irrespective of the type of loop filter used. For this reason the PFD is the preferred phase detector in DPLLs.

All components of the DPLL being known, we can start discussing its dynamic properties.

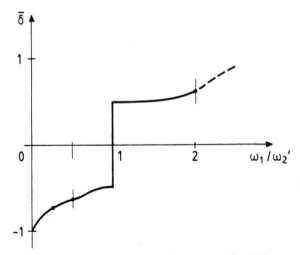

Figure 3.10 Plot of the averaged duty cycle of the PFD output signal $\overline{u_d}$ vs. frequency ratio ω_1/ω_2' (symbols defined in Fig. 3.1). This curve depicts the behavior of the PFD in the unlocked state of the DPLL.

3.2 Dynamic Performance of the DPLL

In Chap. 2 dynamic performance of LPLLs was analyzed in great detail. Behavior in the locked and unlocked states was treated in separate sections (Sec. 2.3 and 2.5, respectively). Because DPLLs perform similarly in many aspects, this discussion can be made more compact, so we treat both locked and unlocked states in this section. When the DPLL has acquired lock and is not pulled out by large phase steps, frequency steps, or phase noise applied to its reference input, its performance can be analyzed by a linear model, as has been done for the LPLL. Figure 3.11 shows the mathematical model of the DPLL in the locked state; it differs from the LPLL model (Fig. 2.7) only by the

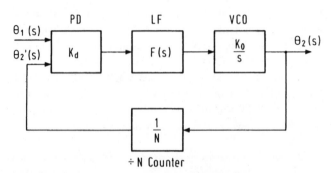

Figure 3.11 Mathematical model of the DPLL. This model is used to analyze tracking performance of the DPLL in the locked state.

divide-by-N counter. Consequently, very similar expressions can be derived for the phase-transfer function $H(s)$ and the error-transfer function $H_e(s)$ of the DPLL. In contrast to the LPLL, the special behavior of the loop filter must be taken into account when driven from the PFD phase detector. Assume for the moment that the loop filter is a passive lag filter (Fig. 2.2a). When it is driven from a conventional signal source, its transfer function $F(s)$ is given by

$$F(s) = \frac{1 + s\tau_2}{1 + s(\tau_1 + \tau_2)}$$

When the PFD is used to drive that filter, no current flows in the resistors R_1 and R_2, however, when the output of the PFD is in the high-impedance state. Under this condition, the voltage across the capacitor remains unchanged. Neglecting leakage currents, the output signal u_f of the passive lag filter can have any voltage level while the input signal is 0 (i.e., while the PFD is in its 0 state). Hence the passive lag filter behaves like an ideal integrator. Its transfer function can be shown to be approximately[15]

$$F(s) = \frac{1 + s\tau_2}{s(\tau_1 + \tau_2)}$$

The combination of the PFD with the passive lag-lead loop filter is commonly referred to as a *"charge pump."* The term *charge pump* is equally used if the PFD drives an active loop filter. Whenever the output signal of the PFD goes high, charge flows into the filter capacitor, i.e. charge is "pumped" into that capacitor. When the PFD output goes low, charge is flowing out of the capacitor, or charge is "pumped" off that capacitor to ground. It will be shown later, that many commercial PLL IC's use such charge pumps, e.g. the popular HC/ HCT744046 circuit. Experiments performed with DPLL systems have shown that the gain of the passive lag filter is not constant but depends on the voltage across the capacitor.[15] Assume that the PFD is powered from a unipolar 5-V supply and that the initial voltage on the capacitor is 2.5 V. When the PFD is in its +1 state, its output signal is pulled up to about 5 V. The voltage drop across the series connection of the resistors R_1 and R_2 is 2.5 V then. After the PFD has delivered a number of positive pulses, the capacitor will have charged to a higher voltage, say to 4 V. Now only 1 V is left across the resistors; i.e., the charging current decreases heavily. If now the PFD switches to its −1 state, its output signal is pulled down to 0 V. At this moment, the voltage drop across the resistors is suddenly increased to 4 V. This phenomenon leads to the fact that the gain of the passive lag filter becomes variable, and the filter dynamics becomes nonlinear. This nonlinearity has an impact on the transient response of the DPLL. If the frequency of the input signal is suddenly increased or decreased in such a way that the output signal u_f of the loop filter will not vary considerably (e.g., from 2.5 to 3.0 V with a supply voltage of 5 V), this effect can be discarded. When larger-frequency steps are applied to the input of the DPLL, the mentioned nonlinearity generally decreases the damping factor ζ of the system. Simulations with the included program show that a predicted damping factor of 0.7 can drop to approximately 0.3. The decrease of ζ can be compensated for by increasing the time constant τ_2 of the loop filter.

A similar effect is observed when the active lag filter is used (Fig. 2.2b). When driven from a normal signal source, its transfer function is known to be

$$F(s) = K_a \frac{1 + s\tau_2}{1 + s\tau_1}$$

where $K_a = -C_1/C_2$. When driven from a PFD phase detector, the voltages across the capacitors of this filter also remain unchanged whenever the PFD is in its high-impedance state. Under this condition the transfer function of the active lag filter can be shown to be approximately

$$F(s) = K_a \frac{1 + s\tau_2}{s\tau_1}$$

This filter then also performs like an ideal integrator.

When the loop is an active PI filter (Fig. 2.2c), it does not matter whether the filter is driven by a "normal" signal source or by a device having a three-state output. In any case, the PI filter acts as an integrator whose transfer function is given by

$$F(s) = -\frac{1 + s\tau_2}{s\tau_1}$$

Knowing the transfer functions of all building blocks of the DPLL, we are able to derive the phase-transfer function $H(s)$, the natural frequency ω_n, and the damping factor ζ. For ω_n and ζ, expressions similar to those for the LPLL are obtained; cf. Eqs. (2.15a) to (2.15c). The formulas for ω_n and ζ are shown in Table 3.1, where the remaining parameters of the DPLL are also listed. The response of the DPLL onto phase steps or frequency steps applied to its reference input is similar to the response of the LPLL; cf. Figs. 2.11 and 2.12. In some aspects, however, the DPLL strongly deviates from the LPLL. This will become evident when we derive formulas for the key parameters such as hold range and lock range. This will be done in the next sections.

3.2.1 The hold range

The hold range $\Delta\omega_H$ is the frequency range within which PLL operation can be statically stable. Under normal operating conditions the PLL never operates at the limits of the hold range. To reach this limit of stability it would be necessary to sweep the reference frequency slowly upward (or downward). If the reference frequency is increased and the dc gain of the loop filter is finite, the phase error increases in proportion. When it attains the maximum value for which the phase detector operates linearly, the hold range is reached.

If an EXOR gate is used as phase detector, the maximum phase error is $\pi/2$. Based on the procedure described in Sec. 2.6.1, we obtain for the hold range

$$\Delta\omega_H = \frac{K_0 K_d F(0)(\pi/2)}{N}$$

where $F(0)$ is the dc gain of the loop filter. As we know from Sec. 2.1, the dc gain is 1 for the passive lag, K_a for the active lag, and ∞ for the passive PI filter. Hence we get the following hold ranges for the EXOR phase detector:

■ Phase detector = EXOR, loop filter = passive lag

$$\Delta\omega_H = \frac{K_0 K_d (\pi/2)}{N} \qquad (3.2a)$$

■ Phase detector = EXOR, loop filter = active lag

$$\Delta\omega_H = \frac{K_0 K_d K_a (\pi/2)}{N} \qquad (3.2b)$$

■ Phase detector = EXOR, loop filter = active PI

$$\Delta\omega_H = \infty \qquad (3.2c)$$

If the JK-flipflop is used for the loop filter, the maximum phase error becomes π, so we get for the hold range

■ Phase detector = JK-flipflop, loop filter = passive lag

$$\Delta\omega_H = \frac{K_0 K_d \pi}{N} \qquad (3.2d)$$

TABLE 3.1 Summary of Parameters and Formulas for Digital PLLs

Parameter category	Symbol	Parameter	Definition
General	ω_0	Center frequency of the VCO	Angular frequency of the VCO at $u_f = 0$
	τ_1, τ_2	Time constants of loop filter	
	ω_n	Natural frequency of the PLL	ω_n is the natural frequency of the PLL system. The PLL responds to an excitation at its input with a transient, normally a damped oscillation with angular frequency ω_n
	ζ	Damping factor	$1/\omega_n\zeta$ = time constant of the damped oscillation
Acquisition	$\Delta\omega_H$	Hold range	Frequency range within which PLL operation can be statically stable
	$\Delta\omega_L$	Lock range	If the frequency offset of the reference signal is smaller than the lock range, the PLL locks within one single-beat note between reference and output frequencies
	T_L	Lock-in time	Time required for the lock-in process
	$\Delta\omega_P$	Pull-in range	If the frequency offset of the reference signal is larger than the lock range but smaller than the pull-in range, the PLL will slowly lock after a number of beat notes between reference and output frequencies
	T_P	Pull-in time	Time required for a pull-in process

	Formulas for second-order PLLs Type of loop filter		
	Passive lag	Active lag	Active PI
	$\omega_n = \sqrt{\dfrac{K_0 K_d}{N(\tau_1 + \tau_2)}}$	$\omega_n = \sqrt{\dfrac{K_0 K_d K_a}{N\tau_1}}$	$\omega_n = \sqrt{\dfrac{K_0 K_d}{N\tau_1}}$
PD is not PFD	$\zeta = \dfrac{\omega_n}{2}\left(\tau_2 + \dfrac{N}{K_0 K_d}\right)$	$\zeta = \dfrac{\omega}{2}\left(\tau_2 + \dfrac{N}{K_0 K_d K_a}\right)$	$\zeta = \dfrac{\omega_n \tau_2}{2}$
PD is PFD	$\zeta = \dfrac{\omega_n \tau_2}{2}$	$\zeta = \dfrac{\omega_n \tau_2}{2}$	$\zeta = \dfrac{\omega_n \tau_2}{2}$
PD = EXOR	$\Delta\omega_H = \dfrac{K_0 K_d \pi/2}{N}$	$\Delta\omega_H = \dfrac{K_0 K_d K_a \pi/2}{N}$	$\Delta\omega_H \to \infty$
PD = JK-flipflop	$\Delta\omega_H = \dfrac{K_0 K_d \pi}{N}$	$\Delta\omega_H = \dfrac{K_0 K_d K_a \pi}{N}$	$\Delta\omega_H \to \infty$
PD = PFD	←	$\Delta\omega_H \to \infty$	→
PD = EXOR	←	$\Delta\omega_L = \pi\zeta\omega_n$	→
PD = JK-flipflop	←	$\Delta\omega_L = 2\pi\zeta\omega_n$	→
PD = PFD	←	$\Delta\omega_L = 4\pi\zeta\omega_n$	→
	←	$T_L \approx \dfrac{2\pi}{\omega_n}$	→
PD = EXOR	Low-gain loops $\Delta\omega_P = \dfrac{\pi}{2}\sqrt{2\zeta\omega_n K_0 K_d - \omega_n^2}$ High-gain loops $\Delta\omega_P = \dfrac{\pi}{\sqrt{2}}\sqrt{\zeta\omega_n K_0 K_d}$	Low-gain loops $\Delta\omega_P = \dfrac{\pi}{2}\sqrt{2\zeta\omega_n K_0 K_d - \omega_n^2/K_a}$ High-gain loops $\Delta\omega_P = \dfrac{\pi}{\sqrt{2}}\sqrt{\zeta\omega_n K_0 K_d}$	$\Delta\omega_P \to \infty$
PD = JK-flipflop	Low-gain loops $\Delta\omega_P = \pi\sqrt{2\zeta\omega_n K_0 K_d - \omega_n^2}$ High-gain loops $\Delta\omega_P = \pi\sqrt{2}\sqrt{\zeta\omega_n K_0 K_d}$	Low-gain loops $\Delta\omega_P = \pi\sqrt{2\zeta\omega_n K_0 K_d - \omega_n^2/K_a}$ High-gain loops $\Delta\omega_P = \pi\sqrt{2}\sqrt{\zeta\omega_n K_0 K_d}$	$\Delta\omega_P \to \infty$
PD = PFD	←	$\Delta\omega_P \to \infty$	→
PD = EXOR	←	$T_P = \dfrac{4}{\pi^2}\dfrac{\Delta\omega_0^2}{\zeta\omega_n^3}$	→
PD = JK-flipflop	←	$T_P = \dfrac{1}{\pi^2}\dfrac{\Delta\omega_0^2}{\zeta\omega_n^3}$	→
PD = PFD	$T_P = 2(\tau_1 + \tau_2)\ln\dfrac{K_0(U_B/2)}{K_0(U_B/2) - \Delta\omega_0}$	$T_P = 2\tau_1 \ln\dfrac{K_0 K_a(U_B/2)}{K_0 K_a(U_B/2) - \Delta\omega_0}$	$T_P = \dfrac{2\tau_1 \Delta\omega_0}{K_0(U_B/2)}$

TABLE 3.1 Summary of Parameters and Formulas for Digital PLLs (*Continued*)

Parameter category	Symbol	Parameter	Definition
Tracking	$\Delta\omega_{PO}$	Pull-out range	Dynamic limit of stable operation of the PLL. The system unlocks if a frequency step larger than $\Delta\omega_{PO}$ is applied to the reference input
	$\Delta\dot{\omega}$	Rate of change of frequency offset	Maximum allowable rate of change of (angular) reference frequency
Noise	P_s, P_n	Signal, noise power	Power of input signal and noise signal applied to the input of a PLL
	B_i	Prefilter bandwidth	Bandwidth of the prefilter (or the input signal source)
	B_L	Noise bandwith	
	$(SNR)_i$	Signal-to-noise ratio of the input signal	
	$(SNR)_L$	Signal-to-noise ratio of the loop	

- Phase detector = JK-flipflop, loop filter = active lag

$$\Delta\omega_H = \frac{K_0 K_d K_a \pi}{N} \tag{3.2e}$$

- Phase detector = JK-flipflop, loop filter = active PI

$$\Delta\omega_H = \infty \tag{3.2f}$$

The situation changes drastically, however, when the PFD is used as phase detector. Because its output is in the high-impedance state when none of the UP or DN outputs is active, the charge on the capacitor(s) of the loop filter remains unchanged when the PFD is in the 0 state. Consequently, the output signal u_f of the loop filter can have a nonzero value even if the average u_d signal is 0. When driven by a tristate source, the loop filter acts like an integrator [i.e., a filter whose transfer function $F(s)$ has a pole at $s = 0$]. The hold range of a DPLL using the PFD becomes infinite, therefore:

- Phase detector = PFD, for all loop filters

$$\Delta\omega_H = \infty \tag{3.2g}$$

All these formulas are also listed in Table 3.1.

	Formulas for second-order PLLs Type of loop filter		
	Passive lag	Active lag	Active PI
PD = EXOR	← ————————	$\Delta\omega_{PO} = 2.46\ \omega_n(\zeta + 0.65)$	——————→
PD = JK-flipflop	← ————————	$\Delta\omega_{PO} = 5.78\ \omega_n(\zeta + 0.5)$	——————→
PD = PFD	← ————————	$\Delta\omega_{PO} = 11.55\ \omega_n(\zeta + 0.5)$	——————→
	← ————————	$\Delta\dot\omega < \omega_n^2$	——————→

$$B_L \approx \frac{\omega_n}{2}\left(\zeta + \frac{1}{4\zeta}\right)$$

$$\mathrm{SNR}_i = \frac{P_s}{P_n}$$

$$\mathrm{SNR}_L = \mathrm{SNR}_i\,\frac{B_i}{2B_L}$$

3.2.2 The lock range

By definition the *lock range* is the offset between reference and (scaled-down) VCO frequency which causes the DPLL to acquire lock within one beat note between reference and (scaled-down) output frequencies. The lock range of the DPLL can be determined by considerations analog to those made in Sec. 2.6.2. We assume that the DPLL is initially out of lock and that the VCO oscillates on its center frequency $N\omega_0$. The reference frequency is offset by $\Delta\omega$ from its center value ω_0, i.e., $\omega_1 = \omega_0 + \Delta\omega$. The signals u_1 and u_2' can then be represented by the Walsh functions

$$u_1(t) = U_{10}w[(\omega_0 + \Delta\omega)t]$$

and

$$u_2'(t) = U_{20}w(\omega_0 t)$$

respectively, where U_{10} and U_{20} are the amplitudes of the square-wave signals. The phase error θ_e is the difference of the phases of these two signals, i.e.,

$$\theta_e(t) = \omega_0 t$$

which is a ramp function.

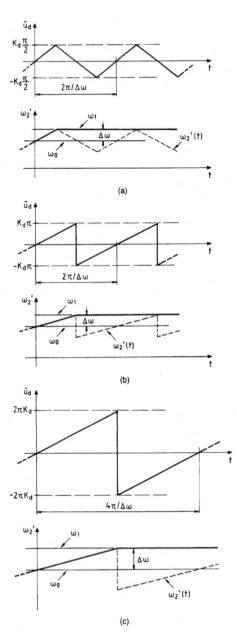

Figure 3.12 This figure explains the determination of the lock-in range of the DPLL. (*a*) The upper curve is the averaged output signal $\overline{u_d}$ of the EXOR phase detector. The lower curve represents the instantaneous output frequency ω_2' of the DPLL. The initial (angular) frequency offset ω_0 has been chosen such that the peak output frequency ω_2' just reaches the input frequency ω_1. Hence the DPLL locks within one cycle of the averaged $\overline{u_d}$ signal. (*b*) Same signals for the JK-flipflop phase detector. The averaged $\overline{u_d}$ signal is a sawtooth in this case. Again, the DPLL locks within one cycle of the averaged $\overline{u_d}$ signal. (*c*) Same signals for the PFD.

Let us calculate first the lock range of a DPLL using an EXOR gate as phase detector. Because the average output signal $\overline{u_d}(t)$ of the EXOR is a triangular function of phase error (Fig. 3.4a), $\overline{u_d}(t)$ becomes a triangular function of time as shown in the upper curve of Fig. 3.12a. When $\Delta\omega$ is higher than the upper corner frequency $1/\tau_2$ of the loop filter (Fig. 2.3), the output signal u_f of the loop filter is given by

$$u_f = u_d F_H$$

where F_H is the gain of the loop filter "at high frequencies," as explained in Sec. 2.6. The shape of the u_f signal is also triangular, therefore, and the frequency of the VCO is modulated by this triangular signal, as shown in the lower curve in Fig. 3.12a. When the frequency offset $\Delta\omega$ is chosen such that the peak of the ω_2' curve just reaches the reference frequency ω_1, $\Delta\omega$ equals the lock range $\Delta\omega_L$. Using the mathematical method developed in Sec. 2.6.2, we obtain for the lock range the approximation

$$\Delta\omega_L \approx \pi\zeta\omega_n \quad \text{(phase detector = EXOR)} \qquad (3.3a)$$

The lock range of the DPLL using the EXOR phase detector is greater than the lock range of the LPLL by a factor of approximately $\pi/2$. This is easily explained by the fact that the maximum output signal of the four-quadrant multiplier is K_d, whereas the maximum output signal of the EXOR is $K_d\pi/2$.

Now the lock range is calculated for the case where the JK-flipflop is used as phase detector. If we assume again that the DPLL is initially out of lock and that the offset between reference frequency ω_1 and center frequency ω_0 is $\Delta\omega$, the phase error becomes a ramp function again. Because the average output signal $\overline{u_d}$ of the JK-flipflop varies in a sawtooth-like fashion with phase error (Fig. 3.6), the average signal $\overline{u_d}(t)$ will also be a sawtooth function, as shown in the upper curve of Fig. 3.12b. Now the frequency of the VCO is modulated in a sawtooth-like manner; see lower curve of Fig. 3.12b. If the frequency offset $\Delta\omega$ is chosen such that the ω_2' curve just touches the ω_1 line, $\Delta\omega$ equals the lock range $\Delta\omega_L$. By an analog consideration we get the approximation

$$\Delta\omega_L \approx 2\pi\zeta\omega_n \quad \text{(phase detector = JK-flipflop)} \qquad (3.3b)$$

A similar procedure can be applied to the PFD. For a DPLL using the PFD the lock range becomes approximately

$$\Delta\omega_L \approx 4\pi\zeta\omega_n \quad \text{(phase detector = PFD)} \qquad (3.3c)$$

The lock-in time T_L can be calculated by an analog consideration as made for the linear PLL. When the DPLL gets locked quickly, the signals u_d and u_f perform a damped oscillation (for $\zeta < 1$) whose angular frequency is approximately ω_n. As can be seen from Fig. 2.12, for example, the lock-in process is completed within one cycle of the damped oscillation at most, so it is a reasonable approximation to state that T_L is one period of the damped oscillation:

$$T_L \approx \frac{2\pi}{\omega_n} \qquad (3.3d)$$

Another useful parameter is the 3 db bandwidth ω_{3db}. It can be computed by the formula given for the LPLL in chap. 2.3 (page 20).

3.2.3 The pull-in range

As stated in Sec. 2.6.3 the pull-in process is a nonlinear phenomenon and is very hard to calculate. The mathematical analysis is treated in greater detail in Appendix A; here we give only the most important results. For the design engineer two parameters are of interest, the *pull-in range* $\Delta\omega_P$ and the *pull-in time* T_P. (Both of these have been defined in Sec. 2.6.3; see also Table 2.1.) Unfortunately, pull-in range and pull-in time depend on the type of phase detector used, so we have to analyze them separately for each type of phase detector.

Let us assume first that the EXOR gate is used as phase detector. Furthermore we assume that the DPLL is out of lock initially, that the VCO operates at its center frequency $N\omega_0$, and that the offset $\Delta\omega$ between reference frequency ω_1 and (down-scaled) VCO frequency ω_0 is large. The signals u_1 and u_2' can then be represented by Walsh functions

$$u_1(t) = U_{10}w[(\omega_0 + \Delta\omega)t]$$

and

$$u_2'(t) = U_{20}w(\omega_0 t)$$

respectively, where U_{10} and U_{20} are the amplitudes of the square-wave signals. The phase error θ_e is the difference of the phases of these two signals, i.e.,

$$\theta_e(t) = \Delta\omega t$$

which is a ramp function. The average output signal $u_d(t)$ is therefore a triangular signal, as shown in the upper trace of Fig. 3.13a. (Let us discard for the moment the asymmetry of the waveform.) The output signal $u_f(t)$ of the loop filter will be some fraction of the $\overline{u_d}$ signal and will modulate the instantaneous frequency ω_2' of the VCO, lower trace

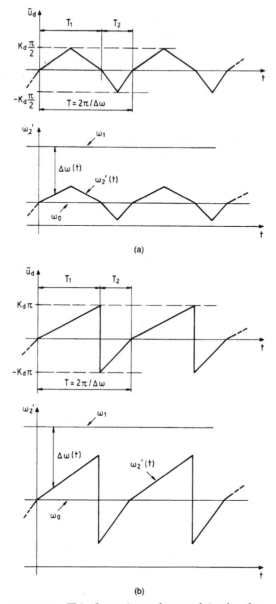

(a)

(b)

Figure 3.13 This figure is used to explain the slow pull-in process of the DPLL. (*a*) Pull-in process of a DPLL using the EXOR as phase detector. The upper trace shows the averaged output signal $\overline{u_d}$ of the EXOR gate, the lower trace the instantaneous output (angular) frequency u_2' of the DPLL. The asymmetry of the waveforms is shown exaggerated. Because the duration of the positive half-wave of $\overline{u_d}$ is larger than the negative, a dc offset is generated. If the loop gain of the DPLL is sufficiently large, the loop is pulled in; i.e., the peak value of ω_2' reaches ω_1 after some time. (*b*) Corresponding waveforms for a DPLL using the JK-flipflop as phase detector.

of Fig. 3.13a. If the triangular waveform were symmetrical (i.e., $T_1 = T_2$), the average frequency ω_2' would remain constant and equal to ω_0. As Fig. 3.13a demonstrates, however, the frequency offset $\Delta\omega$ is not constant but is given by the difference between reference frequency ω_1 and *instantaneous* (scaled-down) VCO frequency ω_2'. Consequently, $\Delta\omega(t)$ becomes smaller during the positive half-wave of the $\overline{u_d}$ signal and larger during the negative half-wave. Therefore, the waveform of $\overline{u_d}$ becomes *asymmetrical,* which is shown exaggerated in Fig. 3.1a. When the $\overline{u_d}$ waveform is asymmetrical, its mean value is no longer zero but becomes slightly positive. This causes the average frequency of the VCO to be *pulled up.* Now different things can happen. If the loop gain, i.e., the product $K_d K_0 F(0)/N$, is very small, the VCO frequency is pulled up by a small amount only and stays stuck at some final value. If the loop gain is larger, however, the pull-in process becomes regenerative: the mean frequency ω_2' is pulled up so much that the $\overline{u_d}$ waveform becomes even more nonharmonic (i.e., the ratio of T_1/T_2 becomes significantly greater). This causes the mean frequency ω_2' to increase even more, etc., and the VCO frequency will be pulled up until it comes close to the reference frequency. Then a locking process will take place. A pull-in process is initiated whenever the initial frequency offset $\Delta\omega$ is smaller than the pull-in range $\Delta\omega_P$. The final results are given here:

■ Phase detector = EXOR, loop filter = passive lag
  Low-gain loops

$$\Delta\omega_P = \frac{\pi}{2} \sqrt{2\zeta\omega_n K_0 K_d - \omega_n^2} \qquad (3.4a)$$

■ Phase detector = EXOR, loop filter = passive lag
  High-gain loops

$$\Delta\omega_P = \frac{\pi}{\sqrt{2}} \sqrt{\zeta\omega_n K_0 K_d} \qquad (3.4b)$$

■ Phase detector = EXOR, loop filter = active lag
  Low-gain loops

$$\Delta\omega_P = \frac{\pi}{2} \sqrt{2\zeta\omega_n K_0 K_d - \frac{\omega_n^2}{K_a}} \qquad (3.4c)$$

■ Phase detector = EXOR, loop filter = active lag
  High-gain loops

$$\Delta\omega_P = \frac{\pi}{\sqrt{2}} \sqrt{\zeta\omega_n K_0 K_d} \qquad (3.4d)$$

If the loop filter is an active PI filter, its dc gain becomes (theoretically) infinite, so the DPLL pulls in under every condition. The pull-in range becomes infinite in this case:

■ Phase detector = EXOR, loop filter = active PI

$$\Delta\omega_P = \infty \tag{3.4e}$$

As demonstrated in Appendix A, it is also possible to calculate an approximate value for the pull-in time T_P. The final result reads

$$T_P = \frac{4}{\pi^2}\frac{\Delta\omega_0^2}{\zeta\omega_n^3} \tag{3.5}$$

for all types of loop filters. $\Delta\omega_0$ is the initial frequency offset, i.e., $\Delta\omega_0 = \omega_1 - \omega_0$. This equation is very similar to that obtained for the LPLL [Eq. (2.46)]; the pull-in time varies with the square of the initial frequency offset ω_0. As we know, however, the pull-in time becomes infinite when the initial frequency offset equals the pull-in range. The approximation of Eq. (3.5) is therefore valid only when $\Delta\omega_0$ is markedly less than $\Delta\omega_P$. Computer simulations have shown that the approximation gives acceptable results when $\Delta\omega_0$ is less than about 0.8 $\Delta\omega_P$. (In practical terms, "acceptable" means that the error of the predicted result is not larger than about 10 percent.)

Now we analyze the pull-in process for the case where the JK-flipflop is used as phase detector. Making the same assumptions as for the EXOR gate, the waveforms of the average $\overline{u_d}(t)$ signal and the instantaneous (down-scaled) output frequency ω_2' look like those drawn in Fig. 3.13b. Instead of triangular waves, we obtain sawtooth waves now. Performing an analog computation as above, we get for the pull-in range

■ Phase detector = JK-flipflop, loop filter = passive lag
Low-gain loops

$$\Delta\omega_P = \pi\sqrt{2\zeta\omega_n K_0 K_d - \omega_n^2} \tag{3.6a}$$

■ Phase detector = JK-flipflop, loop filter = passive lag
High-gain loops

$$\Delta\omega_P = \pi\sqrt{2}\,\sqrt{\zeta\omega_n K_0 K_d} \tag{3.6b}$$

■ Phase detector = JK-flipflop, loop filter = active lag
Low-gain loops

$$\Delta\omega_P = \pi \sqrt{2\zeta\omega_n K_0 K_d - \frac{\omega_n^2}{K_a}} \qquad (3.6c)$$

- Phase detector = JK-flipflop, loop filter = active lag
  High-gain loops

$$\Delta\omega_P = \pi\sqrt{2}\ \sqrt{\zeta\omega_n K_0 K_d} \qquad (3.6d)$$

When the loop filter is an active PI, the pull-in range becomes infinite:

- Phase detector = JK-flipflop, loop filter = active PI

$$\Delta\omega_P = \infty \qquad (3.6e)$$

The pull-in time of this type of DPLL becomes

$$T_P = \frac{1}{\pi^2}\frac{\Delta\omega_0^2}{\zeta\omega_n^2} \qquad (3.7)$$

for all types of loop filters. As above, this formula gives acceptable results when $\Delta\omega_0$ is less than about 0.8 $\Delta\omega_P$.

To conclude this analysis we have to consider the case where the PFD is used as phase detector. As we already noted in Sec. 3.1, the pull-in range becomes infinite now, because the loop filter is driven by a tristate source. The charge on the filter capacitor remains unchanged when the output of the PFD is in the high-impedance state; hence even a passive lag filter works like a real integrator, i.e., like a filter whose transfer function $F(s)$ has a pole at $s = 0$. The pull-in time remains finite, of course, so we need a suitable approximation for T_P.

First we are going to calculate the pull-in time T_P for the case where the passive lag loop filter is used. We assume again that the DPLL is initially out of lock and that the frequency ω_1 of the reference signal u_1 is markedly higher than the (down-scaled) output frequency ω_2'. This situation is sketched by the upper two waveforms in Fig. 3.14. As explained in Sec. 3.1 and Fig. 3.7, the output signal u_d of the PFD then toggles between the states 0 and +1. The third trace shows the waveform of u_d. It was assumed that the PFD is powered by a unipolar supply so that the logical "high" level is U_B, the "low" level 0 V (ground). The centerline of the u_d signal represents the high-impedance state of the PFD (Hi-Z). The average u_d signal is drawn as a dashed line. It has the shape of a sawtooth signal. The average u_d signal is nothing else than the duty cycle of the PFD output. It periodically ramps up from 0 to 1 and is a sawtooth function as well. Obviously, the average duty cycle of u_d is 50 percent. As Fig. 3.10 shows, the average duty cycle δ varies very little with the ratio ω_1/ω_2

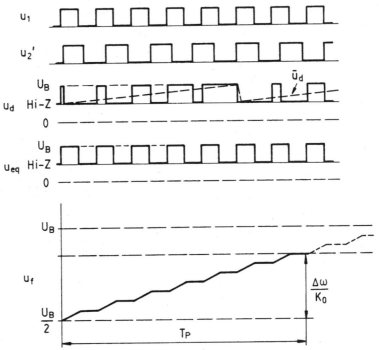

Figure 3.14   The waveforms shown in this figure demonstrate the pull-in process of a DPLL which uses the PFD. The frequency ω_1 of the input signal u_1 is assumed to be higher than the (scaled-down) frequency ω_2' of the output signal u_2' (for symbol definitions, see Fig. 3.1). Consequently, u_d toggles between the states 0 and +1 (third trace). The averaged u_d signal has the same effect as an equivalent signal u_{eq} with constant duty cycle of 50 percent, as explained in the text. This greatly facilitates the computation of the averaged loop filter output signal u_f; see bottom trace.

and can be considered constant for this analysis. Because the time constant τ_1 of the loop filter is much larger than the period of the u_1 signal in Fig. 3.14, an equivalent signal u_{eq} having a constant duty cycle of 50 percent would have the same effect on the loop filter; this equivalent signal is also represented in Fig. 3.14. If the equivalent signal u_{eq} had a duty cycle of 100 percent, capacitor C of the loop filter would simply charge toward the supply voltage U_B with time constant $\tau_1 + \tau_2 = (R_1 + R_2)C$, because C is charged through the series connection of resistors R_1 and R_2 (compare Fig. 2.2a for symbol definitions). Because the duty cycle is only 50 percent, however, the capacitor needs twice as much time to charge. Therefore the loop filter acts like a simple RC filter whose time constant is not $\tau_1 + \tau_2$ but $2(\tau_1 + \tau_2)$. The equivalent model of Fig. 3.15 can now be used to compute the signal u_f across the capacitor C. Since the VCO of the DPLL was assumed to operate at its center frequency $N\omega_0$ at the start of the pull-in process, the initial value of u_f is $U_B/2$. Consequently, the capacitor will try to charge from $U_B/2$ to U_B during the pull-in process. The actual source voltage for the RC filter in Fig. 3.15 is therefore $U_B/2$. After some time, u_f will have reached a level which causes the VCO to generate just the "right" frequency. As shown in Fig. 3.14, this occurs when the voltage on the capacitor has been increased by the amount $\Delta\omega/K_0$. When this happens, the pull-in process terminates, and a lock-in process takes place. As Fig. 3.14 shows, the pull-in time T_P is the time required for the capacitor to increase its voltage by $\Delta\omega/K_0$. This calculation becomes quite easy now. When the passive lag filter is used, the final result reads

■ Loop filter = passive lag

$$T_P = 2(\tau_1 + \tau_2) \ln \frac{K_0(U_B/2)}{K_0(U_B/2) - \Delta\omega_0} \tag{3.8a}$$

where $\Delta\omega_0$ is the initial frequency offset, $\Delta\omega_0 = \omega_1 - \omega_0$. This formula has been derived under the premise that the PFD is driven by a unipolar power supply. In the most general case, the PFD could be driven from a bipolar supply, where the positive and negative supply voltages are U_{B+} and U_{B-}, respectively. Furthermore, the output signal of the PFD output signal could be clipped at the saturation levels $U_{\text{sat+}}$ (positive) and $U_{\text{sat-}}$ (negative), respectively. In the more general case the formula for T_P would read

$$T_P = 2(\tau_1 + \tau_2) \ln \frac{K_0 \dfrac{U_{\text{sat+}} - U_{\text{sat-}}}{2}}{K_0 \dfrac{U_{\text{sat+}} - U_{\text{sat-}}}{2} - \Delta\omega_0} \tag{3.8b}$$

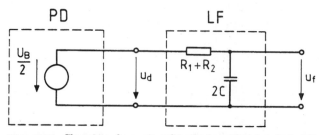

Figure 3.15   Charging of capacitor C in the loop filter is calculated by this equivalent model. Detailed explanations are in the text.

An analog computation can be performed for the cases where the active lag or the active PI is used as a loop filter. The results are:

- Loop filter = active lag

$$T_P = 2\tau_1 \ln \frac{K_0 K_a (U_B/2)}{K_0 K_a (U_B/2) - \Delta\omega_0}$$ (3.9)

- Loop filter = active PI

$$T_P = \frac{2\tau_1 \Delta\omega_0}{K_0 (U_B/2)}$$ (3.10)

Equations (3.9) and (3.10) are valid for unipolar power supplies, too. If a bipolar power supply is used, the same correction must be made as in Eq. (3.8b).

It is worth considering a major difference of the pull-in processes for different types of phase detectors. If the phase detector is an EXOR gate, the instantaneous frequency of the VCO is modulated in both directions around its average value, as seen in the lower trace of Fig. 3.13a. This is to be expected, for the u_d signal is an ac signal. Provided a pull-in process starts, the frequency of the VCO is slowly "pumped up," as has been shown in Fig. 2.18 for the LPLL. A similar "pumping" is observed when the phase detector is a JK-flipflop, cf. Fig. 3.13b. No "pumping" occurs, however, when the PFD is used. Since the driving signal for the loop filter is unipolar here (cf. Figs 3.14 and 3.15), charge is "pumped" into the filter capacitor *in one direction* only, so that the frequency of the VCO is moved in the "right" way at any time. The instantaneous frequency of the VCO approaches the final value from one side only. When the pull-in process is completed, a lock-in process follows. Only then the output frequency performs a damped oscillation; it slightly overshoots the final value and settles after the transient has died out. We will have a closer look at these phenomena when we perform computer simulations with the DPLL (Sec. 3.6).

3.2.4 The pull-out range

As defined in Sec. 2.6.4, the pull-out range is the size of the frequency step applied to the reference input which causes the PLL to lose phase tracking. For the linear PLL, the pull-out range corresponded in the mechanical analogy (Fig. 3.13) with the force impulse which brought the pendulum to tip over. The LPLL lost synchronism when the peak phase error exceeded π. Because the output signal u_d of the phase detector varies with the sine of the phase error θ_e and not with θ_e itself, it was not possible to compute the pull-out range explicitly. Simulations performed on an analog computer were used instead to find the approximation

$$\Delta\omega_{PO} = 1.8 \, \omega_n(\zeta + 1)$$

for the LPLL.[6]

For a DPLL using an EXOR gate as phase detector, we expect that the pull-out range would be about the same. Whereas the output signal of the four-quadrant multiplier varies with the sine of the phase error, it is a triangular function in case of the EXOR (Fig. 3.4a), which is quite similar. In the mechanical analogy, this PLL will also lock out when the peak phase error exceeds π. Since $\overline{u_d}$ is a linear function of θ_e only in the range $-\pi/2 < \theta_e < \pi/2$, an exact calculation is not possible either. The approximation will not be too bad if we simply applied the same formula as that for the LPLL. We expect, however, that the pull-out range would be slightly greater for the DPLL, because the output signal of the EXOR is linear over the full range $-\pi/2 < \theta_e < \pi/2$ and does not flatten out at phase errors approaching $\pi/2$. Using the simulation program distributed with this book, the author performed simulations with this type of DPLL, using damping factors in the range of $0.1 < \zeta < 3$. Then a least-squares fit gave the approximation

$$\Delta\omega_{PO} = 2.46 \, \omega_n(\zeta + 0.65) \tag{3.11}$$

and hence larger values than the LPLL approximation.

A different procedure is used to compute the pull-out range of the DPLLs using a JK-flipflop or a PFD as phase detector. In the case of the JK-flipflop, the pull-out range is the frequency step causing the peak phase error to exceed π; in the case of the PFD, the pull-out range is the frequency step leading to a peak phase error of 2π. Because the average output signal u_d of the JK-flipflop actually *is linear* in the range $-\pi < \theta_e < \pi$ and the average output signal u_d of the PFD *is linear* in the range $-2\pi < \theta_e < 2\pi$, the pull-out range can be computed explicitly. Using the linear model of the DPLL (Figs. 3.1 and 3.7), phase error θ_e is calculated for a frequency step $\Delta\omega$ applied to the reference input. The result is a damped oscillation.[1] Using the rules of differential calculus, it is straightforward to calculate the maximum of the phase error. From there it is quite easy to calculate the size of frequency step which leads to a peak phase error of π (for the JK-flipflop) or 2π (for the PFD). Assuming that the DPLL is a high-gain loop, we get

■ Phase detector = JK-flipflop

$$\Delta\omega_{PO} = \pi\omega_n \exp\left(\frac{\zeta}{\sqrt{1-\zeta^2}} \tan^{-1}\frac{\sqrt{1-\zeta^2}}{\zeta}\right) \quad \zeta < 1 \qquad (3.12a)$$

$$\Delta\omega_{PO} = \pi\omega_n e \quad \zeta = 1 \qquad (3.12b)$$

$$\Delta\omega_{PO} = \pi\omega_n \exp\left(\frac{\zeta}{\sqrt{\zeta^2-1}} \tanh^{-1}\frac{\sqrt{\zeta^2-1}}{\zeta}\right) \quad \zeta > 1 \qquad (3.12c)$$

If $\Delta\omega_{PO}$ is plotted against ζ, we notice that the curve becomes rather flat and could easily be replaced by a linear function. This would ease the computation of the pull-out range considerably, because tables for inverse hyperbolic tangent, for example, are not always at hand. A least-squares fit performed with Eqs. (3.12a) to (3.12c) gave the approximation

$$\Delta\omega_{PO} = 5.78 \; \omega_n(\zeta + 0.5) \qquad (3.12d)$$

An analogous computation was performed for the PFD and gave the accurate results

- Phase detector = PFD

$$\Delta\omega_{PO} = 2\pi\omega_n \exp\left(\frac{\zeta}{\sqrt{1-\zeta^2}} \tan^{-1}\frac{\sqrt{1-\zeta^2}}{\zeta}\right) \quad \zeta < 1 \qquad (3.13a)$$

$$\Delta\omega_{PO} = 2\pi\omega_n e \quad \zeta = 1 \qquad (3.13b)$$

$$\Delta\omega_{PO} = 2\pi\omega_n \exp\left(\frac{\zeta}{\sqrt{\zeta^2-1}} \tanh^{-1}\frac{\sqrt{\zeta^2-1}}{\zeta}\right) \quad \zeta > 1 \qquad (3.13c)$$

Here, the least-squares fit gave the linear approximation

$$\Delta\omega_{PO} = 11.55 \; \omega_n(\zeta + 0.5) \qquad (3.13d)$$

Note that the most important design formulas for the DPLL are summarized in Table 3.1.

3.3 Noise Performance of the DPLL

The noise theory of the PLL evolved mainly during the early, linear era of the PLL. For a long time the four-quadrant multiplier was considered the relevant function block which made it possible to achieve noise suppression. A strong argument for the four-quadrant multiplier has been the addition theorem of the sine function. The principle of noise cancellation in LPLLs is considered by means of Fig. 3.16. Suppose that u_1 is an information signal having superimposed noise, we have

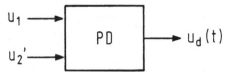

Figure 3.16 Block diagram of the multiplier phase detector as used in LPLLs.

$$u_1(t) = U_{10} \sin[\omega_1 t + \theta_1(t)] + U_n \sin[\omega_3 t + \theta_3(t)]$$

where U_{10} is the amplitude of the information signal and U_n is the amplitude of the noise component. The information is carried by the phase $\theta_1(t)$; ω_3 is the frequency of the noise signal, which also has a time-varying phase $\theta_3(t)$. Normally, the noise signal has a broad frequency spectrum; i.e., we can imagine that not one single but a great many noise terms, (theoretically infinite) which all have different frequencies, are superimposed on the information signal. It is sufficient for this simple study, however, to think of one noise term alone. If the LPLL is locked, the fundamental component of the PLL output signal can be written as

$$u_2(t) = U_{20} \cos(\omega_1 t)$$

As we know, the output signal $u_d(t)$ of the multiplier is the product of $u_1(t)$ and $u_2(t)$. Neglecting higher-frequency terms (which are filtered out by the loop filter), the spectrum of u_d will contain a line at $\omega = 0$, which is the desired output signal of the phase detector and a line at $\omega = \omega_1 - \omega_3$ which stems from the noise signal. Clearly, the second component is an ac term, and if $\omega_1 - \omega_3$ is larger than the lower 3-dB corner frequency of the loop filter, it is filtered out. Generally speaking, only those noise signals contribute to PLL output phase jitter whose frequency is very close to the center frequency of the loop. Noise bandwidth B_L of the LPLL has been shown to be about half the natural frequency ω_n of the LPLL [Eq. (2.58)]. To suppress noise, the noise bandwidth had only to be made sufficiently narrow.

There has been no known theory of noise in DPLLs. Computer simulations on DPLLs by Selle and coworkers[24] have shown, however, that the DPLL offers comparable if not better noise suppression than the LPLL. This becomes plausible if we consider the response of the EXOR phase detector on phase jitter superimposed on a reference signal u_1. Figure 3.17 shows a noiseless reference signal u_1 (second trace). Phase noise is now added to this signal. Normally, phase noise has a broad frequency spectrum. To simplify the analysis, we consider just one spectral component of that noise, i.e., the sinusoidal phase noise signal θ_j, first trace of Fig. 3.17. The superimposed phase noise causes the transitions of the reference signal u_1 to become time-shifted in accordance with phase noise θ_j; the resulting signal u_{1j} is shown by the third trace. Obviously, u_{1j} exhibits *phase jitter*. This phase jitter is immediately reflected in the output signal u_d of the EXOR phase detector, as shown in the second trace from the bottom in Fig. 3.17. Consequently, the average u_d signal is exactly proportional to the superimposed phase noise θ_j, as shown in the bottommost trace. If the frequency of the phase noise is beyond the noise bandwidth B_L of the DPLL, this noise component is suppressed; only the spectral components which are inside the noise bandwidth contribute to phase jitter at the output of the DPLL. The situation is similar to the case with

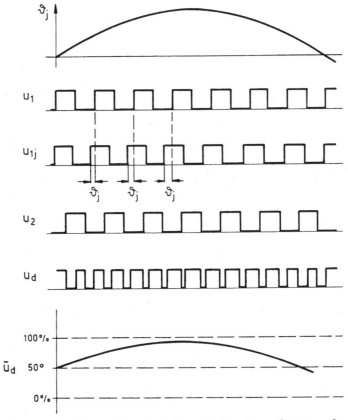

Figure 3.17 This figure is used to demonstrate noise performance of a DPLL which uses an EXOR as phase detector. The topmost trace shows one spectral component of phase jitter superimposed to the input signal u_1 of the DPLL (second trace). The third trace shows the "jittery" input signal u_{1j}. Further explanations are in the text.

the LPLL, so the simplified noise theory of the LPLL can be expected to yield reasonable estimates of phase jitter for the DPLL, too.

The term "signal-to-noise ratio" causes some trouble, however, in the case of the DPLL. With the linear PLL, the information and noise signals simply added up linearly, so we could easily recognize which portion of total power was signal power or noise power, respectively. Unfortunately, this principle cannot be applied immediately to the DPLL. To understand how phase jitter is generated in DPLLs, we consider the simple noise model of Fig. 3.18. The topmost trace shows the waveform of a noiseless, binary information signal u_1. Usually this signal is transmitted over a data link whose bandwidth is limited. If no additional noise were generated, the edges of u_1 would get finite rise time now; see signal u_{1b}. If noise is superimposed, however, the noisy signal u_{1n} appears at the end of the transmission line. To get a "clean" waveform again, the signal is reshaped, normally by a Schmitt trigger having upper and lower threshold levels u_{upper} and u_{lower}, respectively. The reshaped signal is designated u_{1r}; see the bottommost trace in Fig. 3.18. The net effect of finite link bandwidth and added noise is clearly phase jitter on the edges of the reshaped signal. The

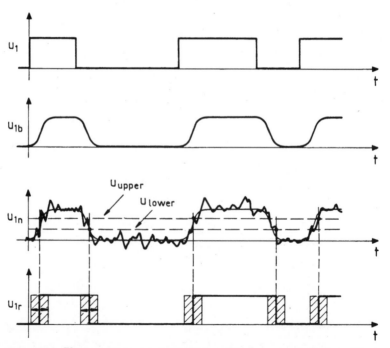

Figure 3.18 Phase noise in a communication signal u_1 causes the transitions of the received signal to become jittered. Further explanations are in the text.

larger the noise level, the wider the regions of time uncertainty become (hatched regions in u_{1r} signal). Because the amplitude of the reshaped signal is constant, we cannot decide which portion of u_{1r} is contributed by signal or noise, respectively, although it becomes possible to define formally a SNR$_i$ of the reference signal by using Eq. (2.50), which reads

$$\overline{\theta_{n1}^2} = \frac{P_n}{2P_s} \tag{3.14}$$

In a practical DPLL application it will be difficult to measure SNR$_i$ directly, but it is not difficult to determine phase noise θ_{n1} experimentally.[17] When a binary reference signal u_1 which has phase jitter is displayed on an oscilloscope and the latter is triggered on the positive edge at say 50 percent of the signal amplitude, an "eye pattern" appears (Fig. 3.19). Because of phase jitter, the edges of the signal are spread over an interval of time. The width w of the transition region is a measure of peak-to-peak phase jitter. In most cases the phase jitter has gaussian distribution; for such a distribution 99.7 percent of the measured phase jitter is within a band of $\pm 3\sigma$, where σ is the square root of the variance σ^2 of phase jitter, which is θ_{n1}^2 in our case. From practical experience it is known that the width w seen by the human eye corresponds to about $\pm 3\sigma$, so we conclude that the square root of phase noise is approximately $w/6$:

$$\sqrt{\overline{\theta_{n1}^2}} \approx \frac{w}{6} \tag{3.15}$$

Having measured the reference phase noise, Eq. (3.14) can be used to calculate the SNR$_i$ at the input of the DPLL. Using the simplified noise theory of the LPLL, SNR$_L$ (signal-to-noise ratio at the output of the DPLL) can be calculated by Eq. (2.63), which reads

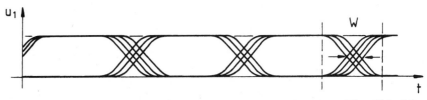

Figure 3.19 Eye pattern of a communication signal having phase jitter. The width of the uncertainty interval is used to estimate phase noise and signal-to-noise ratio of the signal, as explained in the text.

$$(\text{SNR})_L = (\text{SNR})_i \frac{B_i}{2B_L} \qquad (3.16)$$

where B_i is (one-sided) noise bandwidth at the reference input. Equation (3.16) says that the SNR of a DPLL is improved by a factor $B_i/(2B_L)$. This noise analysis was performed for the DPLL using an EXOR gate as phase detector. The question arises now whether these results remain valid if another phase detector is used. The EXOR phase detector differs in one aspect from the JK-flipflop and the PFD. Whereas the output signal u_d of both JK-flipflop and PFD depends on the edges of the u_1 and u_2' waveforms, the output of the EXOR is *level-sensitive*. When a bit stream is transmitted over a transmission link, it becomes possible that a number of succeeding bits are lost by fading. In such a case the reference signal u_{1r} after reshaping stays stuck at either a "low" or "high" logical level for some time. When the EXOR is used as a phase detector, its averaged output signal u_d remains zero during this period of time; i.e., the frequency of the VCO does not run away quickly. The situation is different, however, if a JK-flipflop or a PFD is used. Because these devices are edge-sensitive, their output signal u_d can be stuck at some distinct logical level during the interval when the reference signal fades away. If the output signal u_d of a JK-flopflop phase detector remains stuck at a "high" level, the VCO would run away with maximum speed, which is very undesirable, of course.

We conclude therefore that noise suppression of digital PLLs is about the same for every type of phase detector, as long as no edges of the reference signal get lost by fading. If fading occurs, however, the EXOR provides better noise performance than the other types of phase detectors. The most important design formulas for DPLL noise performance are summarized in Table 3.1.

3.4 DPLL Design

As Sec. 3.2 has shown, the DPLL can be built in many more variants than the LPLL. It is not surprising that the spectrum of DPLL applications is very broad as well. There are DPLL applications in communications where the system is used to extract the clock from a—possibly noisy—information signal. In such an application, noise suppression is of importance. An entirely different application of the DPLL is frequency synthesis. Here, reference noise is not of concern, but the synthesizer should be able to switch rapidly from one frequency to another; hence pull-in time is the most relevant parameter. For these reasons it appears difficult to give a design procedure which yields an optimum solution for every DPLL system. The step-by-step design program presented here should not be considered as a universal tool for the thousand and one uses of the DPLL but rather as a series of design hints. Moreover, in many cases the design of a DPLL will be an iterative process. We may start with some initial assumptions but end up perhaps with a design which is not acceptable, because one or more key parameters (e.g., pull-in time) are outside the planned range. In such a situation we restart with altered premises and repeat the procedure, until the final design appears acceptable. Manufacturers of DPLL ICs have already provided design tools running on the IBM-PC and compatibles. A program distributed by Philips,[15] for example, is used to design DPLL systems using the popular integrated circuits 74HC/HCT4046A and 74HC/HCT7046A; both are based on the old industry standard CD4046 IC (from the 4000 CMOS series) which was originally introduced by RCA. (For details of PLL ICs refer to Table 6.1.)

The design procedure for DPLLs is presented as a flowchart in Fig. 3.20. The individual steps are described in the following. Most of the formulas used to design the DPLL are listed in Table 3.1.

Step 1. In the first step, the input and output frequencies of the DPLL must be specified. There are cases where both input frequency and output frequency are constant but not necessarily identical. In other applications (e.g., frequency synthesizers) the input frequency is always the same, but the output frequency is variable. As a last variant, both input and output frequencies could be variable. Let $f_{1\min}$ and $f_{1\max}$ be the minimum and maximum input frequencies and $f_{2\min}$ and $f_{2\max}$ the minimum and maximum output frequencies, respectively.

Step 2. In this step the scaler ratio must be determined. There are DPLL applications where the output frequency f_2 always equals the reference frequency f_1. Here no down-scaler is needed, i.e., $N = 1$. There are cases where the ratio of output to reference frequency is greater than 1 but remains fixed. Here, a down-scaler with constant divider ratio N is required. When the DPLL is used to build a frequency synthesizer, the ratio of output to reference frequency is variable; thus a range for N must be defined ($N_{\min} \le N \le N_{\max}$). When N is variable, natural frequency ω_n and damping factor ζ will vary with N, as seen from the corresponding equations in Table 3.1. Both of these parameters will vary approximately with $1/\sqrt{N}$. Consequently ω_n will vary in the range $\omega_{n\min} < \omega_n < \omega_{n\max}$, and ζ will vary in the range $\zeta_{\min} < \zeta < \zeta_{\max}$. For these ranges we get approximately

$$\frac{\omega_{n\max}}{\omega_{n\min}} = \sqrt{\frac{N_{\max}}{N_{\min}}} \qquad (3.17)$$

and

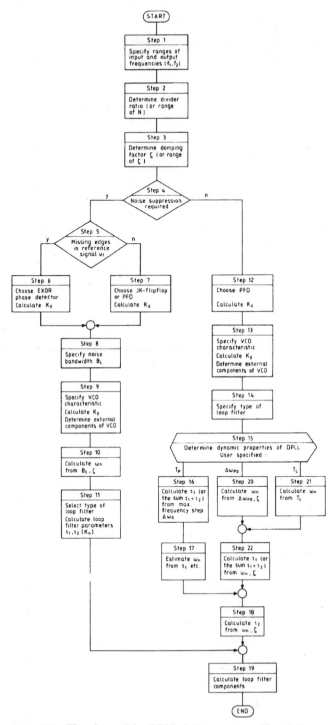

Figure 3.20 Flowchart of the DPLL design program (Sec. 3.4).

$$\frac{\zeta_{\max}}{\zeta_{\min}} = \sqrt{\frac{N_{\max}}{N_{\min}}} \qquad (3.18)$$

respectively. As we know a damping factor between 0.5 and 1 is considered optimum. As long as the ratio $N_{\max}/N_{\min}$ is moderately large, the variation of the damping factor can be accepted; if N varies by a factor of 10, ζ varies by about a factor of 3, which can be tolerated. Much larger variations of ζ, however, have to be avoided, because the loop then would get oscillatory for the smallest and sluggish for the largest damping factor. When N varies over a large range (e.g., 1:100), it is often mandatory to define more than one frequency range for the DPLL and switch the range accordingly. To ease the design in the case of variable N, we specify the parameters of the DPLL such that ζ becomes optimum for a divider ratio N_{mean} which is given by the geometric mean of $N_{\max}$ and $N_{\min}$,

$$N_{\text{mean}} = \sqrt{N_{\min} N_{\max}} \qquad (3.19)$$

(For constant N, $N_{\text{mean}} = N$, of course.) If $N_{\min} = 10$ and $N_{\max} = 100$, for example, N_{mean} would be $31.6 \to 32$. Choosing $\zeta = 0.7$ for $N = 32$ would yield a minimum damping factor of $\zeta_{\min} = 0.4$ and a maximum of $\zeta_{\max} = 1.2$, which is a fair compromise.

Step 3. Determination of damping factor ζ. When N is constant, ζ remains constant too and can be chosen arbitrarily. For constant N, it is optimum to select $\zeta = 0.7$; the DPLL then has Butterworth response, as explained in Sec. 2.3. If N is variable, however, it is recommended to choose $\zeta = 0.7$ for $N = N_{\text{mean}}$, as explained in step 2.

Step 4. In this step the question must be answered whether the DPLL should offer noise suppression or not. If a digital frequency synthesizer has to be built, for example, noise can be discarded, and parameters such as noise bandwidth B_L must not be considered. If noise must be suppressed, however, B_L and related parameters must be taken into account. If noise is of concern, the procedure continues at step 5, otherwise at step 12.

Step 5. Noise must be suppressed by this DPLL. As discussed in Sec. 3.3, the various digital phase detectors behave differently in the presence of noise. In a situation where edges of the reference (input) signal u_1 can get lost, edge-sensitive phase detectors such as the JK-flipflop or the PFD then can hang up in one particular state: the output of the JK-flipflop will switch into the "low" state, and the output of the PFD will switch to the state -1 after a very short time. Consequently, the frequency of the VCO will run away quickly, which is certainly undesirable. The average output signal u_d of the EXOR phase detector, however, will stay at 0 when edges of u_1 are missing. If edges of u_1 are likely to get lost, the procedure continues at step 6, otherwise at step 7.

Step 6. The EXOR phase detector should be selected. The phase detector gain K_d can now be calculated. If the EXOR runs from a unipolar power supply, K_d is given by $K_d = U_B/\pi$, where U_B is the supply voltage. When a bipolar power supply is used, or when the EXOR saturates at levels which differ substantially from the power-supply rails, use the equation given in Sec. 3.1. The procedure continues with step 8.

Step 7. The JK-flipflop or the PFD can be chosen for the phase detector. As explained in Sec. 3.2, the PFD offers superior performance, e.g., infinite pull-in range, so this type of phase detector is normally preferred. The phase detector gain K_d can now be determined. When a unipolar power supply is used, $K_d = U_B/2\pi$ for the JK-flipflop or $K_d = U_B/4\pi$ for the PFD (U_B = supply voltage). When a bipolar power supply is used, or when the phase detector saturates at voltage levels which differ substantially from the power-supply rails, the corresponding equation given in Sec. 3.1 should be used. The procedure continues at step 8.

Step 8. The noise bandwidth B_L must now be specified. As indicated in Table 3.1, B_L is related to the signal-to-noise ratio of the loop $(SNR)_L$ by

$$(SNR)_L = (SNR)_i \frac{B_i}{2B_L} \qquad (3.20)$$

B_L should be chosen such that $(SNR)_L$ becomes larger than some minimum value, typically larger than 4 (which corresponds to 6 dB). In most cases the signal-to-noise ratio at the input of the PLL $(SNR)_i$ is not known a priori. It can be roughly estimated from a measurement of input phase jitter θ_{n1}, as pointed out in Sec. 3.3. Finally, the noise bandwidth B_i at the input of the DPLL must also be known. B_i is nothing else than the bandwidth of the signal source or the bandwidth of an optimal prefilter. After $(SNR)_i$ and B_i are determined, B_L can be calculated from Eq. (3.20).

An additional problem arises when the scaler ratio N must be variable. We know from noise theory that B_L is also related to ω_n and ζ by Eq. (2.58), which reads

$$B_L = \frac{\omega_n}{2}\left(\zeta + \frac{1}{4\zeta}\right) \tag{3.21}$$

When the divider ratio N is variable, both ω_n and ζ vary with N; hence B_L also becomes dependent on N. In such a case we specify B_L for the case $N = N_{\mathrm{mean}}$, as pointed out in step 2. To make sure that B_L does not fall below the minimum acceptable value, we should check its values at the extremes of the scaler ratio N, using Eq. (3.21). If B_L becomes too small at one of the extremes of N, the initial value assumed for $N = N_{\mathrm{mean}}$ should be increased correspondingly.

Step 9. In this step the characteristic of the VCO will be determined. Because the center frequency ω_0 (or the range of ω_0) is known and the range of the divider ratio N is also given, we can calculate the range of output (angular) frequencies which must be generated by the VCO. Let $\omega_{2\mathrm{min}}$ and $\omega_{2\mathrm{max}}$ be the minimum and maximum output frequencies of the VCO, respectively. A suitable VCO must be selected first. In most cases, the VCO will be part of the DPLL system, typically an integrated circuit. Given the supply voltage(s) of the VCO, the data sheet indicates the usable range of control voltage u_f. (For a unipolar power supply with $U_B = 5$ V this range is typically $1 \ldots 4$ V. Let $u_{f\mathrm{min}}$ and $u_{f\mathrm{max}}$ be the lower and upper limit of that range, respectively. The VCO is now designed such that it generates the output frequency $\omega_2 = \omega_{2\mathrm{min}}$ for $u_f = u_{f\mathrm{min}}$ and the output frequency $\omega_2 = \omega_{2\mathrm{max}}$ for $u_f = u_{f\mathrm{max}}$. The corresponding VCO characteristic is plotted in Fig. 3.21. Now the VCO gain K_0 is calculated from the slope of this curve, i.e.,

$$K_0 = \frac{\omega_{2\mathrm{max}} - \omega_{2\mathrm{min}}}{u_{f\mathrm{max}} - u_{f\mathrm{min}}}$$

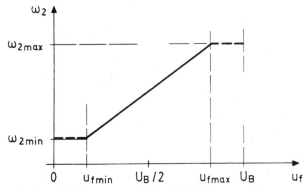

Figure 3.21 Characteristic of the VCO. The curve shows angular frequency ω_2 vs. control voltage u_f. Note that the useful voltage range is less than the supply voltage.

Now the external components of the VCO can be determined also. Usually the data sheets tell you how to calculate the values of these elements.

Step 10. Calculation of natural frequency ω_n. Given noise bandwidth B_L and damping factor ζ, ω_n can be calculated from

$$B_L = \frac{\omega_n}{2} \left(\zeta + \frac{1}{4\zeta} \right)$$

When the divider ratio N is variable, both B_L and ζ have been determined for $N = N_{\mathrm{mean}}$; thus the value obtained for ω_n is valid only for $N = N_{\mathrm{mean}}$ (see the notes in step 2).

Step 11. Specification of the loop filter. The appropriate type of loop filter must be chosen first. The passive loop filter is the simplest; if it is combined with an EXOR or JK-flipflop phase detector, however, the pull-in range becomes limited. If infinite pull-in range is desired, the active PI loop filter should be selected. Given ω_n, ζ, K_0, K_d and N, the equations for ω_n and ζ in Table 3.1 are used to calculate the two time constants τ_1 and τ_2, respectively. When the active lag filter is used, a suitable value for K_a must be chosen in addition. Only values of $K_a > 1$ are reasonable, of course. Typically, K_a is in the range 2 . . . 10. The design proceeds with step 19.

Step 12. (Continued from step 4). Because noise at the reference input can be discarded, the best choice for the phase detector is the PFD. When the PFD runs from a unipolar power supply with supply voltage U_B, its detector gain K_d is given by $K_d = U_B/4\pi$. When a bipolar power supply is used or when the logic levels differ substantially from the voltages of the supply rails, use the general formula given in Sec. 3.1.

Step 13. In this step the characteristic of the VCO will be determined. Because the center frequency ω_0 (or the range of ω_0) is known and the range of the divider ratio N is also given, we can calculate the range of output (angular) frequencies which must be generated by the VCO. Let $\omega_{2\mathrm{min}}$ and $\omega_{2\mathrm{max}}$ be the minimum and maximum output frequencies of the VCO, respectively. A suitable VCO must be selected now. In most cases, the VCO will be part of the DPLL system, typically an integrated circuit. Given the supply voltage(s) of the VCO, the data sheet indicates the usable range of control voltage u_f. (For a unipolar power supply with $U_B = 5$ V this range is typically 1 to 4 V.) Let $u_{f\mathrm{min}}$ and $u_{f\mathrm{max}}$ be the lower and upper limits of that range, respectively. The VCO is now designed such that it generates the output frequency

$\omega_2 = \omega_{2min}$ for $u_f = u_{fmin}$ and the output frequency $\omega_2 = \omega_{2max}$ for $u_f = u_{fmax}$. The corresponding VCO characteristic is plotted in Fig. 3.21. Now the VCO gain K_0 is calculated from the slope of this curve, i.e.,

$$K_0 = \frac{\omega_{2max} - \omega_{2min}}{u_{fmax} - u_{fmin}}$$

Now the external components of the VCO can be determined also. Usually the data sheets tell you how to calculate the values of these elements.

Step 14. Specify the type of loop filter. Because the PFD is used as phase detector, the passive lag filter will be used in most cases. This combination of phase detector and loop filter offers infinite pull-in range and hold range, as explained in Sec. 3.2. Other types of loop filters do not bring additional benefits.

Step 15. Determination of dynamic properties of the DPLL. To select an appropriate value for the natural frequency ω_n, we must have an idea of how the DPLL should react on dynamic events, e.g., on frequency steps applied to the reference input, on a variation of the divider factor N, or the like. Specification of dynamic properties strongly depends on the intended use of the DPLL system. Because different goals can be envisaged, this design step is a decision block having three outputs; i.e., there are (at least) three different ways of specifying dynamic performance of the DPLL.

In the first case, the DPLL is used as a digital frequency synthesizer. Here it could be desirable that the DPLL switches very fast from one output frequency f_{21} to another output frequency f_{22}. If the difference $|f_{21} - f_{22}|$ is large, the DPLL will probably lock out when switching from one frequency to the other. The user then would probably specify a maximum value for the pull-in time T_p the system needs to lock onto the new output frequency. T_p is then used to determine the remaining parameters of the DPLL. If the user decides to specify T_p as a key parameter, the procedure continues at step 16.

In the second case, the DPLL is also used as a digital frequency synthesizer. This synthesizer will generate integer multiples of a reference frequency f_{ref}; i.e., the frequency at the output of the VCO is given by $f_2 = N f_{ref}$, where N is variable. In many synthesizer applications it is desired that the DPLL does not lock out if the output frequency changes from one frequency "channel" to an adjacent "channel," i.e., if f_2 changes from $N_0 f_{ref}$ to $(N_0 + 1) f_{ref}$. In this case, the pull-out range $\Delta\omega_{PO}$ is required to be less than f_{ref}. If the user decides to use $\Delta\omega_{PO}$ as a key parameter, the procedure continues at step 20.

The third case of this decision step represents the more general situation where neither the pull-in time nor the pull-out range are of primary interest. Here the user must resort to a specification which makes as much sense as possible. Probably the simplest way is to make an assumption on the lock-in time T_L (also referred to as settling time) or even to specify the natural frequency immediately. If the third case is chosen, the design proceeds with step 21.

Step 16. Given the maximum pull-in time T_p allowed for the greatest frequency step at the output of the VCO, the time constant τ_1 (or the sum of both time constants $\tau_1 + \tau_2$) of the loop filter is calculated using the formula for T_p given in Table 3.1. When the loop filter is passive, we only can determine the sum $\tau_1 + \tau_2$. For the other types of loop filters τ_1 can be computed directly. For $\Delta\omega_0$ the maximum (angular) frequency step at the VCO output must be entered. The design proceeds with step 17.

Step 17. When the loop filter is passive, we computed the sum $\tau_1 + \tau_2$ in step 16. Hence we can calculate ω_n immediately from the corresponding equation for ω_n in Table 3.1. For the other loop filter types, τ_1 was computed alone in step 16. ω_n can be computed now using the appropriate equation for ω_n in Table 3.1. The design proceeds with step 18.

Step 18. Given ω_n and ζ now, time constant τ_2 has to be calculated using the formula for ζ given in Table 3.1. When the sum $\tau_1 + \tau_2$ has been computed in step 16, τ_1 can now be determined. When a passive loop filter is used, strange things may happen at this point: we perhaps obtained $\tau_1 + \tau_2 = 300$ μs in a previous step and computed $\tau_2 = 400$ μs right now! To realize the intended system we would have to choose $\tau_1 = -100$ μs which is impossible. Whenever we meet that situation, the system cannot be realized with the desired goals (i.e. with the desired values for ω_n and/or ζ). It only becomes realizable if we choose a lower ω_n, e.g., which increases $\tau_1 + \tau_2$, or if we specify a lower ζ, which decreases τ_2.

Step 19. Given τ_1 and τ_2 (plus eventually K_a), the components of the loop filter can be determined. This has been explained in great detail in Sec. 2.8, step 8 and is not repeated here.

Step 20. Given pull-out range $\Delta\omega_{PO}$ and damping factor ζ, the formula for $\Delta\omega_{PO}$ given in Table 3.1 is used to calculate the natural frequency ω_n. The design proceeds with step 22.

Step 21. Given the lock-in time T_L, the natural frequency ω_n is calculated from the formula for T_L in Table 3.1. The procedure continues with step 22.

Step 22. Given ω_n, the time constant τ_1 can be calculated using the formula for ω_n in Table 3.1, when the loop filter is either an active lag or a PI Filter. When the passive lag filter is used, however, only the sum $\tau_1 + \tau_2$ can be computed from the formula for ω_n. In the latter case, the final value for τ_1 will be known after τ_2 has been calculated in step 18. The design proceeds with step 18.

Though the design procedure seems quite complex, it would be premature to consider it universal. Some more special parameters are not at all taken into account, e.g., spectral purity of the output signal of frequency synthesizers and related quantities. We have to consider such effects when discussing the most important DPLL applications. A practical design example which makes use of the design procedure will also be carried out in Sec. 3.5.

3.5 DPLL Applications

The applications of the DPLL are so widespread that we must concentrate on the most important ones. In the following we discuss the usage of DPLLs in frequency synthesis, clock signal recovery, and motor-speed control. This section also includes a case study on the design of a PLL frequency synthesizer which uses the DPLL design procedure pointed out in Sec. 3.4.

3.5.1 Frequency synthesis

Frequency synthesis is one of the major applications of the PLL. In Sec. 3.1 we saw how the digital PLL can be used to generate a frequency that is an integer multiple of a reference frequency. The basic configuration used for frequency synthesis is shown in Fig. 3.22a. This system is capable of generating output frequencies which are an *integer* multiple of the reference frequency f_1. Later in this section it will be demonstrated that even *fractional* multiples of the reference frequency can be generated.

Frequency synthesizers are found in FM receivers, CB transceivers, television receivers, and the like. In these applications there is a need for generating a great number of frequencies with a narrow spacing of 50, 25, 10, 5, or even 1 kHz. If a channel spacing of 10 kHz is desired, a reference frequency of 10 kHz is normally chosen. Most oscillators should be quartz-crystal-stabilized. A quartz crystal oscillating in the kilohertz region is quite a bulky component. It is therefore more convenient to generate a higher frequency, typically in the

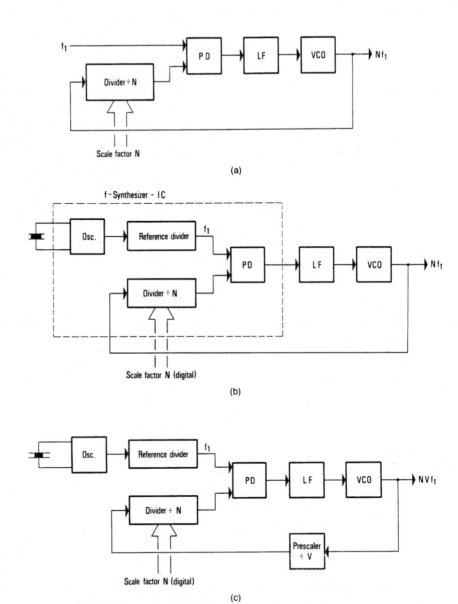

Figure 3.22 Different frequency-synthesizer circuits. (a) Basic frequency-synthesizer system. (b) System equal in performance to (a) but with an additional reference divider that makes it possible to use a higher-frequency reference, normally a quartz oscillator. (c) System extends the upper frequency range by using an additional high-speed prescaler. The channel spacing is increased to Vf_1.

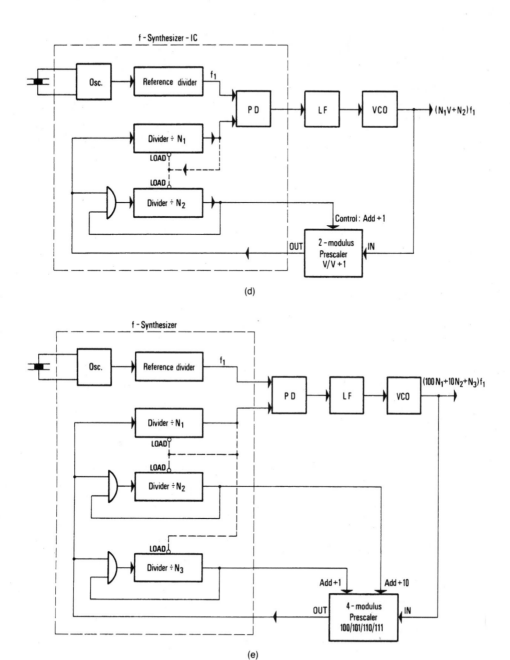

(d)

(e)

Figure 3.22 (*Continued*) (*d*) System similar to (*c*), employing a dual-modulus prescaler. The channel spacing is reduced to f_1. (*e*) System similar to (*d*), but using a four-modulus instead of a dual-modulus prescaler. This not only extends the high-frequency limit, but also allows the generation of lower frequencies.

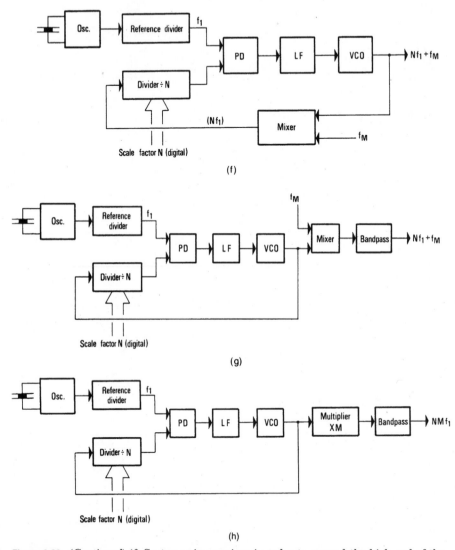

Figure 3.22 *(Continued)* (*f*) System using a mixer in order to expand the high end of the frequency range. (*g*) System similar to (*f*), but with the mixer placed outside of the loop. (*h*) System using an external frequency multiplier for extending the high-frequency end.

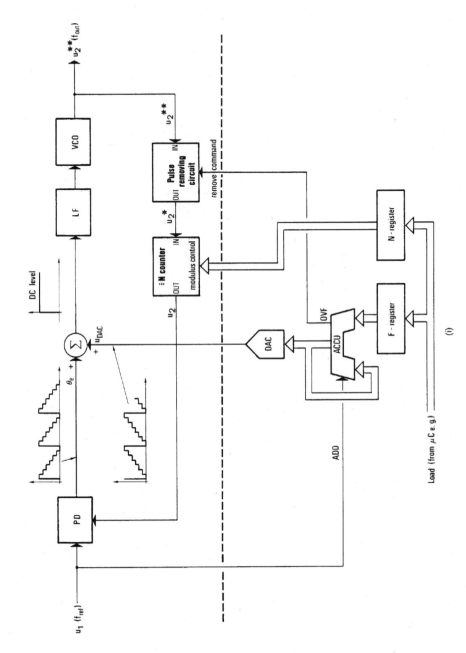

Figure 3.22 (*Continued*) (*i*) Fractional *N*-loop frequency synthesizer. This system allows the generation of output signals whose frequency is a fractional multiple of the reference frequency f_{ref}.

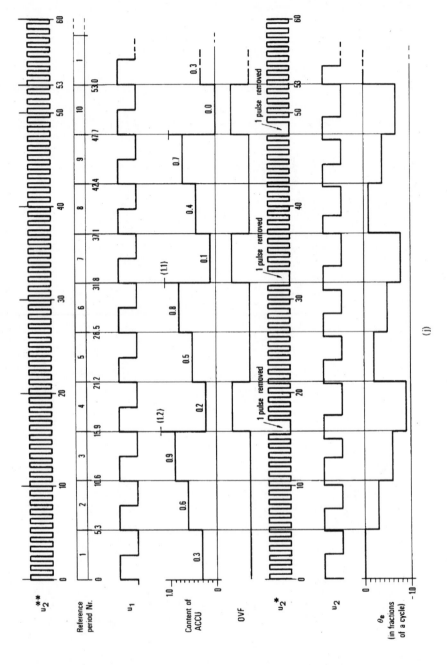

Figure 3.22 (*Continued*) (*j*) Waveforms explaining the operating principle of the fractional *N* loop.

region of 5 to 10 MHz, and to scale it down to the desired reference frequency. In most of the frequency-synthesizer ICs presently available, a reference divider is integrated on the chip, as shown in Fig. 3.22b. The oscillator circuitry is also included on most of these ICs.

One seeks to include as many functions on the chip as possible. It is no major problem to implement all the digital functions on the chip, such as oscillators, phase detectors, frequency dividers, and so on, as indicated by the dashed enclosure in Fig. 3.22b. Due to its low power consumption, high noise immunity, and large range of supply voltages, CMOS is the preferred technology today. The limited speed of CMOS devices precludes their application for *directly* generating frequencies in the range of 100 MHz or more (at least at the time of this writing). To generate higher frequencies, *prescalers* are used; these are built with other IC technologies such as ECL or Schottky TTL (Fig. 3.22c). Such prescalers extend to well beyond 1 GHz the range of frequencies which can be synthesized directly, that is, without mixing techniques.[7,37]

If the scaling factor of the prescaler is V (Fig. 3.22c), the output frequency of the synthesizer becomes

$$f_{\text{out}} = NVf_1$$

Obviously the scaling factor V of the prescaler is much greater than 1 in most cases. This implies that it is no longer possible to generate every desired integer multiple of the reference frequency f_1; if V is say, 10, only output frequencies of $10f_1$, $20f_1$, $30f_1$, . . . can be generated. This disadvantage can be circumvented by using a so-called dual-modulus prescaler, as shown in Fig. 3.22d.[11,38]

A dual-modulus prescaler is a counter whose division ratio can be switched from one value to another by an external control signal. As an example, the prescaler in Fig. 3.22d can divide by a factor of 11 when the applied control signal is HIGH, or by a factor of 10 when the control signal is LOW.

It can be demonstrated that the dual-modulus prescaler makes it possible to generate a number of output frequencies which are spaced only by f_1 and not by a multiple of f_1. The following conventions are used with respect to Fig. 3.22d:

1. Both programmable $\div N_1$ and $\div N_2$ counters are DOWN counters.

2. The output signal of both of these counters is HIGH if the content of the corresponding counters has not yet reached the value 0.

3. When the $\div N_1$ counter has counted down to 0, its output goes LOW and it immediately loads both counters to their preset values N_1 and N_2, respectively.

4. N_1 is always greater than or equal to N_2.

5. As shown by the AND gate in Fig. 3.22d, underflow below 0 is inhibited in the case of the $\div N_2$ counter. If this counter has counted down to 0, further counting pulses are inhibited.

The operation of the system shown in Fig. 3.22d becomes clearer if we assume that the $\div N_1$ counter has just counted down to 0 and both counters have just been loaded with their preset values N_1 and N_2, respectively. We now have to find the number of cycles the VCO must produce until the same logic state is reached again. This number is the overall scaling factor of the arrangement shown in Fig. 3.22d.

As long as the $\div N_2$ counter has not yet counted down to 0, the prescaler is dividing by $V + 1$. Consequently both the $\div N_1$ and the $\div N_2$ counters will step down by one count when the VCO has generated $V + 1$ pulses. The $\div N_2$ counter will therefore step down to 0 when the VCO has generated $N_2(V + 1)$ pulses. At that moment the $\div N_1$ counter has stepped down by N_2 counts, that is, its content is $N_1 - N_2$.

The scaling factor of the dual-modulus prescaler is now switched to the value V. The VCO will have to generate additional $(N_1 - N_2)V$ pulses until the $\div N_1$ counter will step to 0. When the content of N_1 becomes 0, both the $\div N_1$ and the $\div N_2$ counters are reloaded to their preset values, and the cycle is repeated.

How many pulses N_{tot} did the VCO produce in order to run through one full cycle? N_{tot} is given by

$$N_{\text{tot}} = N_2(V + 1) + (N_1 - N_2)V$$

Factoring out yields the simple expression

$$N_{\text{tot}} = N_1 V + N_2 \qquad (3.22)$$

As stated above, N_1 must always be greater than or equal to N_2. If this were not the case, the $\div N_1$ counter would be stepped down to 0 *earlier* than the $\div N_2$ counter, and both counters would then be reloaded to their preset values. The dual-modulus prescaler *never* would be switched from $V + 1$ to V, so the system could not work in the intended way.

If $V = 10$, Eq. (3.22) becomes

$$N_{\text{tot}} = 10N_1 + N_2 \qquad (3.23)$$

In this expression, N_2 represents the units and N_1 the tens of the overall division ratio N_{tot}. Then N_2 must be in the range of 0–9, and N_1 can assume any value greater than or equal to 9, that is, $N_{1\text{min}} = 9$. The smallest realizable division ratio is therefore

$$N_{tot\ min} = N_{1\ min}V = 9 \cdot 10 = 90$$

The synthesizer of Fig. 3.22d is thus able to generate all integer multiples of the reference frequency f_1 starting from $N_{tot} = 90$.

Other factors can of course be chosen for V. If $V = 16$, the dual-modulus prescaler would divide by 16 or 17. The overall division ratio would then be

$$N_{tot} = 16N_1 + N_2$$

Now N_2 would be required to have a range of 0 to 15, and the minimum value of N_1 would be $N_{1min} = 15$. In this case, the smallest realizable division ratio $N_{tot\ min}$ would be 240.

Let us again assume that V is chosen at 10, and that the (scaled-down) reference frequency f_1 of the system in Fig. 3.22d is 10 kHz. The smallest output frequency would then be $90f_1 = 900$ kHz.

For the circuits inside the dashed enclosure CMOS devices are normally used. The counting frequency of older CMOS ICs (such as the old series 74Cxxx) has been limited to approximately 3 MHz, so using these devices a maximum frequency of only about 30 MHz could be realized for a prescaler ratio of $V = 10$. To extend the frequency range, larger prescaler ratios, say $V = 100$, became desirable. Using $V = 100$, ratio N_{tot} would be

$$N_{tot} = 100N_1 + N_2$$

where N_2 must now cover the range of 0 to 99, and N_1 must be at least 99. It should be noted, however, that now the lowest division ratio $N_{tot\ min}$ is no longer 90 but has been increased to

$$N_{tot\ min} = 100N_{1min} = 100 \cdot 99 = 9900$$

If the reference frequency f_1 is still 10 kHz, the lowest frequency to be synthesized is now 99 MHz.

Fortunately, there is another way, which extends the upper frequency range of a frequency synthesizer but still allows the synthesis of lower frequencies. The solution is the *four-modulus prescaler* (Fig. 3.22e). The four-modulus prescaler is a logical extension of the dual-modulus prescaler. It offers four different scaling factors, and two control signals are required to select one of the four available scaling factors.

As an example, the four-modulus prescaler shown in Fig. 3.22e can divide by factors of 100, 101, 110, and 111.[11,36] By definition it scales down by 100 when both control inputs are LOW. The internal logic of the four-modulus prescaler is designed so that the scaling factor is increased by 1 when one of the control signals is HIGH, or increased by 10 when the other control signal is HIGH. If both control signals are HIGH, the scaling factor is increased by $1 + 10 = 11$.

As seen from Fig. 3.22e, there are no longer two programmable $\div N$ counters in the system, but *three*: $\div N_1$, $\div N_2$, and $\div N_3$ dividers. The overall division ratio N_{tot} of this arrangement is given by

$$N_{tot} = 100N_1 + 10N_2 + N_3$$

In this equation N_3 represents the units, N_2 the tens, and N_1 the hundreds of the division ratio N_{tot}. Here N_2 and N_3 must be in the range 0 to 9, and N_1 must be at least as large as both N_2 and N_3 for the reasons explained in the previous example ($N_{1\,min} = 9$).

The smallest realizable division ratio is consequently

$$N_{tot\,min} = 100 \cdot 9 = 900$$

which is lower roughly by a factor of 10 than in the previous example. For a reference frequency f_1 of 10 kHz, the lowest frequency to be synthesized is therefore $900\,f_1 = 9$ MHz.

Let us examine the operation of the system in Fig. 3.22e by giving a numerical example.

Numerical Example. We wish to generate a frequency that is 1023 times the reference frequency. The division ratio N_{tot} is then 1023; hence $N_1 = 10$, $N_2 = 2$, and $N_3 = 3$ are chosen. Furthermore, we assume that the $\div N_1$ counter has just stepped down to 0, so all three counters are now loaded to their preset values. Both outputs of the $\div N_2$ and $\div N_3$ counters are now HIGH, a condition which causes the four-modulus prescaler to divide initially by 111.

Solution After $N_2 \cdot 111 = 2 \cdot 111 = 222$ pulses generated by the VCO, the $\div N_2$ counter steps down to 0. Consequently, the prescaler will divide by 101. At this moment the content of the $\div N_3$ counter is $3 - 2 = 1$. After another 101 pulses have been generated by the VCO, the $\div N_3$ counter also steps down to 0. The division ratio of the four-modulus prescaler is now 100.

The content of the $\div N_1$ counter is now 7. After another 700 pulses have been generated by the VCO, the $\div N_1$ counter also steps down to 0, and the cycle is repeated. To step through an entire cycle, the VCO had to produce a total of

$$N_{tot} = 2 \cdot 111 + 1 \cdot 101 + 7 \cdot 100 = 1023$$

pulses, which is exactly the number desired.

In all frequency synthesizer systems previously considered, multiples of a reference frequency have been generated exclusively by scaling down the VCO output signal by various counter configurations. To produce frequencies in the range of 98.7 to 118.7 MHz with a spacing of 10 kHz, a synthesizer circuit would have had to be designed to offer an overall division ratio of 987 to 1187. As an alternative, one could first generate output frequencies in the range of 8.7 to 18.7 MHz, using a division ratio of 87 to 187, and then *mix up* the obtained frequency band to the desired band. An additional local oscillator operating at a frequency of 90 MHz would be required in this case.

A frequency synthesizer system using an up-mixer is shown in Fig. 3.22*f*. The basic synthesizer circuit employed here corresponds to the simple system shown in Fig. 3.22*b*. Of course, all synthesizer systems using dual- and four-modulus prescalers can be combined with a mixer. In the system of Fig. 3.22*f* the frequency of the local oscillator is f_M. Consequently the synthesizer produces output frequencies given by

$$f_{\text{out}} = N f_1 + f_M$$

The mixer is used here to *mix down* these frequencies in the *baseband* $N f_1$. The mixer also generates a number of further mixing products effectively, generally frequencies given by

$$f_{\text{mix}} = \pm n f_{\text{out}} \pm m f_M$$

where n and m are arbitrary positive integers.

All these mixing products (excluding the baseband $f_{\text{out}} - f_M$) have frequencies that are very much higher than the baseband, so they are filtered easily by either a low-pass filter or even the PLL system itself.

An alternative arrangement using a mixer is shown in Fig. 3.22*g*. In contrast to the previously discussed system, the mixer is not inside but outside the loop. The system generates output frequencies identical with those in Fig. 3.22*f*, but for obvious reasons the desired frequency spectrum has to be filtered out here by a bandpass filter.

Still another way of extending the upper frequency range of frequency synthesizers is given by the *frequency multiplier,* as shown by Fig. 3.22*h*.

Frequency multipliers are normally built from nonlinear elements which produce harmonics, such as varactor diodes, step-recovery diodes, and similar devices. These elements produce a broad spectrum of harmonics. The desired frequency must therefore be filtered out by a bandpass filter. If M is the frequency-multiplying factor, the output frequency of this synthesizer is

$$f_{\text{out}} = MN f_1$$

It should be noted that now the channel spacing is not equal to the reference frequency f_1, but to $M f_1$.

It seems evident that a programmable divide-by-N counter can divide the incoming frequency by an *integer* scaling factor only, and not by a *fractional number*, such as 10.5. Dividing by 10.5 becomes possible, however, if the divide-by-N counter is made to scale down alternately first by 10 and then by 11. On the average this counter effectively divides the input frequency by 10.5. Fractional division ratios of any complexity can be realized (at least theoretically). A ratio of 5.3 is obtained if a $\div N$ counter is forced to divide by 5 in 7 cycles of each group of 10 cycles and by 6 in the remaining 3 cycles. A ratio of 27.35 is obtained if a $\div N$ counter is forced to divide by 28 in 35 cycles of each group of 100 cycles and by 27 in the remaining 65 cycles.

A frequency synthesizer capable of generating output frequencies which are a fractional multiple of a reference frequency is called a *fractional N loop*.[10,39] Figure 3.22i shows the block diagram of a fractional N loop. Its division ratio is the number $N{:}F$, where N is the integer part and F is the fractional part. For example, if a division ratio of 26.47 is desired, then $N = 26$ and $F = 0.47$.

The upper portion of Fig. 3.22i, separated by a dashed line, shows an ordinary frequency-synthesizer system generating an output signal whose frequency is N times the reference frequency f_{ref}. The block labeled "pulse-removing circuit" and the summing block Σ should be disregarded for the moment. The integer and fractional parts (N and F, respectively) of the division ratio $N{:}F$ are stored in a buffer register as shown in the lower part of Fig. 3.22i. The numbers N and F can be loaded via a serial or parallel data link from a microprocessor.

The F register stores a fraction number; F can be stored in any code (binary, hexadecimal, BCD, and so on). If F is represented as a two-digit fractional BCD number, for example, the F register is an 8-bit register, where the individual states are assigned weights of 0.8, 0.4, 0.2, and 0.1 (tenths register) and 0.08, 0.04, 0.02, and 0.01 (hundredths register).

We now investigate how a frequency synthesizer can divide its output frequency f_{out} by fractional numbers. Let us see, for example, how the system can generate an output frequency $f_{\text{out}} = 5.3 f_{\text{ref}}$. We can understand the operation of the fractional N loop by examining the waveforms shown in Fig. 3.22j. Assume that the output frequency is already 5.3 times the reference frequency, and refer to the waveforms u_2^{**} (VCO output signal) and u_1 (reference signal) in Fig. 3.22j.

The signal u_2^{**} shows 53 cycles during the time interval when the reference signal u_1 is executing 10 reference cycles. (In the following, the term *reference cycle* or *reference period* will be used to designate one full oscillation of the reference signal u_1.) Let the system start at the time $t = 0$.

During the time interval where u_1 generates its first reference cycle, the $\div N$ counter is required to divide the signal u_2^{**} by a factor of 5.3. This is impossible, of course, so the $\div N$ counter will divide by 5 only.

In the first reference period, 0.3 pulse is then "missing." This error (0.3) has to be memorized somewhere in the system; an accumulator (ACCU) is used for this purpose.

The ACCU uses the same code as the F register. If a two-digit fractional BCD format is used, the ACCU is capable of storing fractional BCD numbers within a range of 0.00 to 0.99. As seen from Fig. 3.22i, the ACCU adds the fractional number 0.F supplied by the F register to its original content whenever the ADD signal performs a positive transition, such as at the beginning of each reference period. If we assume that the initial content of the ACCU was zero at $t = 0$, the ACCU will accumulate an error of 0.3 cycle during the first reference period, indicating that the synthesizer has "missed" 0.3 pulse during the first reference period.

In the second reference period the $\div N$ counter is again required to divide by 5.3. Because this is not possible, it will continue to divide by 5 instead. Since it has already missed 0.3 cycle in the first reference period, the total error has now accumulated to 0.6 cycle. In the third reference period the accumulated error is 0.9 cycle, and in the fourth reference period, it is 1.2 cycles. However, the ACCU cannot store numbers greater than 1; consequently it overflows and generates an OVF signal (Fig. 3.22i). The content of the ACCU is now 0.2, as seen in Fig. 3.22j. The OVF pulse generated by the ACCU causes the pulse-removing circuit to become active, and the next pulse generated by the VCO is removed from the $\div N$ counter. This pulse removal has the same effect as if the $\div N$ counter divided by 6 instead of 5.

As Fig. 3.22j shows, the ACCU overflows again in the seventh and tenth reference periods. Three pulses will therefore be removed from the $\div N$ counter in a sequence of 10 reference periods. Because the $\div N$ counter divides by 5 forever, $10 \cdot 5 + 3 = 53$ pulses are produced by the VCO during 10 reference periods. This is exactly what was intended. However, one problem has been overlooked. If the VCO oscillates at 5.3 times the reference frequency f_{ref}, it produces 5.3 cycles during one reference period. The PD in Fig. 3.22i will consequently measure a phase error of -0.3 cycle (or $-0.3 \cdot 2\pi$ rad) after the first reference period in Fig. 3.22j. Thus the phase error has *negative polarity* because the reference signal u_1 *lags* the signal u_2.

After the second reference cycle the phase error has increased to -0.6 cycle, and so on. The phase error θ_e is plotted versus time in Fig. 3.22j; its waveform looks like a staircase.

This phase error is applied to the input of the loop filter and will modulate the frequency of the VCO. Such a staircase-shaped modulation of the VCO frequency is not desired, however, because the pulse-removing technique just discussed has already compensated for the phase error. There is an elegant way to avoid this undesired frequency modulation: The waveforms of Fig. 3.22j show that the content of the ACCU has the same amplitude as the phase error θ_e but opposite polarity. The content of the ACCU is therefore converted to an analog signal by a DAC (digital-to-analog converter); the output signal u_{DAC} is added to the output signal of the phase detector. Since the two staircase signals cancel each other, the input signal to the loop filter is a dc level when the fractional N loop has reached a stable operating point.

3.5.2 Spectral purity of synthesized signals

The frequency synthesizer shown in Fig. 3.22a generates an output signal whose frequency is exactly N times the reference frequency. In the ideal case the VCO would operate at its center frequency. The output signal of the loop filter would then be exactly zero. Under these conditions the output signal of the VCO would be a "pure" square wave, i.e., a signal without any phase jitter. The spectrum of this signal would consist of a line at $f = f_0$ and—as is normal for a rectangular signal—a number of harmonics.

Under normal PLL operation, however, the phase detector delivers correction signals as soon as the phase of the VCO output signal deviates from the ideal phase. If an EXOR gate is used as phase detector, its output signal is a square-wave signal whose frequency is twice the reference frequency, as shown in Fig. 3.3. Because the gain of every loop filter is nonzero at that frequency, a *ripple signal* appears at the input of the VCO. The ripple signal modulates the output signal of the VCO. As we will see, this can lead to spurious sidebands in the VCO output signal, which affects its spectral purity, of course. The effects of the ripple are different for the various types of phase detectors and must therefore be treated separately.

Let us look first at what can happen when the EXOR phase detector is used. In a DPLL frequency synthesizer (Fig. 3.22) with divider ratio $N = 1$, the instantaneous phase of the VCO output signal would be modulated by a signal whose frequency is twice that frequency. Obviously this cannot have an impact on the VCO frequency. When $N = 2$, nothing happens again, because now the instantaneous phase of the VCO output signal is modulated by a signal whose frequency is identical with the VCO frequency. But adverse effects start as soon as N becomes 3 or larger. For general N, the frequency of the VCO output signal is $N f_{\mathrm{ref}}$, and the ripple signal has a fundamental frequency of $2 f_{\mathrm{ref}}$. For $N \geq 3$ spurious sidebands appear at the VCO output whose frequencies are $N f_{\mathrm{ref}} \pm 2 f_{\mathrm{ref}}$, $N f_{\mathrm{ref}} \pm 4 f_{\mathrm{ref}}$, etc.

The situation is similar when the JK-flipflop is used as phase detector. In the locked state the JK-flipflop generates an output signal whose frequency is identical with the reference frequency (Fig. 3.5). If the divider ratio of the frequency synthesizer is 1, the ripple on the VCO input signal only alters the duty cycle of the VCO output signal but does not generate spurious sidebands. Spurious signals appear, however, when N becomes 2 and greater. Then the spectrum of the output signal contains lines at frequencies $N f_{ref} \pm f_{ref}$, $N f_{ref} \pm 2 f_{ref}$, etc.

When the PFD is used as phase detector, its output signal is theoretically in the 0 state, when the VCO operates exactly on its center frequency, as shown in Fig. 3.8a. Under normal DPLL operation, however, the PFD will output correction pulses, as shown in Fig. 3.8b and c. The polarity of the correction pulses will change periodically, so positive and negative correction pulses will appear in succession. Fortunately the duration of the correction is short when the DPLL is locked, so the ripple on the VCO input signal will be much smaller than in the two former cases.

It is of vital interest, of course, to know the power of such sidebands. Because the amplitude of the first pair of sidebands is always much greater than the amplitude of the higher ones, it is sufficient for practical purposes to have an approximation for the first sideband. In analogy to noise signals, sideband suppression S_1 has been defined as the quotient of signal power to the first sideband power:

$$S_1 = \frac{P_s}{P_{sb1}} \tag{3.24a}$$

where P_s is the power of the synthesized signal and P_{sb1} is the power of the first spurious sideband. Mostly, a logarithmic quantity $S_{1(dB)}$ is specified:

$$S_{1(dB)} = 10 \log S_1 \tag{3.24b}$$

If the EXOR or the JK-flipflop is used, an approximation for $S_{1(dB)}$ is given by[15]

$$S_{1(dB)} = 20 \log \frac{K_0 U_B}{2\pi^2 f_r} |F(2\pi f_r)| \qquad \text{(dB)} \tag{3.25}$$

Here f_r denotes ripple frequency. $F(2\pi f_r)$ is the gain of the loop filter at the ripple frequency, and U_B is the supply voltage of the phase detector. In case of the EXOR gate, f_r equals twice the reference frequency. When the JK-flipflop is used, however, f_r is identical with the reference frequency. For the PFD another approximation has been found:[15]

$$S_{1(\mathrm{dB})} = 20 \log (K_0 U_B \tau)|F(2\pi f_r)| \quad (\mathrm{dB}) \qquad (3.26)$$

Here the ripple frequency f_r is not necessarily identical with the reference frequency but can be considerably lower, as will be discussed in the following. τ is the width of the correction pulses. The following consideration yields an approximation for the quantity τ. When a DPLL using the PFD as phase detector operates exactly on its center frequency, the output signal of the PFD is theoretically in the 0 state all the time. The reference signal u_1 and the (scaled-down) output signal u_2' (refer to Fig. 3.23a) would then be exactly in phase. In reality, the frequency of the VCO output signal will slowly drift away, which causes a time lag between these two signals. When this time lag is 10 ps, for example, the PFD will theoretically generate a correction pulse whose duration is 10 ps as well. Because the logical circuits inside the PFD have nonzero propagation delays and rise times, the PFD is never capable of generating such short pulses. It will produce a correction pulse only when the delay between the u_1 and u_2' signals has become greater than a time interval called *backlash*. For high-speed CMOS circuits, the backlash is typically in the range of 2 to 3 ns. The PFD never generates an output pulse shorter than the

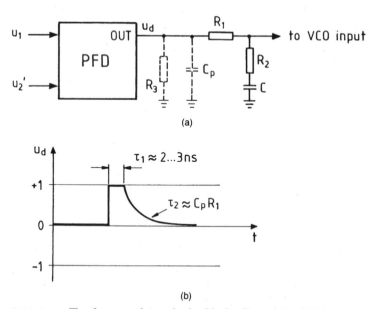

(a)

(b)

Figure 3.23 The figure explains the backlash effect of the PFD and the effect of parasitic capacitance at the phase detector output. (a) Schematic of the PFD including parasitic capacitor C_p. The loop filter is also shown. (b) Waveform of the u_d signal. The duration τ_1 of the output pulse cannot be less than the backlash interval.

backlash interval. Figure 3.23b shows the PFD output signal u_d which as been generated just at the instant where the positive edge of the u_1 signal led the positive edge of u_2' by an amount equal to the backlash, which is denoted τ_1 here. The width of the correction pulse is nearly equal to τ_1 in this case. Unfortunately, each real device contains parasitic capacitances. In our example, parasitic capacitance C_p (Fig. 3.23a) will charge to the supply voltage by the correction pulse, so the decay of that pulse will slow down. Typically, C_p is in the order of 5 to 10 pF. The time constant of the decay is approximately $\tau_2 \approx R_1 C_p$. It can be made small by selecting a low value for resistor R_1, but R_1 cannot be chosen arbitrarily low, because this would overload the PFD output. Typically, R_1 must be higher than about 500 Ω. Thus τ_2 will also be in the order of 5 ns or more. This means that the value of τ in Eq. (3.26) is approximately given by the sum $\tau_1 + \tau_2$ and is typically around 10 ns. Because the correction pulses cannot be arbitrarily narrow, the pulse as depicted in Fig. 3.23b alters the instantaneous frequency of the VCO more than would normally be required. In the example of Fig. 3.23b the frequency of the VCO is increased. Therefore, the time delay between the edges of the u_1 and u_2' becomes shorter in succeeding cycles of the reference signal. After some cycles the delay crosses zero and becomes negative, so that u_2' leads u_1 now. The PFD will only produce a correction pulse (of negative polarity), however, when the time lag exceeds the backlash again. This leads to the unhappy situation that the PFD generates a positive correction pulse at some instant, stays in the zero state during a number of succeeding cycles (perhaps 10), produces a negative correction pulse then, and so forth. The frequency of the ripple signal is therefore a *subharmonic* of the reference signal, and if the frequency of this subharmonic is very low, the ripple is not attenuated by the loop filter, which is undesirable, of course. (It is possible to calculate approximately the subharmonic frequency, using the formulas given in ref. 15. Usually the subharmonic frequency is about 1/20 the reference frequency.)

Subharmonic frequency modulation of the VCO can be eliminated by the addition of resistor R_3 in Fig. 3.23a. Normally, R_3 is a high-value resistor which slowly but steadily discharges filter capacitor C toward ground. If the discharging current is higher than the charging current produced by the parasitic capacitance, the PFD is forced to generate positive correction pulses *in every cycle* of the reference signal. The ripple frequency is now identical with the reference frequency. Because the cutoff frequency of the loop filter is lower than the reference frequency in most cases, the ripple signal will be attenuated by the loop filter, which decreases the level of the spurious sidebands. When resistor R_3 is used, the spurious sidebands occur at frequencies $N f_{\mathrm{ref}} \pm f_{\mathrm{ref}}$, $N f_{\mathrm{ref}} \pm 2 f_{\mathrm{ref}}$, etc.

In high-quality frequency synthesizers, the level of the spurious sidebands is still unacceptable, even if the mentioned tricks are used. To reduce the sidebands further, we must lower the gain of the loop filter at the ripple frequency [cf. Eqs. (3.25) and (3.26)]. The simplest way to do this would be by introducing an additional pole in the loop filter, as shown in Fig. 3.24a to c. Capacitor C_3 introduces an additional time constant $\tau_3 = R_2 C_3$. Normally τ_3 is chosen about $\tau_2/10$.[7] Under this condition the loop filter has an additional break frequency which is approximately at $\omega = 1/\tau_3$. Plotting the Bode diagram of the modified loop filter enables us to estimate the improved sideband suppression.[15] The loop filters shown in Fig. 3.24 are second-order low-

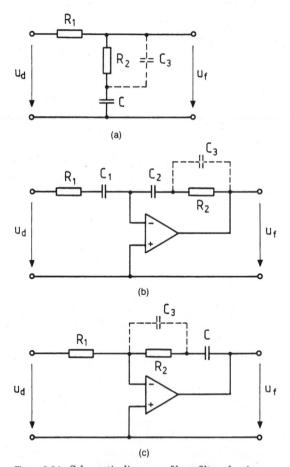

Figure 3.24 Schematic diagram of loop filters having an additional pole to suppress spurious sidebands in digital frequency synthesizers. (a) Passive lag filter. Capacitor C_3 introduces the additional pole. (b) Active lag filter. (c) Active PI filter.

pass filters. To increase ripple rejection even more, loop filters of higher order, typically up to order 5, are used in many designs.[11]

3.5.3 Case study on frequency synthesis

We now use the procedure described in Sec. 3.4 to design a frequency synthesizer which is based on a digital PLL device. As an example the synthesizer should be able to produce a set of frequencies in the range from 1 to 2 MHz with a channel spacing of 10 kHz; i.e., frequencies of 1000, 1010, 1020, . . . , 2000 kHz will be generated. For this design we use the popular 74HC/HCT4076 CMOS device which is based on the old industry standard CD 4046 originally introduced by RCA. This circuit contains three different phase detectors, an EXOR gate, a JK-flipflop, and a PFD. Because noise must not be considered in this design, we use the PFD as phase detector. Because this detector offers infinite pull-in range for any type of loop filter, we use the simplest of these, the passive lag-lead. Here again the PFD forms a *charge pump* in combination with the passive loop filter. The supply voltage U_B is chosen as 5 V. With these assumptions we are ready to start the design, following the procedure shown in Fig. 3.20 and described in Sec. 3.4.

Step 1. Determine ranges of input and output frequencies. The input frequency is constant, $f_1 = 10$ kHz. The output frequency is in the range from 1 to 2 MHz; thus $f_{2min} = 1$ MHz, $f_{2max} = 2$ MHz.

Step 2. The divider ratio must be variable in the range $N = 100 \ldots 200$. The DPLL will be optimized ($\zeta = 0.7$) for the divider ratio $N_{mean} = \sqrt{N_{min}N_{max}} = 141$, as will be shown in step 3.

Step 3. Determination of damping factor ζ. Selecting $\zeta = 0.7$ at $N = N_{mean}$ yields the following minimum and maximum values for ζ:

$$\zeta_{min} = 0.59 \text{ for } N = 200$$

$$\zeta_{max} = 0.83 \text{ for } N = 100$$

This range is acceptable.

Step 4. Noise is not of concern in this DPLL design, so the procedure continues with step 12.

Step 12. Selection of the phase detector type. The PFD is chosen, as noted in the introductory remarks. The phase detector gain becomes $K_d = 5/4\pi = 0.4$ V/rad.

Step 13. VCO layout. According to the data sheet of the 74HC4046A IC, the VCO operates linearly in the voltage range of u_f = 1.1 to 3.9 V approximately. Therefore, the characteristic of Fig. 3.25 can be plotted. If the VCO input voltage exceeds about 3.9 V, the VCO generates a very high frequency (around 30 MHz); if it falls below 1.1 V, the VCO frequency is extremely low, i.e., some Hz. From this characteristic, the VCO gain becomes $K_0 = 2.24 \cdot 10^6$ rad/s/V. Following the rules indicated in the data sheet, resistors R_1, R_2, and capacitor C_1 (Fig. 3.26) are found from graphs. The resistors must be chosen to be in the range 3 . . . 300 kΩ. The parallel connection of R_1 and R_2 should furthermore yield an equivalent resistance of more than 2.7 kΩ. We obtain

$$R_1 = 47 \text{ k}\Omega$$

$$R_2 = 130 \text{ k}\Omega$$

$$C_1 = 100 \text{ pF}$$

Step 14. Loop filter selection. As indicated above, we choose the passive lag filter.

Step 15. Here we must make some assumptions on dynamic behavior of the DPLL. It is reasonable to postulate that the DPLL should lock within a sufficiently short time, e.g., within 1 ms. Hence we set T_L = 1 ms. The procedure continues with step 21.

Step 21. Given T_L, the natural frequency ω_n is calculated:

$$\omega_n = \frac{2\pi}{T_L} \approx 6300 \text{ s}^{-1}$$

The procedure continues with step 22.

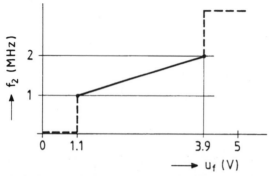

Figure 3.25 Characteristic of the VCO for the CMOS IC type 74HC/HCT4046.

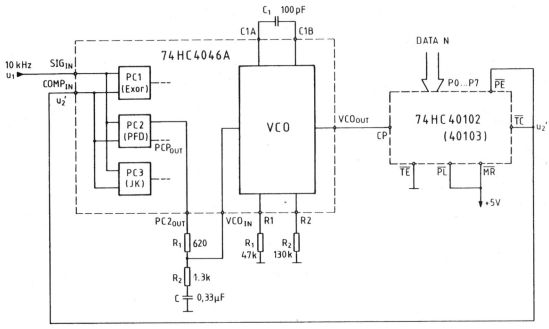

Figure 3.26 Schematic diagram of the DPLL frequency synthesizer designed in Sec. 3.5.3.

Step 22. Because a passive loop filter is used, the formula for ω_n in Table 3.1 can only be used to calculate the sum $\tau_1 + \tau_2$. We obtain

$$\tau_1 + \tau_2 = 644 \ \mu\text{s}$$

The procedure continues with step 18.

Step 18. Given τ_1 and ω_n, we use the formula for ζ in Table 3.1 to calculate τ_2. We obtain

$$\tau_2 = 445 \ \mu\text{s}$$

Now the time constant τ_1 can be computed. Because $\tau_1 + \tau_2 = 644 \ \mu\text{s}$ (from step 22) and $\tau_2 = 455 \ \mu\text{s}$, τ_1 becomes

$$\tau_1 = 199 \ \mu\text{s}$$

Step 19. Calculation of loop filter components. Given τ_1 and τ_2, the loop filter components R_1, R_2, and C can be determined. For optimum sideband suppression, capacitor C should be chosen as large and resistors R_1 and R_2 as low as possible. Selecting $C = 0.33 \ \mu\text{F}$ gives the resistors (rounded to the next values of the R24 series)

$$R_1 = 620 \ \Omega$$

$$R_2 = 1.3 \ \text{k}\Omega$$

The sum of R_1 and R_2 is higher than the minimum allowable load resistance (470 Ω).

The final design is shown in Fig. 3.26. The 74HC4046A DPLL contains three phase detectors, PC1 (EXOR), PC2 (PFD), and PC3 (JK-flipflop). PC2 has two outputs, $PC2_{OUT}$ and PCP_{OUT}. $PC2_{OUT}$ is the phase PFD output, and $PC2_{OUT}$ is an in-lock detection signal, i.e., a logical signal which becomes "high" when the PLL has acquired lock. Only the $PC2_{OUT}$ is used in this application. For the down-scaler, an 8-bit presettable down counter of type 74HC40102 or 74HC40103 is used. The 74HC40102 consists of two cascaded BDC counters, whereas the 74HC40103 is a binary counter. When the PE (preset enable) input is pulled "low," the data on input port P0 . . . P7 are loaded into the counter. The counter counts down on every positive transition at the CP input (count pulse). If the counter has counted down to 0, the TC output (terminal count) goes low. Connecting TC with PE forces the counter to reload the data on the next counting pulse. If N is the number represented by the data bus, the counter divides by $N + 1$ (and not by N). To scale down by a factor of 100, for example, we must therefore apply $N = 99$ to the input port.

As mentioned earlier, the Philips company provides a diskette with a design program for this type of DPLL IC.[15] The author repeated the design using this program and got design parameters very similar to those obtained by his own procedure.

3.5.4   Clock signal recovery

The DPLL is used in almost every digital communication link. Digital data can be transmitted over serial or parallel data channels. In any case, the data are clocked, i.e., they are sent in synchronism with a clock signal, which is normally not transmitted. Thus the receiver is forced to extract the clock information out of the received signal.

There are basically two different principles of transmitting digital data, namely, *baseband* and *carrier-based* transmission. In baseband transmission, the digital signal is directly sent over the link; when a carrier system is used, however, the digital signal is first modulated onto a carrier, usually a high-frequency signal. Amplitude (AM), frequency (FM), and phase (PM) modulation are used here, or it is even possible to combine different modulation techniques, such as AM + PM.[20] In carrier systems, a number of digital signals can be modulated onto different carriers; hence the bandwidth of carrier systems can be much larger than the bandwidth of baseband systems.

In this section we deal only with baseband systems. Let us assume that a digital signal, i.e., an arbitrary sequence of 0s and 1s, has to be transmitted over a serial link, using, for example, the familiar RS-232, RS-422, or the RS-485 standard.[28] As we will immediately see, this problem is more serious than would be expected at a first glance. First of all, it would seem obvious to transmit the signal such that the 1s are represented as a "high" voltage level and the 0s as a "low" voltage level, e.g., 5 V and 0 V, respectively. This simplest data format is sketched in the first row of Fig. 3.27 and is referred to as "NRZ-Code" (NRZ = nonreturn to zero). In effect, this is the most commonly used code in *asynchronous* communications. In asynchronous communication, only one single character (an ASCII character in most cases) is transmitted in one message. Such a character mostly consists of a series of 8 data bits. The data bits are headed by a start bit; a stop bit is transmitted after the data bits; eventually a parity bit is used to enable the receiver to detect single-bit errors. The voltage level on the transmission line is usually "high" when no data are transmitted. The start bit is always a "low" level. The data bits can be "high" or "low," but the stop bit is always "high." This ensures that every message starts with a high-to-low transition of the data signal, so the receiver "knows" exactly when the transmission of a character starts. The situation becomes more serious, however, when not a few but hundreds or thousands of bits have to be transmitted in succession. If the NRZ code were used and say 1000 1s were transmitted in succession, no clocking information would be available during that interval. If the clock generator within the data receiver drifts only slightly away, the receiver will not be able to decide whether 999, 1000, or 1001 1s have been sent.

To provide more clocking information, a number of other codes have been devised, as shown in Fig. 3.27. To start with the simplest, the NRZ-mark code produces a level transition whenever a 1 is transmitted. On the other hand, the NRZ-space code generates an edge on the transmission of a 0. As is easily seen, the NRZ-mark code easily fails when a long sequence of 0s is sent; the same problem occurs with the NRZ-space code when a series of 1s are transmitted. The RZ code (return to zero) generates a sequence of "high-low" whenever a 1 is transmitted. This code is prone to failure if long sequences of 0s occur in the data stream. A better choice is the biphase-level code, which produces a sequence "high-low" for a 1 and a sequence "low-high" for a 0. This code shows up at least one edge in every bit cell. (A bit cell is the time interval in which one bit is transmitted.) Because clock recovery is very simple with this code, the biphase code is the most often used in high-speed data communications. A drawback of this code is in the fact that the required bandwidth to transmit a given number of bits per second is twice the bandwidth of the NRZ code.[20]

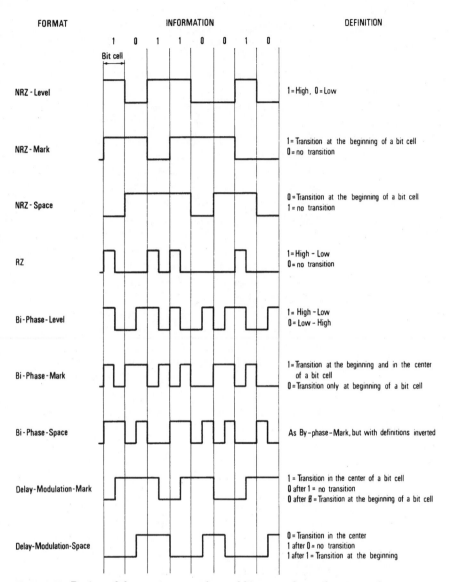

Figure 3.27 Review of the most commonly used binary and pseudo-ternary formats.

There is another code called *delay-modulation code* which combines the advantages of the biphase and NRZ codes. In the delay-modulation-mark code, a transition in the center of a bit cell is generated on the transmission of a 1. If a 0 follows a 1, no transition takes place in the next bit cell. If another 0 follows the first 0, however, a transition at the start of the next bit cell is generated. The bandwidth of this code is almost the same as for the NRZ code.

From this discussion it becomes evident that the methods of extracting clock information must depend on the particular code used. Let us have a look at the most important recovery techniques.

Let us start with recovering the clock signal for the RZ format (Fig. 3.27). The corresponding block diagram is shown in Fig. 3.28. Clock-signal recovery is accomplished by one DPLL. The center frequency of this PLL is chosen to be approximately equal to the baud rate of the data signal. The PLL is synchronized by the transitions of the demodulated data signal u_z^*, as shown in Fig. 3.28. If an EXOR PD is used, the VCO generates a square-wave signal which is in quadrature to the demodulated data signal, as seen from the waveforms in Fig. 3.28. This phase relationship is advantageous, for the positive transitions of the VCO output signal can be used to strobe the data signal.

Synchronization of the clock-recovery PLL takes place on every logic 1 contained in the data signal. During a succession of logic 0s the VCO continues to oscillate at its instantaneous frequency. Extended sequences of 0s have to be avoided since the frequency of the VCO could drift away to the extent that synchronization gets lost.

Longer sequences of 0s are avoided if *parity checking* is used. In parity checking an additional bit of information is added to each group of, say, eight successive data bits. When odd parity is chosen, the total number of 1s, including the parity bit, must be *odd*. The choice of odd parity then ensures that in every sequence of nine bits there is *at least one bit* that is not a 0.

One problem still needs attention: the problem of *initialization*. Every message is finished and a pause will follow the message. During

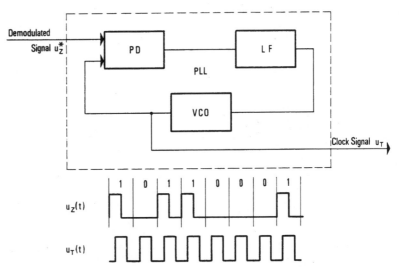

Figure 3.28 Operating principle for clock recovery in the RZ format.

the pause no signal is transmitted, a situation which is identical to a long sequence of 0s in the case of the RZ code. When a new message is started, synchronization is likely to be lost. Synchronization must be reestablished; this is done by a fixed preamble which precedes every message. In the case of the RZ code, a typical preamble consists of a series of 1s.

As stated earlier, the major drawback of the RZ format is the large bandwidth required. The NRZ format needs about half that bandwidth, but clock-signal recovery is slightly more difficult because the frequency spectrum of the NRZ does not necessarily contain a spectral line at the clock frequency.[1,20]

One method of extracting the clock frequency from the NRZ signal is shown in Fig. 3.29. The demodulated NRZ signal u_z^* is first differentiated. Then, the differentiated signal u_{diff} is rectified by an absolute value circuit.[20,34] As seen in the waveforms in Fig. 3.29, the rectified signal u_{diff}^* contains a frequency component synchronous with the clock. Hence it can be used to synchronize a PLL; the output signal of its VCO is the recovered clock signal u_T. Many analog-differentiator and absolute-value circuits have been developed; both operations can be performed alternatively by one single digital circuit.

Figure 3.30 shows a so-called *edge-detector* circuit, in which propagation delays of gates are used to produce a pulse on each transition of the input signal. The rise and fall times of the input signal must be shorter than the cumulative propagation delay of the four cascaded

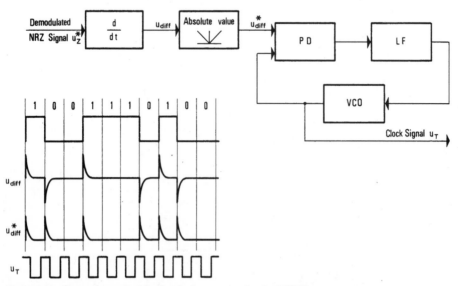

Figure 3.29 Operating principle for clock recovery in the NRZ format.

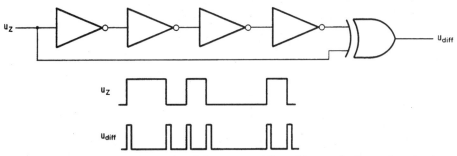

Figure 3.30 The function of an analog differentiator followed by an absolute-value circuit can be alternatively performed by a digital edge detector.

inverters. If this condition is not met, a Schmitt trigger must be used to clean up the transition.

If the biphase or the delay-modulation format is utilized, the demodulated data signal u_z^* can have transitions at the beginning and in the center of a bit cell, as pointed out in Fig. 3.27. As in the circuits in Figs. 3.28 and 3.29, the demodulated signal u_z^* can be used here to synchronize a PLL operating at twice the clock frequency. A circuit recovering the clock signal for the delay-modulation (DM) format is shown in Fig. 3.31.[27] The waveforms produced by this device are depicted in Fig. 3.32.

An edge-detector circuit produces a short positive pulse u_{f1} on every transition (positive and negative) of the demodulated signal u_z^*. The signal u_{f1} is used to synchronize a PLL which operates at twice the clock frequency $2f$ (Fig. 3.32). The output signal of the VCO ($2f$) is scaled down in frequency by a factor of 2; the rightmost JK-flipflop of Fig. 3.31 is used for this purpose.

The clock signal u_T is now defined to be LOW in the first half of every bit cell and HIGH in the second half. We can see from Fig. 3.31 that the circuit could also settle at the *opposite phase* (u_T being HIGH in the first half of the bit cells). This would result in a faulty interpretation of the received data, because the start and the center of every bit cell are erroneously exchanged.

To establish the correct phase of the recovered clock signal, it is necessary initially to reset the JK-flipflop at the right time. For clarity assume that the recovered clock signal u_T really has the *wrong phase* at the beginning, as shown in Fig. 3.32. An additional circuit is needed, which takes corrective action. Consider again the signal u_{f1} in Fig. 3.32. The time interval between any two consecutive pulses of u_{f1} is not constant, but can show the values of 1, 1½, or 2 periods of a bit cell. An interval longer than 2 periods is impossible. If a three-stage binary UP counter (labeled as the 101 detector in Fig. 3.31) is

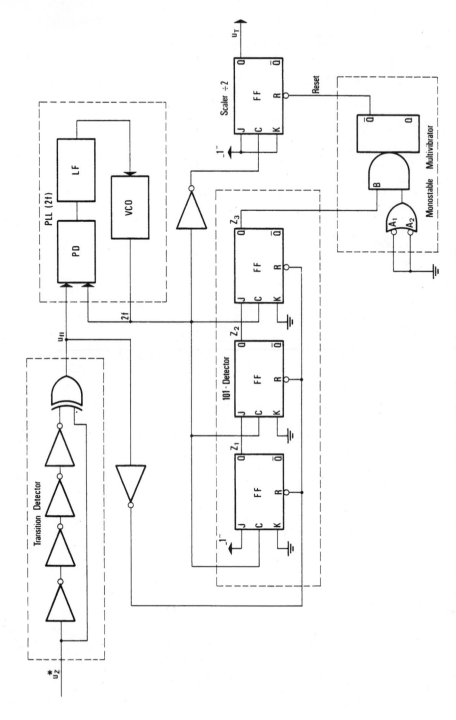

Figure 3.31 Circuit for clock signal recovery with the DM (delay modulation) format. All flipflops are triggered onto the negative-going edge of the clock signal C. Input B of the monostable multivibrator triggers on the positive-going edge; inputs A_1 and A_2 trigger on the negative-going edge. Inputs A_1 and A_2 are unused here.

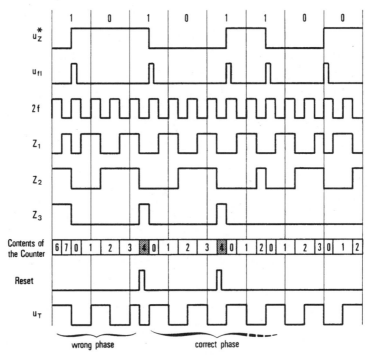

Figure 3.32 Waveforms of the circuit of Fig. 3.31.

used to count the (negative) transitions of the signal $2f$ and if this counter is reset by every u_{f1} pulse, its content never exceeds 4. Moreover, as we can clearly see in Fig. 3.32 the content of 4 can only be obtained in the first half of a bit cell, and never in the second half.

This fact is used to properly reset the divide-by-2-flipflop in Fig. 3.31. Whenever the 3-bit counter reaches 4, a monostable multivibrator (single shot) is triggered. This resets the JK-flipflop. As shown by the waveforms, the phase of the recovered clock is set to its correct state. To enable the corrective action, a preamble of the form 101 ... should precede every message.

It is no major problem to conceive clock-recovering circuits for other formats. As seen from the three examples discussed, the PLL is the key element in each of the applications.

3.5.5 Motor-speed control

Very precise motor-speed controls at low cost are possible using the PLL. The advantages offered by the PLL technique become evident if the PLL motor-speed controls are first compared with conventional

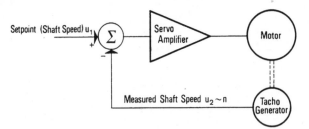

Figure 3.33 Block diagram of a conventional motor-speed control system.

motor-speed controls. A classical scheme for motor-speed control is sketched in Fig. 3.33. The set point for the motor speed is given by the signal u_1. The shaft speed of the motor is measured with a tachometer; its output signal u_2 is proportional to the motor speed n. Any deviation of the actual speed from the set point is amplified by the servo amplifier whose output stage drives the motor. The gain of the servo amplifier is usually high but finite; to drive the motor, a nonzero error must exist.

Other sources of errors are nonlinearities of the tachometer and drift of the servo amplifier. A further drawback is the relatively high cost of the tachometer itself.

The PLL technique offers a much more elegant solution. Figure 3.34 is the block diagram of a PLL-based motor-speed control system. The entire control system is just a DPLL in which the VCO is replaced by a combination of a motor and optical tachometer, as shown in Fig. 3.34. The tachometer signal is generated by a fork-shaped optocoupler in which the light beam is chopped by a sector disk; the detailed circuit is shown in Fig. 3.35. The optocoupler is usually fabricated from a

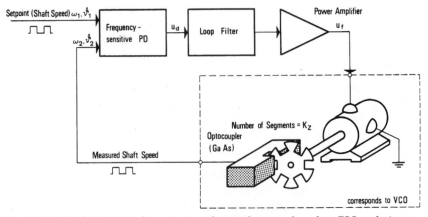

Figure 3.34 Block diagram of a motor-speed control system based on PLL techniques.

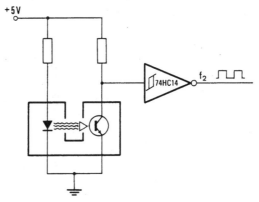

Figure 3.35 Circuit of the optical tachometer generator shown in Fig. 3.34.

light-emitting diode (LED) and a silicon phototransistor. In order to obtain a clean square-wave output, the phototransistor stage is normally followed by a Schmitt trigger, such as 74HC/HCT14-type CMOS device.

The signal generated by the optocoupler has a frequency proportional to the speed of the motor. Because the phase detector compares not only the frequencies ω_1 and ω_2 of the reference and the tachometer signals but also *their phases,* the system settles at *zero-velocity error.* To enable looking at every initial condition, the PFD is used as phase detector.

To analyze the stability of the system, the transfer functions of all blocks in Fig. 3.34 must be known. The transfer functions of the PD and of the loop filter are known, but the transfer function of the motor-tachometer combination still must be determined.

If the motor is excited by a voltage step of amplitude u_f, its angular speed $\omega(t)$ will be given by

$$\omega(t) = K_m u_f \left[1 - \exp\left(-\frac{t}{T_m}\right) \right] \tag{3.27}$$

where K_m is the proportional gain and T_m is the mechanical time constant of the motor. The step response of the motor is plotted on the right in Fig. 3.36. Equation (3.27) indicates that ω will settle at a value proportional to u_f after some time. Applying the Laplace transform to Eq. (3.27) yields

$$\Omega(s) = U_f(s) \frac{K_m}{1 + sT_m} \tag{3.28}$$

The phase angle ϕ of the motor is the time integral of the angular

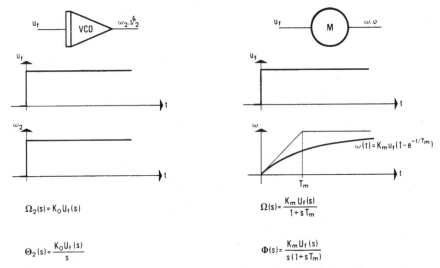

Figure 3.36 The step response of an ordinary VCO compared with that of a motor. Note that the VCO is a first- and the motor a second-order system.

speed ω. Therefore, we get for its Laplace transform $\Phi(s)$ the expression

$$\Phi(s) = U_f(s) \frac{K_m}{s(1 + sT_m)} \tag{3.29}$$

The sector disk shown in Fig. 3.34 has K_Z teeth. This implies that the phase of the tachometer signal is equal to phase ϕ multiplied by K_Z. Consequently we obtain for $\Theta_2(s)$

$$\Theta_2(s) = \frac{K_m K_Z}{s(1 + sT_m)} U_f(s) \tag{3.30}$$

The transfer function $H_m(s)$ of the motor is therefore given by

$$H_m(s) = \frac{K_m K_Z}{s(1 + sT_m)} \tag{3.31}$$

The motor is evidently a second-order system, whereas the VCO [according to Eq. 2.11] was a first-order system only. In Fig. 3.36 the transient response of the motor is compared with that of an ordinary VCO. The motor-speed control system of Fig. 3.34 is therefore a third-order system. The mathematical model of the control system can now be plotted (Fig. 3.37). The servo amplifier is supposed to be a zero-order gain block with proportional gain K_a. The poles of this amplifier

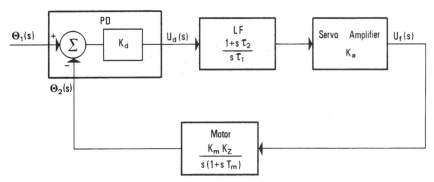

Figure 3.37 Mathematical model of the motor-speed control system of Fig. 3.34.

normally can be neglected because they are at much higher frequencies than the poles of the motor. The closed system has three poles. Therefore, a filter with a zero must to be specified for the loop filter; otherwise the phase of the closed-loop transfer function would exceed 180° at higher frequencies and the system would become unstable. The active PI-filter is chosen here for the loop filter.

The individual blocks in Fig. 3.37 can be combined into fewer blocks, a result which yields the simpler block diagram of Fig. 3.38. In this system the transfer function of the forward path is defined by $G(s)$, whereas the transfer function of the feedback network (motor) is given by $H(s)$.

When a motor-speed control system is designed practically, some parameters are initially given, such as the motor parameters K_m and T_m or the number of teeth K_Z of the sector disk. The remaining parameters (K_a and τ_2) then have to be chosen for best dynamic performance and maximum stability of the system. There are many ways to solve this problem. It is possible to calculate the hitherto unspecified

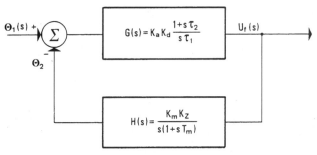

Figure 3.38 Condensed mathematical model of the motor-speed control system.

parameters by purely mathematical methods. In design engineering, however, more practical methods are preferred. To use the simplest one, we optimize DPLL performance by the Bode diagram.[3] In the Bode diagram, amplitude and phase response of the open-loop gain of the system are plotted. From Fig. 3.39, the open-loop gain $G_0(s)$ is given by

$$G_0(s) = G(s)H(s) = K \frac{1 + s\tau_2}{s^2(1 + sT_m)} \qquad (3.32)$$

where K contains the gain factors of the individual blocks,

$$K = \frac{K_d K_a K_m K_z}{\tau_1} \qquad (3.33)$$

In a practical design, phase detector gain K_d, number of teeth K_z, motor gain K_m, and motor time constant T_m are given. The remaining parameters K_a (amplifier gain) and the two time constants τ_1 and τ_2 can be chosen freely. We assume here that

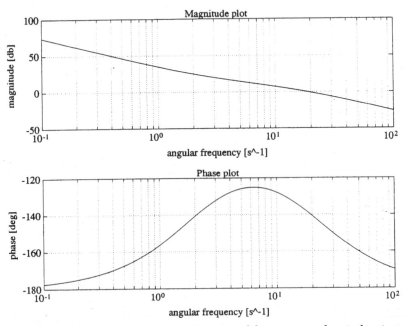

Figure 3.39 Bode diagram for the open-loop gain of the motor-speed control system according to Fig. 3.34.

$$T_m = 0.05 \text{ s}$$

$$K_m = 1$$

We use a math tool called "Control Kit"[26] to plot and optimize the Bode diagram of our control system. The "Control Kit" is a software package which runs under the well-known math program MATLAB.[25] Looking at Eq. (3.32) we see that the magnitude response will start with a slope of -40 dB/decade at low frequencies, due to the term s^2 in the denominator. Above angular frequency $\omega_m = 1/T_m$, the slope will become -60 dB/decade, so at higher frequencies the phase would approach 270°, if there were no zero in the transfer function. The zero of the transfer function must be placed such that the phase stays well below 180° at the frequency where the magnitude curve goes through the 0-dB line. To get acceptable stability of the loop, we require a *phase margin*[3] of at least 30°. We conclude that the break frequency of the zero term $(1 + s\tau_2)$ must be well below the break frequency $1/T_m$, say by a factor of 10. Therefore, τ_2 is tentatively set to 0.5 s. Using the "Control Kit," we vary the overall gain such that the magnitude is just 0 dB at the break frequency $\omega = 1/T_m$. A value of $K = 50$ was required to attain this goal. The resulting Bode diagram is shown in Fig. 3.39. The magnitude curve starts with -40 dB/decade at low frequencies. The transfer function zero at $\omega = 1/\tau_2 = 2$ s^{-1} causes the slope to change to -20 dB/decade; i.e., the magnitude curve "bends up." This causes the phase curve to increase from $-$ 180° toward $-$ 90°, as is seen from the phase plot. At the break frequency $\omega = 1/T_m = 20$ s^{-1}, the slope of the magnitude curve becomes $-$ 40 dB/ decade again. The magnitude crosses the 0-dB line at about $\omega = 20$ s^{-1}, and the phase margin is about 40°, which is sufficient for good transient response. Knowing total gain K, the amplifier gain K_a and the remaining time constant τ_1 can be specified as soon as the number of teeth K_z is given.

3.6   Simulating the DPLL on the PC

Let us perform some computer simulations with the DPLL now.

Case study 1: Influence of the phase detector type.   In a first case study, we will analyze the influence of the phase detector type on the dynamics of the DPLL. After starting the simulation program, click at the *CONFIG* menu and select menu item *Params*. In the *PLL Type Selection* dialog box, click at the *DPLL* radio button. Pressing the *OK* button brings the *DPLL Circuit Selection* dialog box. Select *Passive Lag* for the loop filter, *VCO* (without scaler) for oscillator type, and *EXOR* for phase detector. Press the *OK* button. In the *PLL Parameter Selection* dialog box, enter the following values:

PLL supply voltages:	
Supply+	5.0
Supply−	0
Phase detector:	
$U_{\text{sat}+}$	4.5 V
$U_{\text{sat}-}$	0.5 V
Loop filter:	
τ_1	500 μs
τ_2	50 μs
Oscillator:	
K_0	130,000 $\text{s}^{-1}\,\text{V}^{-1}$
$U_{\text{sat}+}$	4.5 V
$U_{\text{sat}-}$	0.5 V

Pressing the *OK* button brings the message box which lists the predicted values $\omega_n = 17{,}347\ \text{s}^{-1}$ and $\zeta = 0.486$. The formulas in Table 3.1 predict a pull-out range $\Delta f_{PO} = 7719$ Hz and a pull-in range Δf_P of 13,192 Hz. The phase detector gain did not have to be entered in the simulation, because the program calculated it from the exact formula $K_d = (U_{\text{sat}+} - U_{\text{sat}-})/\pi = (4.5 - 1.5)/\pi = 1.27$ V/rad (cf. Sec. 3.1). Now press the *Yes* button in the message box. Select the *SIMULATION* menu and then the *RUN* menu item. In the *DPLL Simulation* dialog box, enter the following parameters:

f-Step radio button	Checked
f_0	100,000
df	2000
dphi	Any value
nSamp	4
T	0.000724 s (set by default)
Filter checkbox	Checked
ZoomFact	1

Press the *Run* button; this brings the picture of Fig. 3.40. Because the frequency step applied to the reference input is much less than the pull-out range, the lock-in process of the DPLL looks like the transient response of a linear second-order system. Now let us check the pull-out frequency. Increasing the frequency step to $df = 8000$ Hz shows the response of Fig. 3.41. The dip in the u_d waveform indicates that the phase error became larger than 90° but less than 180°, hence the frequency would have to be increased very little to pull out the loop. Indeed the loop pulls out for $df = 8100$ Hz (Fig. 3.42), thus Δf_{PO} is around 8000 Hz, which is not far away from the predicted value of 7719 Hz (less than 5 percent error). In the next simulation we will check the pull-in range. A couple of trials with different *df* values show that the loop still pulls in for $df = 14{,}100$ Hz (Fig. 3.43) but no longer

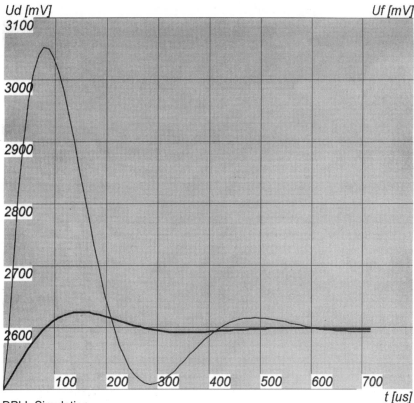

Ud [mV]   *Uf [mV]*

DPLL Simulation
Tue Jan 30 08:48:34 1996 •
PD = EXOR
LF = passive lag
OSC = VCO
Center frequency = 100000 Hz
Response on frequency step 2000 Hz

Figure 3.40 DPLL simulation. Phase detector = EXOR, loop filter = passive lag. Response of the DPLL onto a frequency step Δf_1 = 2000 Hz applied to the reference input.

for df = 14,200 Hz, thus $\Delta f_P \approx$ 14,100 Hz. This is about 6 percent higher than the predicted value. For df = 14,100 Hz, the pull-in time became about 20 ms, which is quite slow. If we tried to compute T_p by the formula given in Table 3.1, we would get $T_p \approx$ 1 ms. What's the matter here? As we stated in Sec. 3.2.3, the formula for T_p only delivers useful values when the initial frequency offset Δf_0 is smaller than about 80 percent of the pull-in range Δf_0. This is clearly not the case here. If you lower the frequency step to say 10 kHz, you will find the pull-in time of the simulation in good agreement with the predicted value.

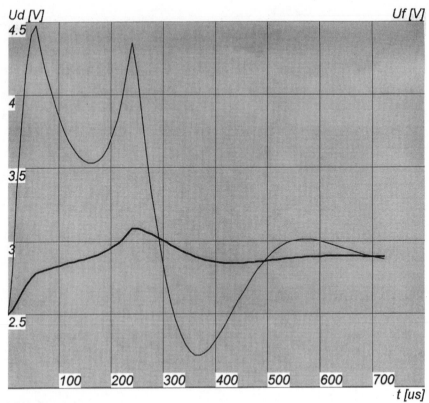

LPLL Simulation
Tue Jan 30 08:49:21 1996 •
PD = EXOR
LF = passive lag
OSC = VCO
Center frequency = 100000 Hz
Response on frequency step 8000 Hz

Figure 3.41 Same as Fig. 3.39, but frequency step $\Delta f_1 = 8$ kHz. A slightly larger frequency step would cause the PLL to unlock temporarily.

As we know from theory, the pull-in range should become "infinite," if we used the PFD phase detector instead of the EXOR gate. We therefore go back to the *CONFIG* menu and replace the EXOR by the PFD. The passive lag is still used for the loop filter. The VCO gain is again chosen as 130,000 rad s^{-1} V^{-1}. Because the VCO is assumed to saturate at an input voltage of $u_f = 4.5$ V, the maximum excursion of u_f from its center value (2.5 V) is 2 V. Hence the VCO is only able to increase the frequency f_2 by $(130,000/2\pi) \cdot 2 \approx 40$ kHz. To avoid driving the VCO into saturation, we apply a frequency step which is slightly less than 40 kHz, say 35 kHz to the reference input of the

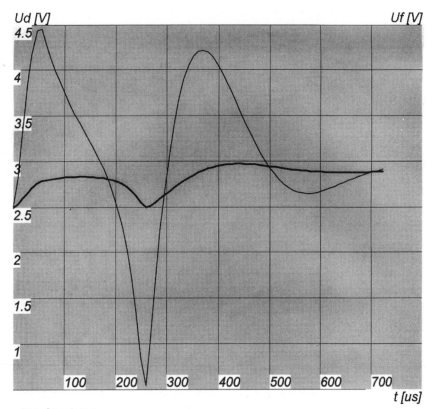

Ud [V]   Uf [V]

LPLL Simulation
Tue Jan 30 08:49:49 1996•
PD = EXOR
LF = passive lag
OSC = VCO
Center frequency = 100000 Hz
Response on frequency step 8100 Hz

Figure 3.42 Same as Fig. 3.40, but the frequency step has been chosen such that the PLL just locks out ($\Delta f_1 = 8.1$ kHz).

DPLL. The result of the simulation is shown in Fig. 3.44. The pull-in process is completed in about 1.5 ms and is followed by a lock-in process. It is interesting to compare the waveform for u_d with the waveform in the previous simulation, where the EXOR phase detector was used. When the EXOR was used, the u_d signal oscillated almost symmetrically around the center line (2.5 V). In case of the PFD, however, u_d is always larger than 2.5 V during the entire pull-in process. The charge pump formed by the series connection of PFD and loop filter pumps charge into the filter capacitor in one direction only, so the pull-in process becomes much faster.

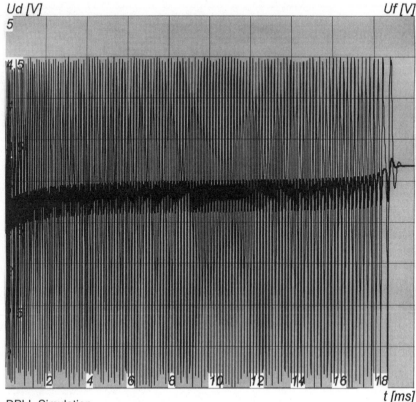

DPLL Simulation
Tue Jan 30 08:51:15 1996 •
PD = EXOR
LF = passive lag
OSC = VCO
Center frequency = 100000 Hz
Response on frequency step 14100 Hz

Figure 3.43 Same as Fig. 3.38, but the frequency step is slightly less than the pull-in range ($\Delta f_1 = 14.1$ kHz).

Case study 2: Influence of the loop filter type.   It has been shown in Sec. 3.2.3 that the pull-in range of the DPLL using the EXOR phase detector can also be infinite when the active PI is used as loop filter. Let us therefore simulate this situation, too. In the *CONFIG* menu we choose the *EXOR* and the *active PI* filter. All other parameters are left unchanged. As Fig. 3.45 shows, the DPLL pulls in in roughly 5 ms. This DPLL is markedly slower than the DPLL simulated previously. This stems from the fact that with a PFD the output frequency of the VCO is pulled in a positive direction during the entire pull-in process,

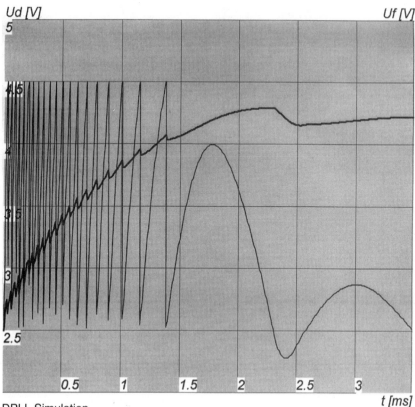

DPLL Simulation
Fri Feb 09 13:22:29 1996 •
PD = Phase-Frequency Detector
LF = passive lag
OSC = VCO
Center frequency = 100000 Hz
Response on frequency step 35000 Hz

Figure 3.44 DPLL simulation. Phase detector = PFD; loop filter = passive lag. The pull-in range is virtually infinite. In practice, the pull-in range equals the frequency range of the VCO ($\Delta f_p = \pm 40$ kHz). To avoid clipping of the u_f signal, a slightly smaller frequency step of $\Delta f_1 = 35$ kHz is simulated here.

whereas with the EXOR phase detector the frequency of the VCO "pumps" up slowly.

Many more things could be simulated using this program. The reader is asked to use his or her own creativity here. Some hints for further studies could be the following:

- Compare the performance of a DPLL using the EXOR with that of DPLL using the JK-flipflop as phase detector.

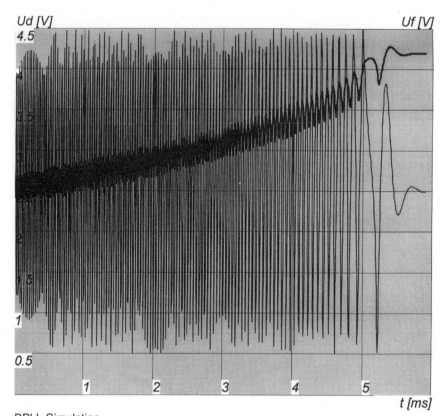

DPLL Simulation
Fri Feb 09 13:24:45 1996 •
PD = EXOR
LF = active PI
OSC = VCO
Center frequency = 100000 Hz
Response on frequency step 35000 Hz

Figure 3.45 This simulation shows that infinite pull-in range can also be achieved by using the EXOR as phase detector and the active PI as loop filter. Response of the DPLL onto a frequency step $\Delta f_1 = 35$ kHz.

■ Study the acquisition and tracking performance of the DPLL frequency synthesizer as designed in Sec. 3.5.3.

4

The All-Digital PLL (ADPLL)

4.1 ADPLL Components

As we have seen in Chap. 3, the classical DPLL is a semianalog circuit. Because it always needs a couple of external components, its key parameters will vary because of parts spread. Even worse, the center frequency of a DPLL is influenced by parasitic capacitors on the DPLL chip. Its variations can be so large that trimming can become necessary in critical applications. Many parameters are also subject to temperature drift.

The all-digital PLL does away with these analog-circuitry headaches. In contrast to the DPLL, it is an entirely *digital* system. Let us know first that the term "digital" is used here for a number of different things. First of all, "digital" means that the system consists exclusively of logical devices. But "digital" also signifies that the signals within the system are digital too. Hence it can be a binary signal (or "bit" signal), as was the case with the classical DPLL, but it can as well be a "word" signal, i.e., a digital code word coming from a data register, from the parallel outputs of a counter, and the like. When discussing the various types of ADPLL, we find the whole palette of such digital signals.

To realize an ADPLL, all function blocks of the system must be implemented by purely digital circuits. Digital versions of the phase detector are already known, but we now have to find digital circuits for the loop filter and for the VCO, too. As we will see in Sec. 4.1.3, the digital counterpart of the VCO is the digital-controlled oscillator (DCO). There are an almost unlimited number of purely digital function blocks for the ADPLL; to save space, we concentrate on the most frequently used.

4.1.1 All-digital phase detectors

The three most important digital phase detectors have already been discussed in Sec. 3.1. When digital word signals instead of bit signals are used, a number of additional phase-detector circuits become available.

A logical evolution of the simple JK-flipflop PD is the FF-counter phase detector illustrated in Fig. 4.1a. The corresponding waveforms are shown in Fig. 4.1b.

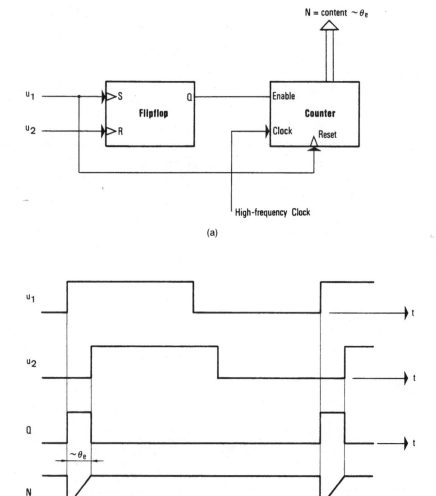

(a)

(b)

Figure 4.1 Flipflop-counter PD. (a) Block diagram. (b) Corresponding waveforms.

The reference (input) signal u_1 and the output (or scaled-down output) signal u_2 of the DCO (or VCO) are binary-valued signals. They are used to set or reset an edge-triggered *RS* flipflop. The time period in which the Q output of the flipflop is a logic 1 is proportional to the phase error θ_e. The Q signal is used to gate the high-frequency clock signal into the (upward) counter. Note that the counter is reset on every positive edge of the u_1 signal.

The content N of the counter is also proportional to the phase error θ_e, where N is the n-bit output of this type of phase detector. The frequency of the high-frequency clock is usually Mf_0, where f_0 is the frequency of the reference signal and M is a large positive integer.

Another all-digital phase-detector circuit has become known by the name Nyquist-rate phase detector (NRPD).[8] The name stems from the well-known Nyquist theorem which states that a sampled signal can be constructed only when the sampling rate is at least twice the highest-frequency component of the signal. The block diagram of the NRPD is shown in Fig. 4.2a, and the corresponding waveforms are seen in Fig. 4.2b.

The input signal should be an analog signal, such as a signal transmitted over a data line. It is periodically sampled and digitized at the clock rate. In the example shown, the clock rate chosen is 16 times the signal frequency, which is 8 times the Nyquist rate. The signal u_2 is an N-bit digital word, generated by a DCO. (Refer also to Sec. 4.1.3.) Furthermore, the signal u_2 has been drawn as a sine wave in Fig. 4.2b; another waveform (such as a square wave) could be used as well. The digitized signal u_1 and the signal u_2 are multiplied together by a software multiplying program. Thus the NRPD operates similarly to the linear PD introduced in Chap. 2.

The resulting phase error signal θ_e is also shown in Fig. 4.2b. Its average value $\overline{\theta}_e$ will have to be filtered out by a succeeding digital loop filter, as will be demonstrated in Sec. 4.1.2.

Still another method of measuring the phase error is the *zero-crossing technique*.[8] The simplest zero-crossing phase detector is illustrated in Fig. 4.3a; its waveforms are shown in Fig. 4.3b. The reference signal u_1 is supposed to be analog; u_2 is a binary signal. The positive transitions of u_2 are used to clock the analog-to-digital converter (ADC), so u_1 is sampled once during every reference period. The digital output signal of the ADC is then proportional to the phase error. Usually this signal is held in a buffer register until the next conversion is completed; thus the phase-error signal θ_e is a quasicontinuous signal, as shown by the dashed line in Fig. 4.3b.

Figure 4.4 shows another variant of digital PD, the so-called *Hilbert-transform PD*. The key element of this PD is a Hilbert transformer. This is a special digital filter which shifts the phase of a sinusoidal

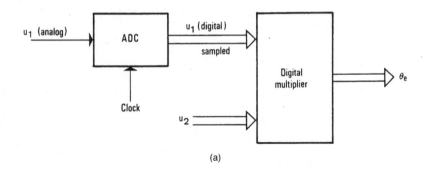

(a)

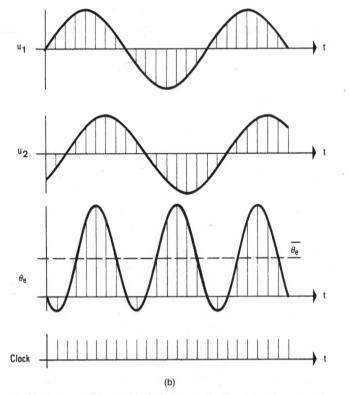

(b)

Figure 4.2 Nyquist-rate PD. (*a*) Block diagram. (*b*) Corresponding waveforms.

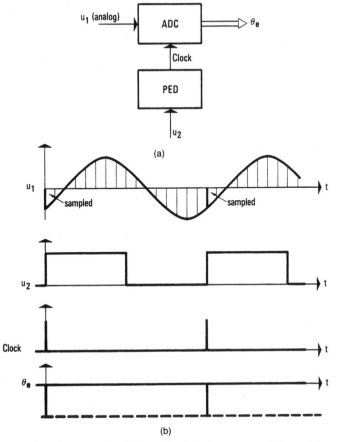

Figure 4.3 Zero-crossing PD sampling the phase error at the positive transitions of the reference signal. (a) Block diagram (PED = positive-edge detector). (b) Corresponding waveforms.

input signal by exactly $-90°$ irrespective of its frequency.[13] Moreover, the gain of the Hilbert transformer is 1 at all frequencies. In the block diagram of Fig. 4.4a, the Hilbert transformer is shown in the top left corner and is marked by the symbol $-\pi/2$. Assuming the input of the device is a digital word signal of the form

$$u_1(t) = \cos(\omega_0 t + \theta_e)$$

the output signal of the Hilbert transformer is given by

$$u_1(t) = \cos\left(\omega_0 t + \theta_e - \frac{\pi}{2}\right) = \sin(\omega_0 t + \theta_e)$$

The Hilbert-transform PD extracts the phase error θ_e by trigonometric

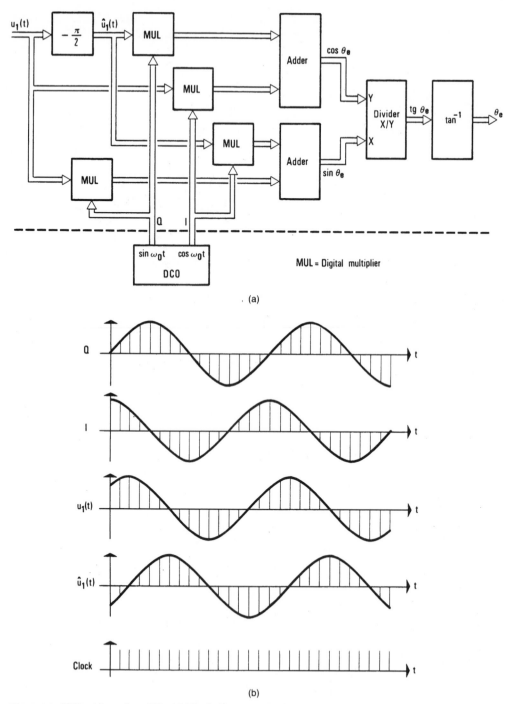

Figure 4.4 Hilbert-transform PD. (*a*) Block diagram. (*b*) Corresponding waveforms.

computations, which are shown in the following. The DCO used in this type of phase detector is supposed to generate two output signals, an *in-phase* signal $I = \cos(\omega_0 t)$ and a *quadrature* signal $Q = \sin(\omega_0 t)$. By the trigonometric operations

$$\cos \theta_e = Iu_1 + Q\hat{u}_1$$

$$\sin \theta_e = I\hat{u}_1 - Qu_1$$

the sine and cosine of the phase error are computed. Dividing the sine by the cosine term yields the tangent of phase error; by using a digital algorithm for the inverse tangent ($\tan^{-1}$), the phase error is obtained. The double lines in Fig. 4.4a signify that all signals within this device are digital word signals. The signals of the Hilbert-transform PD are shown in Fig. 4.4b.

As in the NRPD, all mathematical computations are performed under the control of a clock signal whose frequency is usually M times the signal frequency $f_0 = \omega_0/2\pi$.

The variety of mathematical operations required by the Hilbert transform strongly suggests implementation of this method by software.

A similar but simpler way to calculate the phase error is given by the digital-averaging phase detector (Fig. 4.5).

As in the method discussed previously, the DCO is also required to generate in-phase and quadrature signals I and Q, respectively. These are again multiplied by the digital reference signal $u_1(t)$, but the signals $\cos \theta_e$ and $\sin \theta_e$ are obtained by simply averaging (or integrating) the output signals of the multipliers over an appropriate period of time.[29] Note that this arrangement already includes a filtering func-

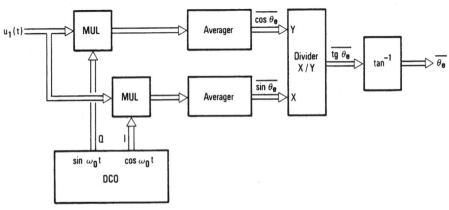

Figure 4.5 Digital-averaging PD.

tion, defined by the impulse transfer function of the averaging filter used.[14] This method, too, lends itself particularly to implementation by software.

4.1.2 All-digital loop filters

As seen in the preceding section, different all-digital PDs generate different types of output signals. The PDs discussed at the end of Sec. 4.1.1 produce N-bit digital output signals, whereas simpler types, such as the XOR or the phase/frequency detector deliver one or two binary-valued output signals. It becomes evident that not every all-digital loop filter is compatible with all types of all-digital PDs. We have to consider which types of loop filters can be matched to the various PDs discussed previously.

Probably the simplest loop filter is built from an ordinary UP/DOWN counter (Fig. 4.6a). The UP/DOWN counter loop filter preferably operates in combination with a phase detector delivering UP or DN (DOWN) pulses, such as the PFD. It is easily adapted, however, to operate in conjunction with the XOR or JK-flipflop phase detectors and others. As shown in Fig. 4.6a, a pulse-forming network is first needed which converts the incoming UP and DN pulses into a counting clock and a direction ($\overline{\text{UP}}$/DN) signal (as explained by the waveforms in Fig. 4.6b).

On each UP pulse generated by the phase detector, the content N of the UP/DOWN counter is incremented by 1. A DOWN pulse will decrement N in the same manner. The content N is given by the n-bit parallel output signal u_f of the loop filter. Because the content N is the weighted sum of the UP and DN pulses—the UP pulses have an assigned weight of $+1$, the DN pulses, -1—this filter can roughly be considered an *integrator* having the transfer function

$$H(s) = \frac{1}{sT_i}$$

where T_i is the integrator time constant. This is, however, a very crude approximation, since the UP and DN pulses do not carry any information (in this application at least) about the actual size of the phase error; they only tell whether the phase of u_1 is leading or lagging u_2.

One of the most important digital loop filters is the K counter (Fig. 4.7). This loop filter always works together with the EXOR or the JK-flipflop phase detector. As Fig. 4.7a shows, the K counter consists of two independent counters, which are usually referred to as "UP counter" and "DOWN counter." In reality, however, both counters are always counting upward. K is the modulus of both counters; i.e., the contents of both counters is in a range from $0 \ldots K - 1$. K can be

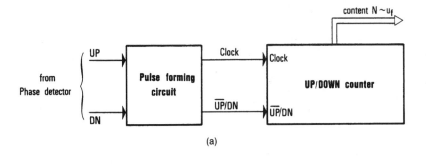

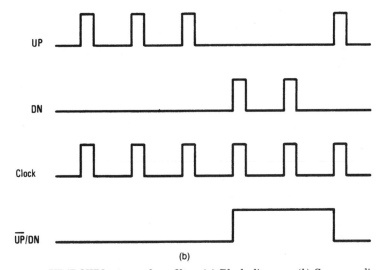

Figure 4.6 UP/DOWN counter loop filter. (*a*) Block diagram. (*b*) Corresponding waveforms.

controlled by the K modulus control input and is always an integer power of 2. The frequency of the clock signal (K clock) is by definition M times the center frequency f_0 of the ADPLL, where M is typically 8, 16, 32, ... The operation of the K counter is controlled by the DN/$\overline{\text{UP}}$ signal. If this signal is high, the "DN counter" is active, while the contents of the UP counter stays frozen. In the opposite case, the "UP counter" counts up but the DN counter stays frozen.

Both counters recycle to 0 when the contents exceeds $K - 1$. The most significant bit of the "UP counter" is used as a "carry" output, and the most significant bit of the "DN counter" is used as a "borrow" output. Consequently, the carry is high when the contents of the UP counter is equal to or more than $K/2$. In analogy, the borrow output

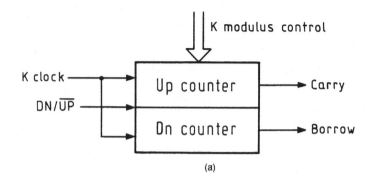

(a)

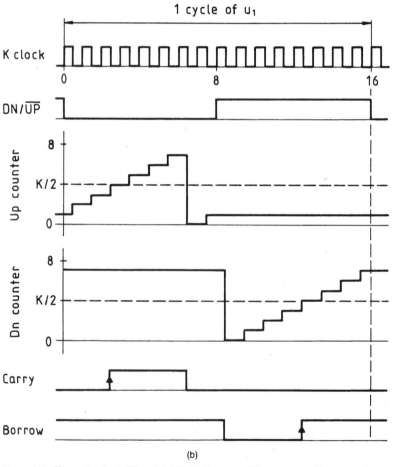

(b)

Figure 4.7 K counter loop filter. (a) Block diagram. (b) Corresponding waveforms.

gets high when the contents of the DN counter is equal to or more than $K/2$. As will be shown in Sec. 4.3, the positive-going edges of the carry and borrow signals are used to control the frequency of a digitally controlled oscillator.

Figure 4.7b shows the signals of the K counter. The DN/$\overline{\text{UP}}$ input is controlled by the output of a phase detector. In this example it was assumed that a JK-flipflop is used and that the ADPLL operates on its center frequency. As explained by Fig. 3.5, input signal u_1 and output signal u_2' of the PLL are in antiphase then, and the output signal u_d of the phase detector is a square wave having exactly 50 percent duty cycle. Hence the DN/$\overline{\text{UP}}$ signal is high in one half-cycle of the u_1 signal and low in the other. The frequency of the K clock is assumed to be 16 times the center frequency ($M = 16$). The counter modulus K has been arbitrarily set to 8. Looking at the waveforms in Fig. 4.7b we see that the UP counter counts on the first 8 K clock pulses, and the DN counter counts on the next 8 pulses. Under these conditions, the UP counter generates exactly one carry pulse within each cycle of the u_1 signal, and the DN counter generates exactly one borrow pulse in the same interval. As will be seen later, the carry and borrow pulses then cancel. We assume now that there exists a phase error in the loop; thus the duty cycle of the DN/$\overline{\text{UP}}$ signal becomes asymmetric. When this signal is low during a longer fraction of one u_1 cycle than it is high, the UP counter gets more clock pulses on average than the DN counter. The average number of carries then becomes larger than the average number of borrows per unit of time. When the DN/$\overline{\text{UP}}$ signal is permanently low, the UP counter is active all the time. When the DN/$\overline{\text{UP}}$ signal is permanently high, however, the DN counter is working continually. The K counter is part of the popular type 74xx297 ADPLL, which will be treated in Sec. 4.3.

Another digital loop filter is the so-called N-before-M-counter (Fig. 4.8). The performance of this filter is very nonlinear. In Fig. 4.8 it is suggested that the N-before-M filter operates in conjunction with a phase detector generating UP and DN pulses, as was the case with the PFD. The N-before-M filter uses two frequency counters scaling down the input signal by a factor N and one counter scaling down by M, where $M > N$ always. The $\div M$ counter counts the incoming UP and DN pulses, s shown in Fig. 4.8. As also seen in the diagram, the upper $\div N$ counter will produce one CARRY output when it has received N UP pulses. But it will generate this CARRY only when the $\div M$ counter does not receive M pulses. Otherwise the $\div N$ counter would have been *reset*. We can say that the upper $\div N$ counter will produce a CARRY pulse whenever more than N pulses of an ensemble of M pulses have been UP pulses. A similar statement can be made for the lower $\div N$ counter in Fig. 4.8, which will output BORROW pulses only when the majority of incoming pulses are DN pulses.

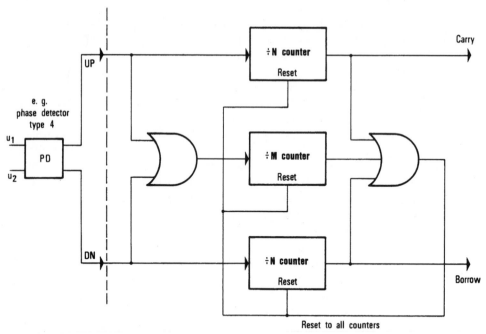

Figure 4.8 Block diagram of the N-before-M loop filter.

The outputs of the N-before-M filter can be used in a similar way to control a DCO, as indicated for the K counter.

We will now deal with digital loop filters compatible with an N-bit parallel input signal. The obvious solution for this case is the *digital filter*, which operates by itself with N-bit input and N-bit output signals. With digital loop filters, any desired transfer function performed by an analog loop filter (and many additional ones) can be reproduced. As we know, the performance of an analog loop filter is described by its transfer function $F(s)$

$$F(s) = \frac{U_f(s)}{U_d(s)}$$

which is the ratio of the Laplace transforms of the signals u_f and u_d (for signal definitions, refer to Fig. 3.1). When the filter action has to be performed by digital means, the transfer function $F(s)$ is normally transformed into z-domain, yielding the so-called z-transfer function $F(z)$, where z is the z-operator. An introduction to digital filtering is given in Appendix C. $F(z)$ is the ratio of the z-transforms of the signals u_f and u_d, i.e.,

$$F(z) = \frac{U_f(z)}{U_d(z)}$$

When implementing the digital filter, this equation is transformed back into time domain, which yields a recursion of the form[12,13,14,19]

$$u_f(nT) = b_0 u_d(nT) + b_1 u_d([n - 1]T) + b_2 u_d([n - 2]T) + \cdots$$
$$- a_1 u_f([n - 1]T) + a_2 u_f([n - 2]T) + \cdots \quad (4.1)$$

The signals u_f and u_d are sampled signals now, which means that they exist only at discrete time instants $t = 0, T, 2T \ldots nT$, where T is the sampling interval. The a_i and b_i terms are called *filter coefficients*. The sampling frequency $f_s = 1/T$ must be chosen much larger than the 3-dB corner frequency of the filter, which has to be realized, typically 10 to 20 times the corner frequency.[13,14] $u_f(nT)$, $u_f([n - 1]T) \ldots$ denote the values of the sampled signal u_f at sampling instants $t = nT$, $t = (n - 1)T. \ldots$ The recursion formula calculates the output signal $u_f(nT)$ in the nth sampling instant from the value of u_d sampled at this instant and from one or more previously sampled values of u_d. Furthermore, in a recursive digital filter, $u_f(nT)$ depends on values of u_f calculated in previous sampling instants. The number of "delayed" samples of u_f and u_d which have to be taken into account is equal to the order of the digital filter. (For a first-order digital filter, for example, only the filter coefficients a_1, b_0, and b_1 do not vanish.)

4.1.3 Digital-controlled oscillators

A variety of DCOs can be designed; they can be implemented by hardware or by software. We consider the most obvious solutions here.

Probably the simplest solution is the $\div N$ counter DCO (Fig. 4.9). A $\div N$ counter is used to scale down the signal generated by a high-frequency oscillator operating at a fixed frequency. The N-bit parallel

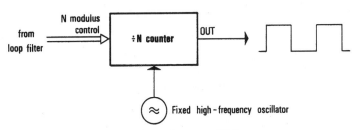

Figure 4.9 Block diagram of a $\div N$ counter DCO.

output signal of a digital loop filter is used to control the scaling factor N of the $\div N$ counter.

Another DCO type is the so-called *increment-decrement* (ID) counter shown in Fig. 4.10*a*.[9,16] This DCO is intended to operate in conjunction with those loop filters that generate CARRY and BORROW pulses, such as the K counter or the N-before-M filter discussed in Sec. 4.1.2. The operation of the ID counter follows from the waveforms shown in Fig. 4.10*b*. As Fig. 4.10*a* shows, the ID counter has three inputs, a clock input (ID clock), an increment (INC), and a decrement (DEC) input. Carry pulses (as delivered, for example, by a K counter, Fig. 4.7) are fed to the INC, borrow pulses to the DEC input. The ID counter is sensitive on the positive-going edges of the carry and borrow inputs; the duration of these signals is not of concern here. In the absence of carry and borrow pulses, the ID counter simply divides the ID clock frequency by 2; it produces an output pulse (IDout) on every second ID clock; see waveforms in Fig. 4.10*b*. To understand the function of the ID counter, one must know that this circuit contains a toggle flip-flop, which is not shown in the schematic diagram of Fig. 4.10*a*. As the waveform "Toggle-FF" in Fig. 4.10*b* shows, the toggle flipflop toggles on every positive edge of the ID clock if no carries and borrows are present. The output of the ID counter (IDout) is obtained by the logical function

$$IDout = \overline{IDclock} \cdot \overline{Toggle\text{-}FF}$$

Now we assume that a carry pulse appears at the INC input of the ID counter. The carry signal is processed only in the period where the toggle flipflop is set "high." If the carry gets "true" when the toggle flipflop is in the "low" state (Fig. 4.10*c*), the toggle goes high onto the next positive edge of the ID clock. It stays low, however, during two ID clock intervals thereafter. This means that the next IDout pulse is advanced in time by one ID clock period. If the carry becomes "true" when the toggle flipflop is set "high," this flipflop is set "low" onto the next two ID clocks, as shown in Fig. 4.10*d*. Because the carry can only be processed when the toggle flipflop is in the high state, the maximum frequency of the IDout signal is reached when the toggle flipflop follows the pattern "high low low high low low. . . ." Consequently, the output frequency of the ID counter cannot be as high as the frequency of the ID clock, but at most two-thirds of that value. This of course limits the hold range of the ADPLL, as will be shown in Sec. 4.3. Figure 4.10*e* demonstrates what happens when a borrow pulse is generated. In analogy, a borrow is processed only when the toggle flipflop is in the low state. As soon as a borrow is sensed, the toggle flipflop is set high onto the succeeding two positive edges of the ID clock. The

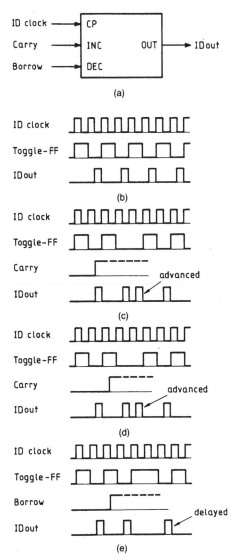

Figure 4.10 ID counter DCO. (*a*) Block diagram. (*b*) Waveforms for the case where no carries and no borrows are applied to the INC and DEC inputs, respectively. (*c*) Waveforms for the case where a carry input is applied when the toggle flipflop is in the 0 state. (*d*) Waveforms for the case where a carry input is applied when the toggle flipflop is in the 1 state. (*e*) Waveforms for the case where a borrow input is applied to the DEC input.

next IDout pulse is therefore delayed by one ID clock period. The ID counter delivers the minimum output frequency when the toggle flipflop shows the pattern "low high high low high high. . . ." Thus the minimum output frequency of the ID counter is one-third the ID clock frequency. As we will see in Sec. 4.3, the limited output frequency range of the ID counter restricts the realizable hold range of the ADPLL. (Note that the explanation of ID counter performance has been slightly simplified; the actual ID counter circuit not only consists of the mentioned toggle flipflop, but also contains eight more flipflops and a number of gates. The exact operation of the ID counter can be deduced from the data sheet of the 74HC/HCT297.)

Because the ID counter needs three ID clock periods to process one carry or one borrow, the maximum frequency of carry or borrow pulses must not be higher than one-third the frequency of the ID clock. If more carries or borrows are delivered, some are "overslept." When the average frequency of the carries is such that all are processed, the instantaneous frequency of the IDout signal is increased by $n/2$ Hertz when n carries are detected in 1 second. This is most easily understood if we assume that the frequency of the ID clock is 32 Hz, for example. Without any carry, the output frequency would be 16 Hz. If 8 carries are detected within 1 second, the "next" IDout pulse is advanced eight times in 1 second by 1/32 second. The number of output pulses is therefore increased from 16 to 20 Hz during that period, and not from 16 to 24 Hz. Generally, one carry pulse causes 1/2 cycle to be added to the IDout signal, and one borrow pulse causes 1/2 cycle to be deleted correspondingly.

The two DCO circuits discussed earlier are better suited for hardware than for software implementations. The waveform synthesizer DCO—the third and last DCO to be considered here—lends itself almost ideally to implementation by software. We will discuss software implementations of the PLL in Chap. 5. This type of DCO generates sine and/or cosine waveforms by looking up tables stored in read only memory (ROM).[30] The block diagram of a waveform-synthesizer DCO is shown in Fig. 4.11a.

The waveforms in Fig. 4.11b demonstrate how the synthesizer generates sine waves of different frequencies (1 Hz and 0.5 Hz in this example). It operates at a fixed clock rate (i.e., it calculates a sample of the synthesized signal at the sampling instants $t = 0, T, 2T, \ldots,$ nT, irrespective of the desired frequency). Lower-frequency signals are generated with higher resolution than higher-frequency signals.

In the example of Fig. 4.11b it has been arbitrarily assumed that the sampling period is 50 ms; most actual waveform synthesizers operate much faster, of course. It shows that a 1-Hz sine wave is generated with a resolution of 20 samples for a full period. In the case of

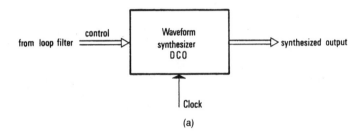

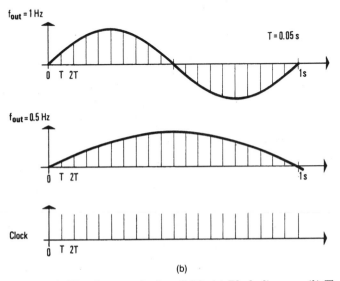

Figure 4.11 Waveform-synthesizer DCO. (*a*) Block diagram. (*b*) The waveforms show how sine waves with frequencies of 1 Hz and 0.5 Hz, respectively, are synthesized.

the 0.5-Hz sine wave, twice as many samples (40) are produced within a full period. When generating a 1-Hz sine wave, the synthesizer calculates the sine function for the phase angles $\phi = 0$ (initial value), $2\pi/20$, $2(2\pi/20)$, $3(2\pi/20)$, . . . , and the phase angle ϕ is incremented by an amount $\Delta\phi = 2\pi/20$ at every clock period.

If an arbitrary frequency f is to be produced, the increment $\Delta\phi$ is given by

$$\Delta\phi = 2\pi f T$$

where T is the sampling period.

The synthesizer calculates $\Delta\phi$ from the N-bit parallel f-control signal delivered by the loop filter (shown in Fig. 4.11a). The waveform synthesizer is capable of generating the appropriate signal to the f-

control input. It is no problem to generate "simultaneously" a sine and a cosine function, as required, when a Hilbert-transform phase detector is used with an ADPLL system (refer to Fig. 4.4a).

Digital waveform synthesizers are easily implemented using single-chip microcomputers.[30] The speed of trigonometric computations can be greatly enhanced by using table-lookup techniques rather than by calculating a sine function with a Taylor series or Chebyshev polynomials. An example of the application of a waveform-synthesizer DCO is presented in Sec. 4.2.

4.2 Examples of Implemented ADPLLs

Based on the numerous variants of all-digital PDs, loop filters, and DCOs, an almost unlimited number of ADPLLs can be built by combining compatible functional blocks. There is an extended literature on this subject, and a review of the most important systems is found in ref. 8. A detailed discussion of every possible ADPLL system would go beyond the scope of this book; we therefore consider only three typical ADPLL implementations.

The first two are hardware implementations. There is no reason why they could not be designed using software as well. The last example to be discussed is a typical software-based system, which encompasses a large variety of mathematical operations. A hardware implementation of this type of PLL is certainly not impossible, but the hardware would be very complex.

The first example of an ADPLL is depicted in Fig. 4.12a.[39] In this circuit the input signal u_1 is first preprocessed by a pulse-forming network (cf. dashed box on the left). The generated signals are shown in Fig. 4.12b. First, D-flipflop FF1 scales down the frequency of the input signal by a factor of 2. The down-scaled signal is denoted u_1^*. By ANDing u_1 and u_1^*, a clock signal CK is generated that is applied to the counting input of an up/down counter. The signal u_1^* is used to set the state of D-flipflop FF3, which serves as a phase detector. The duration of CK is one-quarter of the period of u_1^*. Moreover, the negative going edges of u_1^* trigger a monoflop which generates very short pulses. The pulses are labeled *Start*. Their duration must be shorter than the period of the high-frequency clock f_c which will be discussed in the following. The dashed enclosure on bottom right of Fig. 4.12a represents a DCO (digital-controlled oscillator). It is built from a cascade of two counters, a variable modulo-N divider and a fixed modulo-M counter. The variable modulo-N divider is a DOWN counter. Its content starts with the number N which is loaded in parallel by the Load input. The clock signal f_c causes the divider to count down. When it reaches the terminal count (TC)—which is 0 in this case—a pulse is delivered at

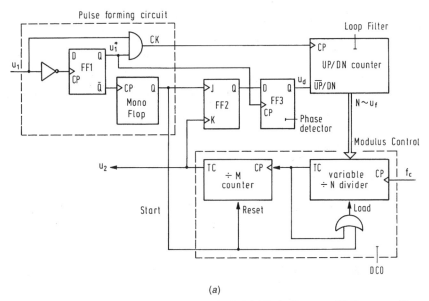

(a)

Figure 4.12 All-digital PLL system, example 1. (a) Block diagram. (b) Corresponding waveforms. Two cases are shown: (1) divider ratio N too small; (2) divider ratio N too large.

the TC output. This immediately reloads the content N (see the OR gate at the Load input), and the modulo-N divider continues counting down. The fixed modulo-M counter is an UP counter. Its content is reset to 0 upon applying a Reset signal. The pulses applied to the CP input ramp up its content until it reaches the terminal count, which is a positive number here. As soon as the terminal count is reached, a pulse is delivered at the TC output, and the content wraps around to zero again. The counter then continues counting up.

When the circuit is locked, the clock frequency f_c should be

$$f_c = NMf_1$$

where f_1 is the frequency of the input signal u_1. If this condition is met, the high-frequency clock delivers exactly $N \cdot M$ pulses during one cycle of the input signal u_1, whose period is marked T_0 in Fig. 4.12b. In this case the frequency of u_2 equals the frequency of u_1^*, i.e. $f_1/2$. When the locking process starts, the modulus N of the variable modulo-N divider can have an arbitrary value, thus N is either too high or too low. The phase detector must adjust the value of N by increasing or decreasing the content of the UP/DN counter which is shown on right top of Fig. 4.12a, until N has the correct value. Because the UP/DN counter immediately controls the frequency of the DCO, this counter

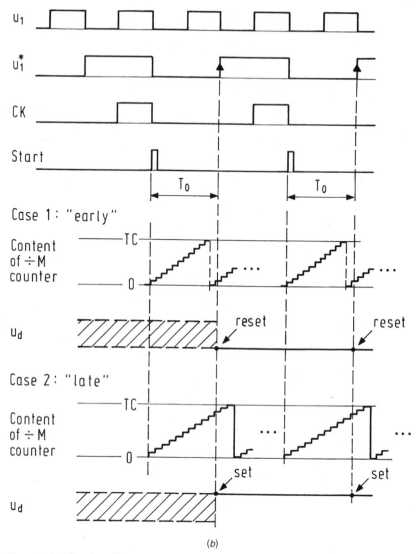

Figure 4.12 *(Continued)*

acts as a loop filter. To see how the loop becomes locked, we want to check the waveforms in Fig. 4.12*b*.

The START pulse resets the modulo-M counter on each high-to-low transition of signal u_1^* (see the waveform "Content of $\div M$ counter"). As mentioned above, the duration of the START pulse must be shorter than the period of the f_c clock. If its pulse width were chosen larger, the LOAD pulse of the modulo-N divider would last during several cycles of the f_c clock, thus inhibiting the counter to change its content. If N already had its correct value, the terminal count of the modulo-M counter would be reached exactly T_0 seconds after reset. In a first case named *"Case 1: early"* it was assumed that N is smaller than required, thus the clock frequency at the CP of the modulo-M counter is too high, and the terminal count is reached in shorter time, i.e., before the next low-to-high transition of u_1^*. The positive-edge-triggered JK-flipflop FF2 was initially set to its 1 state by the START pulse. Now the TC output of the modulo-M counter resets the FF2 before the low-to-high transition of u_1^* occurs. Consequently the next positive edge of u_1^* resets D-flipflop FF3, whose output is labeled u_d (phase detector output) in Figs. 4.12a and 4.12b. Note that the state of u_d has been indeterminate at the start of the locking process, hence its waveform is drawn by the shaded area in Fig. 4.12b. Because u_d is low now, the next CK (clock) pulse causes the UP/DN counter to increase its content (N) by 1. This lowers the instantaneous frequency of the TC output of the modulo-N divider, and the terminal count of the modulo-M counter will be reached later in the next cycle of u_1^*. This process repeats until the modulo-M counter reaches TC after occurrence of the positive edge of u_1^*.

In the other case named *"Case 2: late,"* N was too large initially so that the modulo-M counter required a period longer than T_0 to reach TC. Under this condition, D-flipflop FF3 will be set 1 on the next low-to-high transition of u_1^*. This causes the next CK pulse to decrease the content N of the UP/DN counter. Consequently, the instantaneous output frequency of the modulo-N divider will become higher. When the lock process has been completed, the content N of the UP/DN counter will usually toggle between two adjacent values N and $N+1$ in successive cycles of u_1^*. As a numerical example, assume that the high-frequency clock is $f_c = 10$ MHz and the input frequency is $f_1 = 10.1$ kHz. The overall divider ratio of the cascade of both counters (modulo-M and modulo-N) then should be $N \cdot M = 990.1$. Supposing that $M = 100$ (i.e., two cascaded decade counters are used for the modulo-M counter), N will toggle between 9 and 10 in alternative cycles. This obviously results in a phase jitter of the ADPLL's output signal u_2. The phase jitter can be made arbitrarily small by increasing the frequency f_c of the clock signal. N will then settle at higher values. Note, however, that the lock-in process will be become slower then, because the UP/DN counter would probably have to change its initial content much more but can increase or decrease it only in increments of 1 in one cycle of u_1^*.

Basically, this circuit is highly nonlinear, but it turns out that its inherent stability is just a consequence of this nonlinearity. When trying to model the circuit we become aware that the UP/DN counter which acts as a loop filter acts like an integrator. If a phase error persists for an extended period of time, the content of the UP/DN ramps up or down depending on the sign of the phase error, hence behaving like an integrator. As we know from the theory of the LPLL, a VCO is also modeled as an integrator, because its output phase θ_2 is proportional to the integral of applied control signal u_f, cf. Eq. (2.10). An analogous statement can be made for a DCO; hence our circuit contains a cascade of two integrators, which implies that its phase transfer function $H(s)$ has two poles at $s = 0$. Because there are no compensating "zeros" in this system, it would get unstable. Most happily, this does not occur, because the contents of the counters within the DCO are reset on every START pulse, as described above. In other words, the "integral" term at the output of the DCO is not allowed to "wind up." Due to the nonlinearities of the circuit, it becomes very difficult to establish a mathematical model. No such model has been developed to the author's knowledge.

The second ADPLL system described here is shown in Fig. 4.13. This is the most often used ADPLL configuration; it is available as an integrated circuit with the designation 74xx297, where xx stands for the family specification (HC, HCT, LS, S, etc.). We analyze this ADPLL in greater detail in Sec. 4.3. The IC contains two phase detectors: an EXOR gate and a JK-flipflop. In the schematic of Fig. 4.13, the EXOR is used. The loop filter is formed by the previously discussed K counter (Fig. 4.7), and the already known ID counter (Fig. 4.10) is used as DCO. This ADPLL system requires an external divide-by-N-counter.

The system is supposed to operate at a center frequency f_0 referred to the input u_1. The K counter and the ID counter are driven by clock signals having frequencies of M times and $2N$ times the center frequency f_0, respectively. Normally both M and $2N$ are integer powers of 2 and are mostly derived from the same oscillator. In many cases, $M = 2N$, so both clock inputs can be tied together.

Assume for the moment that the EXOR phase detector is used and that the ADPLL operates on its center frequency. Then the ID counter is required to scale down the ID clock precisely by 2. The average number of carry and borrow pulses delivered by the K counter must therefore be the same, too. This is possible only when the phase difference between the signals u_1 and u_2' is 90°. In this case, the output signal of the EXOR gate is a symmmetrical square wave whose frequency is twice the center frequency. Consequently, the UP counter will count during two quarters of the reference cycle and the DOWN counter will count in the remaining two quarter periods. Because the average number of carries and borrows precisely matches, no cycles are added to or deleted from the ID counter. If the reference frequency is increased, however, the output signal of the EXOR phase detector must

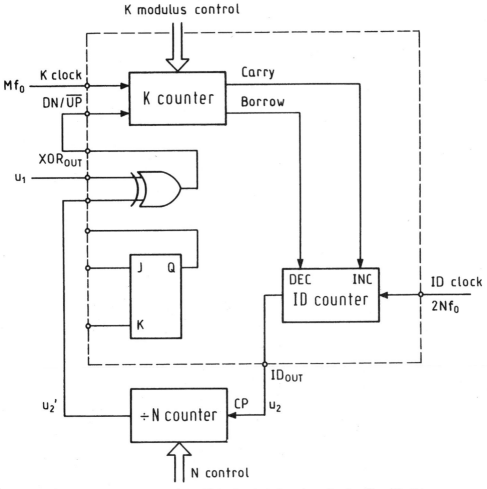

Figure 4.13 All-digital PLL, example 2. This circuit is based on the familiar IC of type 74HC/HCT297. The EXOR phase detector is used here. For the external $\div N$ counter, an IC of the type 74HC/HCT4040 can be used.

become asymmetrical in order to allow the K counter to produce more carries than borrows on average.

The last example of an ADPLL is illustrated in Fig. 4.14. A similar system has been implemented by software on a TMS320 single-chip microcomputer (Texas Instruments).[30] The system of Fig. 4.14 is built from functional blocks introduced previously.

1. A Hilbert-transform phase detector (Fig. 4.4)

2. A first-order digital loop filter

3. A waveform-synthesizer DCO (Fig. 4.11)

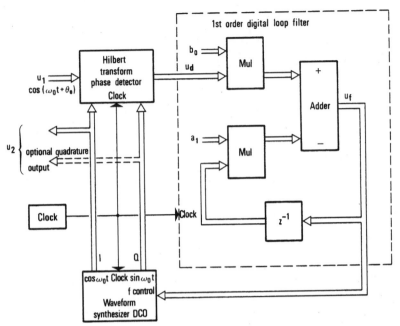

Figure 4.14 All-digital PLL system, example 3. This system is best implemented by software.

As indicated in the block diagram, the arithmetic and logic operations within the functional blocks are performed under control of a clock. This means that all routines calculating the output variables of the blocks are executed periodically. The DCO generates the in-phase and quadrature signals I and Q required by the Hilbert-transform PD to calculate the phase error, $u_d \sim \theta_e$. The output signal u_d is digitally filtered by the loop filter, which performs the operation

$$u_f(nT) = b_0 u_d(nT) - a_1 u_f[(n-1)T]$$

where a_1 and b_0 are filter coefficients as defined in Eq. (4.1). The mathematical operations performed by the digital filter are represented by the dashed block in Fig. 4.14.

One of the major benefits of the software implementation is the simplicity of changing the structure of the ADPLL system. With only minor program modifications the first-order loop filter could be turned into a second-order one.[30] This would yield a third-order PLL.

4.3   Theory of a Selected Type of ADPLL

Because there are so many variants of purely digital phase detectors, loop filters, and controlled oscillators, an enormous number of different ADPLL systems can be built. Among these variants, some will perform similarly to LPLLs. Others will operate like classical DPLLs, but the functioning of many ADPLLs will have almost nothing in common with LPLLs and DPLLs. For this reason it is absolutely impossible to create a generalized "theory of the ADPLL." To investigate the behavior of a particular ADPLL type, the user is forced to look for appropriate models of the corresponding function blocks and then try to get a reasonable description in the form of transfer functions (e.g., phase-transfer functions), Bode diagrams, or the like. In many cases, application of standard tools (like linear control theory) will fail, because the systems to be analyzed are mostly nonlinear.

To demonstrate that analyzing an ADPLL is not an entirely hopeless job, we investigate the dynamic performance of the most popular ADPLL type, the familiar 74xx297, which was already shown in Fig. 4.13.

4.3.1 Effects of discrete-time operation

It is our aim to investigate the most important key parameters such as hold range, lock range, and lock-in time. We assume for the moment that the EXOR is used as phase detector, as shown in Fig. 4.13. Performance of the ADPLL is most conveniently analyzed by the waveforms of the circuit, which are shown in Fig. 4.15. The signals are plotted for the simple case that the reference frequency f_1 equals the center frequency f_0. The frequency of the K clock has been chosen 16 times the clock frequency ($M = 16$). The K counter modulus is assumed to be 4 ($K = 4$). (Note, however, that the minimum value of K for the 74HC/HCT297 is 8). The divider ratio of the $\div N$ counter is 8 in this example ($N = 8$), so the K clock and the ID clock can be taken from the same generator. One cycle of the reference signal u_1 consists therefore of 16 cycles of the K clock. The contents of the UP and DOWN counters are denoted Kup and Kdn, respectively. As can be seen from the data sheet of the 74HC/HCT297, these two counters are reset when power is applied to the circuit. When it has been operating for an undefined period of time, the contents will be arbitrary at a given time. At the instant where the 0th K clock occurs (Fig. 4.15), Kup and Kdn can therefore be assigned arbitrary numbers. For the polarities of the clock pulses the following conventions are made:

- Both counters of the K counter count onto the negative edges of the K clock.

- The toggle flipflop within the ID counter toggles onto the positive edge of the ID clock.

- All flipflops of the $\div N$ counter count onto the negative edge of the corresponding clock signal; i.e., the first stage of the $\div N$ counter

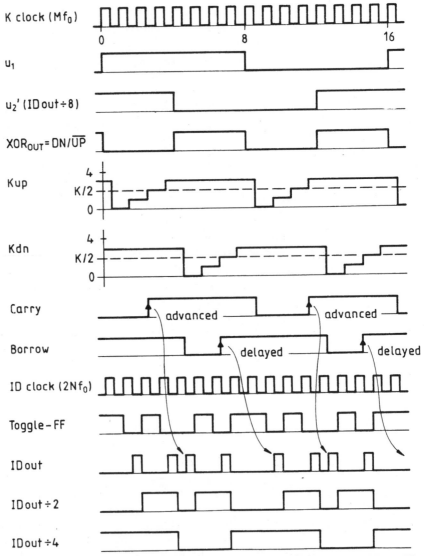

Figure 4.15 Waveforms of an ADPLL system using the IC type 74HC/HCT297. The ADPLL operates at its center frequency. The EXOR PD is used, and the parameters of the circuit are $M = 16$, $K = 4$, $N = 8$. Note that K has been chosen small to simplify the drawing; in a real application, K cannot be less than 8, however.

(denoted IDout ÷ 2 in Fig. 4.15) counts onto the negative edge of IDout, the second stage (denoted IDout ÷ 4) counts onto the negative edge of the IDout ÷ 2 output signal, etc.

As we see from the Kup and Kdn waveforms, the UP counter is active during the first and third quarters of the reference cycle, whereas the DOWN counter is active during the second and fourth. Hence carries appear on ID clocks 2 and 10, while borrows are detected on ID clocks 6 and 14. (Remember that the carry and borrow signals depicted in Fig. 4.15 are simply the outputs of the most significant bit of the corresponding counter. Hence the carry becomes high when the content of the UP counter has reached $K/2$.) As the IDout waveform shows, its pulses are periodically advanced and delayed by one cycle of the ID clock. The bottommost two signals represent the scaled-down IDout signal, where the scaling factors are 2 and 4, respectively.

Because of the carry and borrow pulses, the output of the toggle flipflop becomes asymmetric. Therefore, the IDout signal does not have constant frequency but rather exhibits phase jitter. The output signals IDout ÷ 2 and IDout ÷ 4 are also asymmetrical, but the output signal u_2' (which corresponds to IDout ÷ 8) is symmetrical again. The reason for this is that there is exactly one carry and one borrow in the period where u_2' is high, and there is exactly one carry and one borrow in the period where u_2' is low. The ripple introduced by delaying and advancing the IDout pulses is therefore canceled at the output of the ÷N counter.

It would be premature, however, to conclude that there is never ripple in the output signal u_2' of this type of ADPLL. Let us choose a larger value for K in the next example, e.g., $K = 8$. All other parameters remain unchanged. Figure 4.16 shows what happens now. When the ADPLL operates at its center frequency again, the UP counter counts up by 4 in one quarter period of the reference signal u_1 on average, and the DOWN counter counts up by 4 on one quarter period of u_1 on average. Carries and borrows are now generated only in each second UP-counting or DOWN-counting intervals. Figure 4.16 shows two periods of the reference signal, which corresponds to 32 K clock cycles. Carries are produced onto K clocks 1 and 18, and borrows are produced onto K clocks 11 and 23. Because advancing and delaying of the IDout pulses no longer cancel in a particular half-cycle of u_2', this signal shows ripple; note the half-cycle of u_2', for example, which goes from K clock 4 to 11. Its duration is 7 K clock pulses instead of 8. In succeeding reference periods (not shown in this figure) there must be half-cycles of u_2' which have a duration of 9 K clock pulses.

We conclude that there is no ripple on the output signal if the K modulus is chosen such that it produces exactly one carry (or one borrow) in a quarter period of the reference signal. Choosing

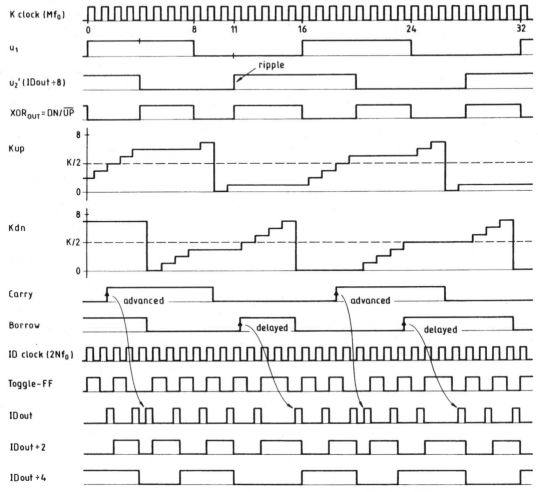

Figure 4.16 Like Fig. 4.15, but the parameters of the ADPLL are $M = 16$, $K = 8$, and $N = 8$. The ADPLL operates at its center frequency. Note that the UP counter and the DOWN counter of the K counter recycle only on every second UP-counting or DOWN-counting period, respectively.

$$K = \frac{M}{4} \qquad (4.2)$$

provides zero ripple when the EXOR phase detector is used. An ADPLL having $K = M/4$ is therefore called a "minimum ripple" configuration. As we will see later, every ADPLL exhibits some ripple if it operates at other frequencies.

If $K > M/4$, ripple is produced. It is easy to calculate the amount of ripple to be expected under this condition. Ideally, the duty factor δ of the output signal u_2' is $\delta = \delta_0 = 0.5$. If carries and borrows in succeeding UP- and DOWN-counting periods do not cancel, and edges of the u_2' signal can be advanced or delayed by at most one ID clock cycle, i.e., by a time interval of $1/(2Nf_0)$. Consequently, the actual duty factor can vary in the range

$$0.5 \left(1 - \frac{1}{N} \right) < \delta < 0.5 \left(1 + \frac{1}{N} \right) \qquad (4.3)$$

i.e., the relative duty factor deviation is $1/N$ at worst. It will be demonstrated later that ripple can be suppressed by the addition of only a few components.

Let us check what happens if K is chosen smaller than $M/4$. If we consider Fig. 4.15 again and assume $K = 2$ now (which is not possible, however, with the 74HC/HCT297 IC), the UP counter would generate 2 carries in one UP-counting period, and the DOWN counter would generate 2 borrows in one DOWN-counting period. Because 2 carries and 2 borrows would cancel in succeeding UP-counting and DOWN-counting periods, there should theoretically be no ripple in the u_2' waveform. This is true only, however, when the ID clock frequency is chosen large enough so that the ID counter is able to process all carries and borrows. When the UP counter is counting up for an extended period of time, it produces a carry every $K/(Mf_0)$ seconds. If a number of carries have to be processed in succession by the ID counter, the delay between any two carries should be larger than 3 ID clock periods, as shown in Sec. 4.2.3 (see also Fig. 4.10). Because the duration of one ID clock cycle is $1/(2Nf_0)$ seconds, we have no overslept carries (or borrows) when the condition

$$N > N_{\mathrm{min}} = \frac{3M}{2K} \qquad (4.4a)$$

is met. Now M, K, and N are mostly integer powers of 2, so in practical applications the minimum N will be chosen

$$N > N_{\mathrm{pract}} = \frac{2M}{K} \qquad (4.4b)$$

where N_{pract} is also an integer power of 2. In this example, we have $N_{\mathrm{pract}} = 4 \cdot 16/2 = 16$. Because we have chosen $N = 8$, the frequency of the carries and borrows would clearly be too high, so the ADPLL would not work properly.

Generally we can conclude that for an ADPLL using the EXOR phase detector we have no ripple when M is chosen to produce at least one carry in a quarter cycle of u_1 ($M \geq K/4$) and when N is chosen such that all carries and borrows can be processed ($N \geq 2M/K$). When

M is chosen smaller than $K/4$, less than one carry (or borrow) is generated within one quarter period of u_1, so ripple will be created. The amount of ripple is given by Eq. (4.3).

Now we have to investigate how the ADPLL performs when the JK-flipflop phase detector is used. We suppose that the signals u_1 and u_2' are connected to the J and K inputs of the flipflop in Fig. 4.13, respectively, and that the Q output of this flipflop is tied with the $\overline{\mathrm{DN/UP}}$ input of the K counter. As we know from theory of the DPLL, the signals u_1 and u_2' should be in antiphase when the PLL operates at its center frequency. The waveforms for this example are shown in Fig. 4.17 for the parameters $M = 16$, $K = 8$, and $N = 8$. If both signals

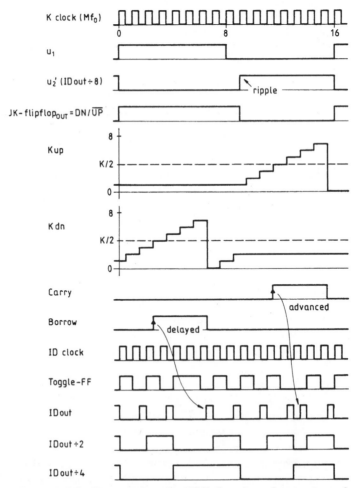

Figure 4.17 Like Fig. 4.15, but the JK-flipflop phase detector is used. The ADPLL operates at its center frequency. The parameters of the circuit are $M = 16$, $K = 8$, and $N = 8$.

u_1 and u_2' were precisely symmetrical, the DOWN counter would be active in the first half of the reference cycle, and the DOWN counter would be active in the second half. On average, both UP and DOWN counters would overflow once in a counting period. As the waveforms show, the DOWN counter is active in the first half of the cycle. Because the borrow causes one of the IDout pulses to be delayed by one ID clock, the waveform for u_2' becomes asymmetric. The asymmetry is easily calculated in this case, too. In one upward-counting period, the UP counter overflows $M/(2K)$ times, hence produces $M/(2K)$ carries. The same number of borrows are generated by the DOWN counter as well. The $M(2K)$ carries cause the next positive edge of the u_2' signal to be advanced by $M/(2K)$ ID clock cycles, where one ID clock cycle lasts for $1/(2Nf_0)$ seconds. It follows that the duty cycle of u_2' can vary in the range

$$0.5 \left(1 - \frac{M}{2KN}\right) < \delta < 0.5 \left(1 + \frac{M}{2KN}\right) \qquad (4.5)$$

When K is chosen smaller than $M/2$, the UP counter produces more than one carry in one UP-counting period, which of course increases the ripple on the output signal. To get minimum ripple, we should choose

$$K = \frac{M}{2}$$

for the JK-flipflop PD. In contrast to the EXOR phase detector, the waveform of u_2' is unsymmetrical even if K is specified for minimum ripple. In many cases, $M = 2N$; i.e., the K clock and the ID clock are taken from the same generator. Then the duty cycle is in the range

$$0.5 \left(1 - \frac{1}{2K}\right) < \delta < 0.5 \left(1 + \frac{1}{2K}\right) \qquad (4.6)$$

Selecting a large value for K reduces the ripple accordingly.

4.3.2   The hold range of the ADPLL

It is time now to see what happens if the frequency f_1 of the reference signal deviates from the center frequency f_0. We assume first that an EXOR is used as phase detector. Furthermore, let the reference frequency be $f_1 = 1.25\, f_0$. Figure 4.18 shows the waveforms for the case $M = 16$, $K = 4$, and $N = 8$. For the ADPLL to generate a higher frequency, the UP counter must operate during longer time intervals than the DOWN counter. The phase error must therefore become posi-

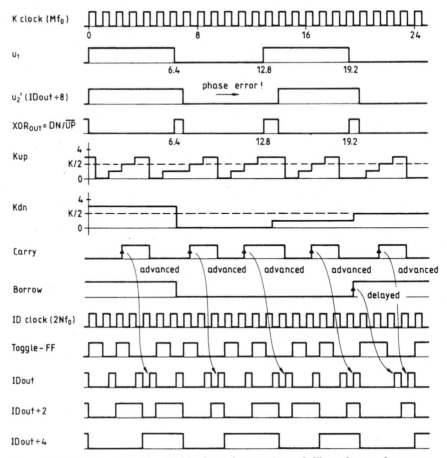

Figure 4.18 Like Fig. 4.15. The EXOR phase detector is used. The reference frequency is 1.25 times the center frequency here. Consequently, the UP-counting section of the K counter is active most of the time, and more carries than borrows are generated on average.

tive. As Fig. 4.18 shows, the signals u_1 and u_2' are nearly in phase now, which corresponds to a phase error near 90°. The UP counter is active most of the time, and many more carries than borrows are produced. This forces the ID counter to increase its output frequency.

A simple consideration yields the range of frequencies the ADPLL can work with. It is clear that the ADPLL generates the maximum output frequency when the K counter is continually counting up. The frequency of carry pulses then is given by

$$f_{max} = f_0 \frac{M}{K} \qquad (4.7)$$

Because each carry applied to the increment input of the ID counter causes 1/2 cycle to be added to the IDout signal (refer to Sec. 4.1.3 and Fig. 4.10), the frequency at the output of the ID counter is increased by

$$\Delta f_{\mathrm{IDout}} = f_0 \, \frac{M}{2K} \qquad (4.8)$$

Because the $\div N$ counter scales down that frequency by N, the maximum frequency deviation from the center frequency the ADPLL can handle is

$$\Delta f_H = f_0 \, \frac{M}{2KN} \qquad (4.9a)$$

This is nothing else than the *hold range* of the ADPLL. The hold range given by Eq. (4.9a) is only realizable, however, if N is chosen larger than N_{min} defined in Eq. (4.4). If N is smaller than N_{min}, some of the carries and borrows are overslept, and the hold range is limited to

$$\Delta f_H = \frac{f_0}{3} \qquad (4.9b)$$

A simple consideration shows that the hold range given by Eq. (4.9a) or (4.9b) is a theoretical limit, which is not realized in practice. For the parameters $M = 16, K = 4, N = 8$ and for an EXOR phase detector, we obtain a hold range of $\Delta f_H = 1.25 \, f_0$. This is exactly the situation depicted in Fig. 4.18. We should note, however, that the duration of one cycle of the reference signal is now 4/5 of the duration of the reference period $1/f_0$, or in other words, the duration of one reference cycle is 12.6 cycles of the K clock. Assuming that the reference signal performs a positive edge on the 0th K clock, the next edges occur after 6.4, 12.8, 19.2, . . . , K clocks. Thus the edges of u_1 generally do not coincide with the edges of the output signal u_2'. As a consequence, the K counter is not continually counting up, but there are short intervals where the DOWN counter becomes active. Borrow pulses occur from time to time, as seen from Fig. 4.18. We see very clearly from the waveform of u_2' that the phase error increases from cycle to cycle. Hence the ADPLL is not able to operate at the limit of the hold range. The author does not know an exact method to calculate the usable frequency range of an ADPLL, but computer simulations have shown that the maximum frequency deviation comes close to the hold range, say to about 90 percent of it.

From the LPLLs and DPLLs it is well known, however, that the useful frequency range is not given by the hold range but rather by the lock-in or pull-in ranges. The question arises here, whether such parameters can also be defined for the ADPLL under concern. Also here, the author cannot give an answer which is theoretically justified. As the following analysis of phase-transfer function indicates, this type of ADPLL is a first-order loop whose time constant is in the order of the period of the reference signal, hence a very fast system. Simulations show that even for very large frequency steps applied to the reference input the ADPLL does not lock out, so for the practitioner it seems affordable to state that lock-in range, pull-out range, pull-in range, and hold range are all about the same for this circuit. We perform some ADPLL simulations in Sec. 4.6.

It can be observed that the output signal u_2' exhibits ripple or phase jitter whenever the reference frequency f_1 deviates from the center frequency f_0. If the quotient f_1/f_0 is a rational fraction

$$\frac{f_1}{f_0} = \frac{m}{n}$$

where m, n = integer, then we have $mT_0 = nT_1$, where $T_0 = 1/f_0$ and $T_1 = 1/f_1$; i.e., n cycles of the reference signal have exactly the same duration as m cycles of a signal whose frequency equals the center frequency. In this case the ripple pattern of the u_2' signal becomes periodic. If f_1/f_0 is not rational, however, the ripple pattern does not repeat itself.

4.3.3 Frequency-domain analysis of the ADPLL

As with the LPLL and the DPLL, it is possible to derive the phase-transfer function and the error-transfer function for the ADPLL. To get the phase-transfer function, a mathematical model of the ADPLL (Fig. 4.13) must be found. The model is shown in Fig. 4.19. The phase

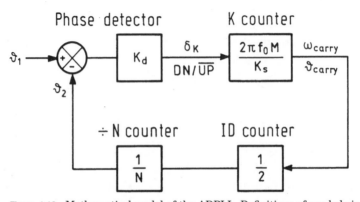

Figure 4.19 Mathematical model of the ADPLL. Definitions of symbols in text.

detector represents a zero-order block with gain K_d. The phase detector output signal controls the duty factor δ_K of the K counter. This duty factor is defined by the average fraction of time the UP counter is active. Thus for $\delta_k = 1$ the UP counter is permanently active, whereas for $\delta_K = -1$ the DOWN counter is permanently active. If the EXOR phase detector is used, δ_K will be 1 for a phase error of $\theta_e = \pi/2$ and -1 for $\theta_e = -\pi/2$. For the EXOR we then have

$$K_d = \frac{2}{\pi} \tag{4.10a}$$

For the JK-flipflop, the phase detector gain is

$$K_d = \frac{1}{\pi} \tag{4.10b}$$

In a number of other texts, the phase error is alternatively specified in cycles (of the reference period $1/f_1$) and not in radians. Here, the phase detector gains become 4 for the EXOR and 2 for the JK-flipflop. We prefer, however, to specify all phase signals in radians in order to get the required phase-transfer function. Now the mathematical model of the K counter must be found. As explained in Sec. 4.3.2, the number of carry pulses generated per second is given by

$$f_{\text{carry}} = \delta_K \frac{Mf_0}{K}$$

The corresponding angular frequency ω_{carry} is therefore

$$\omega_{\text{carry}} = \delta_K 2\pi \frac{Mf_0}{K}$$

Because we are looking for the phase-transfer function, we must know the phase θ_{carry} of the K counter output signal. Because the phase is simply the integral of angular frequency over time, the phase-transfer function of the K counter becomes

$$K_K(s) = \frac{\Theta_{\text{carry}}(s)}{\Delta_K(s)} = \frac{2\pi Mf_0}{Ks} \tag{4.11}$$

where $\Delta_K(s)$ and $\Theta_{\text{carry}}(s)$ are the Laplace transforms of the signal δ_K and θ_{carry}, respectively. Because each carry pulse applied to the increment input of the ID counter causes $1/2$ cycle to be added to the IDout signal, the ID counter can be modeled simply by a zero-order block having the gain $1/2$. Clearly, the $\div N$ counter is a block with gain

$1/N$. Having the model, the phase-transfer function $H(s)$ is now found to be

$$H(s) = \frac{\omega_0}{s + \omega_0} \tag{4.12}$$

where ω_0 is given by

$$\omega_0 = \frac{K_d \pi M f_0}{KN}$$

Moreover, the error-transfer function $H_e(s)$ is

$$H_e(s) = \frac{s}{s + \omega_0} \tag{4.13}$$

Clearly, this ADPLL is a first-order system. Its time constant is given by

$$\tau = \frac{1}{\omega_0} = \frac{KN}{K_d \pi M f_0} \tag{4.14}$$

Thus for the EXOR phase detector the time constant of the ADPLL becomes

$$\tau(\text{EXOR}) = \frac{KN}{2M f_0} \tag{4.15a}$$

and for the JK-flipflop phase detector

$$\tau(JK) = \frac{KN}{M f_0} \tag{4.15b}$$

If the K modulus is chosen for minimum ripple, i.e., $K = M/4$ for the EXOR and $K = M/2$ for the JK-flipflop PD, we obtain $\tau = (N/8)T_0$ for the EXOR and $\tau = (N/2)T_0$ for the JK-flipflop PD, where $T_0 = 1/f_0$. This shows that for small divider ratio N the ADPLL settles extremely fast. Only for large N its response onto phase or frequency steps becomes slower. Some numerical examples will be presented in Sec. 4.6.

4.3.4 Ripple reduction techniques

We saw in Sec. 4.3.1 that ripple in ADPLLs can be minimized by choosing an optimum value for the K counter modulus K; i.e., $K = M/4$ when the EXOR PD is used or $K = M/2$ when the JK-flipflop PD is used. There are other simple ways to reduce ripple. Figure 4.20a

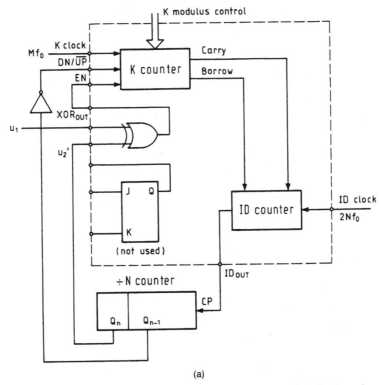

(a)

Figure 4.20 Ripple cancellation scheme for an ADPLL. (*a*) This circuit makes use of the ENABLE input of the *K* counter. For details, refer to text. (*b*) The ADPLL operates on its center frequency. (*c*) The reference frequency is higher than the center frequency.

shows a ripple cancellation scheme which uses a feature of the *K* counter that has not yet been mentioned: the ENABLE input of the *K* counter. Only one additional inverter is required in this circuit. The *K* counter is in operation only when the ENABLE input *EN* is high. To suppress ripple, the second most significant bit of the $\div N$ counter is used in addition (denoted Q_{n-1}). The frequency at the Q_{n-1} output is twice the frequency of the output signal u_2'. In this circuit, the EXOR phase detector is utilized. Its output is not connected, however, to the DN/$\overline{\text{UP}}$ input of the *K* counter, but rather to its ENABLE input *EN*. The DN/$\overline{\text{UP}}$ input is driven by the Q_{n-1} signal now. This forces the signals u_1 and u_2' to be nearly in phase when the ADPLL operates at its center frequency (Fig. 4.20*b*). If this is the case, the EN input is FALSE most of the time, so neither the UP counter nor the DOWN counter are active. Only when the reference frequency deviates from the center frequency is a phase difference between u_1 and u_2' established. Figure 4.20*c*

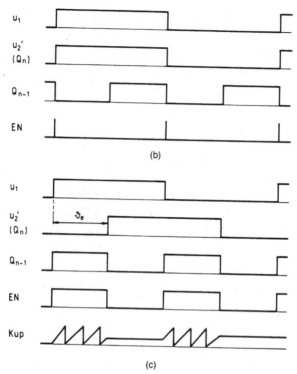

Figure 4.20 *(Continued)*

shows the case for maximum phase error θ_e. Now, the *EN* signal is high during about 50 percent of the time. At the same time, the UP counter becomes active, so the *K* counter generates the maximum number of carries. As we easily recognize from the waveforms, the average number of carries produced is only about half that number for a normal ADPLL circuit (Fig. 4.13). Consequently, the hold range is reduced roughly by a factor of 2. It has been shown[16] that the hold range of the circuit becomes

$$\Delta f_H = \frac{M f_0}{2N(2K + 1)} \tag{4.16}$$

If $M = 2N$ and K is large, this reduces to

$$\Delta f_H \approx \frac{f_0}{2K} \tag{4.17}$$

Still other ripple cancellation schemes are discussed in ref. 16.

4.3.5 Higher-order ADPLLs

The ADPLL considered in this section is a first-order system. As we have seen in Eqs. (4.15a and b), its settling time can be made extremely short. This can be an advantage in some applications; there are other cases, however, where a slower response—combined with better noise-suppression capability—is desired. If two ADPLLs are cascaded, a second-order ADPLL is obtained. This arrangement has been considered by Rosink[16] in some more detail.

4.4 Typical ADPLL Applications

Because of the availability of low-cost ADPLL ICs this type of PLL can replace the classical DPLL in many applications today. The ADPLL is mainly used in the field of digital communications. A good example is the FSK decoder, which will be considered in the following. FSK is the abbreviation for "frequency shift keying." In FSK data transmission, serial binary data are transmitted using two different frequencies, one of which represents a logical 0, the other a logical 1.

An extremely simple FSK decoder circuit is shown in Fig. 4.21. The center frequency f_0 of the ADPLL is chosen such that it is between the two frequencies used by the FSK transmitter. Because the JK-flipflop phase detector is utilized here, the two signals u_1 and u_2' would be exactly in antiphase if the ADPLL were operating at its center frequency (see also Fig. 4.17). Therefore, the contents of the divided-by-N counter would be just $N/2$ at the time where the reference signal u_1 performs a positive transition. If the reference frequency is greater than the center frequency, the output signal u_2' must lead the reference signal u_1 in order to enable the K counter to produce more carries than borrows on the average. Consequently, the contents of the divide-by-N counter will be greater than $N/2$ at the positive transient of u_1. If the reference frequency is lower than the center frequency, however, the reverse is true, and the contents of the divide-by-N counter is less than $N/2$ at that instant of time. Because the output signal u_2' is low for a content less than $N/2$ and high for a content equal to or more than $N/2$, the information transmitted by the FSK signal can be recovered simply by storing the u_2' signal in a D-flipflop every time the reference signal goes high. This D-flipflop is shown at the bottom of Fig. 4.21. In Sec. 4.5 we discuss a design procedure for ADPLLs and use it to specify the parameters of this application. Many other ADPLL applications are discussed on the data sheets of the 74HC/HCT297[31] and in various application notes delivered by the IC suppliers.[9,16]

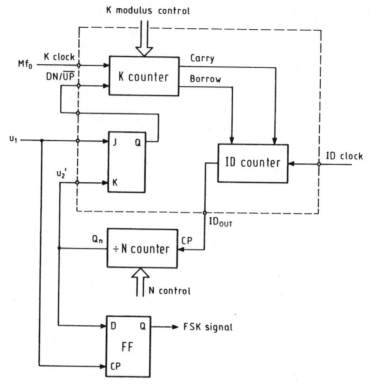

Figure 4.21 FSK decoder using the 74HC/HCT297 ADPLL IC. An additional D-flipflop is used to deliver the demodulated FSK signal.

4.5 Designing an ADPLL

Designing an ADPLL is much easier than designing an LPLL or DPLL, because we have to specify only the three parameters M, K, and N. For the 74HC/HCT297, K must always be an integer power of 2 and can be selected in the range 2^3 to 2^{17}. For the divide-by-N counter (refer to Fig. 4.13, for example), a straight-binary counter is used in most cases, so N will usually be an integer power of 2, too. Moreover, to use the same signal generator for the K clock and for the ID clock, we set $M = 2N$ whenever possible. Consequently all ADPLL parameters are integer powers of 2 in most cases.

The selection of the phase detector represents a further degree of freedom. Because the JK-flipflop phase detector is edge-sensitive, it can only be used in applications where no cycles of the reference signal are missing. Otherwise, the output of the JK-flipflop would hang up in the 1 or 0 state all the time where the reference signal disappears, which inevitably would lock out the loop.

As we have seen in Sec. 4.3, there are a number of interdependences among the quantities M, K, and N. If minimum ripple is a goal, M should be chosen to be $4K$ when the EXOR PD is used, or $2K$ when the JK-flipflop is used. Given the ratio M/K, the only parameter which can be freely selected is N. To avoid "oversleeping" of carries and borrows, N should be greater than a minimum value N_{min}, which is given by Eq. (4.4). As indicated by Eq. (4.9a), the resulting hold range is a function of M, K, and N. Furthermore, the settling time τ of the ADPLL also depends on these parameters, as seen from Eqs. (4.15a and b).

On the base of these general considerations, we can derive the design procedure given in Fig. 4.22. Like the procedures worked out for the LPLL and the DPLL, this program should not be considered as a universal recipe for every kind of ADPLL but rather as a checklist. To demonstrate the design procedure, we now implement the FSK decoder described in Sec. 4.4; refer also to Fig. 4.21.

Case study: Designing an ADPLL FSK decoder

Step 1. To specify the parameters of this ADPLL system, we must define first its center frequency and hold range. Let us assume that the FSK transmitter uses the frequencies $f_{11} = 2100$ Hz and $f_{12} = 2700$ Hz to encode the binary informations 0 and 1, respectively. Following the description in Sec. 4.4, we choose a center frequency of $f_0 = 2400$ Hz. To ensure that both frequencies of the FSK transmitter are within the hold range of the ADPLL, we specify a hold range of $\Delta f_H = 600$ Hz. We do not specify a settling time τ for the moment but will check at the end of the design whether the resulting settling time is appropriate or not.

Step 2. The decision of whether the EXOR or the JK-flipflop will be used as phase detector has been made in advance. Let us keep in mind, however, that this ADPLL relies on an input signal which does not show up missing pulses.

Step 3b. The JK-flipflop is used as phase detector.

Step 4b. For minimum ripple, we choose $M = 2K$.

Step 5. To get the simplest circuit, we choose $M = 2N$. The clocks for the K counter and for the ID counter are identical then and can be taken from the same signal source.

Step 6. This step reveals the first flop in our design, because K should be chosen 4 to get the desired hold range of 600 Hz. The 74HC/HCT297 circuit does not allow $K = 4$, but only $K = 8, 16, \ldots$. The design proceeds with step 9.

Step 9. To get a hold range of 600 Hz, the assumption $M = 2N$ cannot be made. If we will keep $M = 2K$ (the condition for minimum ripple), the hold range can be expressed as

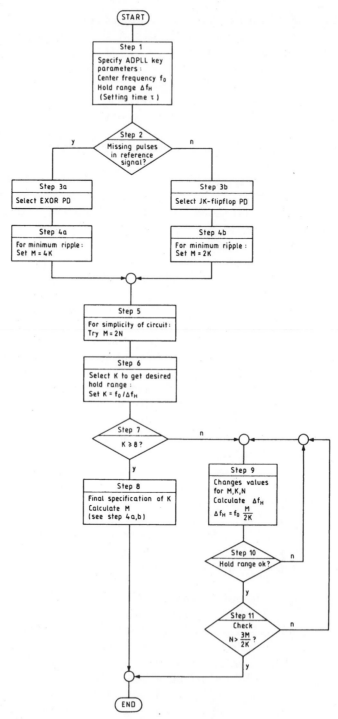

Figure 4.22 Design procedure for the ADPLL.

$$\Delta f_H = f_0 \cdot \frac{M}{2KN} = f_0 \cdot \frac{1}{N}$$

The hold range becomes 600 Hz if we specify $N = 4$. Choosing the minimum value for K, $K = 8$, we get $M = 16$. Hence we must insert an additional JK-flipflop between the K clock and the ID clock which scales down the K clock frequency by a factor of 2.

Step 10. The hold range is okay now (600 Hz).

Step 11. To avoid overslept carries and borrows, N must be larger than the minimum value $N_{min} = 3M/2K$. Because $N_{min} = 3$ in our case, $N = 4$ is a valid choice.

The design is completed now. Finally, we check the settling time τ of this circuit. Using Eq. (4.15b), we get $\tau = 2/f_0 = 2T_0$, where T_0 is the duration of one reference cycle. This indicates that the ADPLL settles within two cycles of the reference signal, which is certainly acceptable in this application. When simulating the ADPLL on the PC, we have a look at this circuit and see how it performs.

4.6   Simulating the ADPLL on a PC

The simulation program distributed with the book also performs ADPLL simulations. The following case studies will reveal the substantial differences in the behavior of all-digital PLLs when compared with the classical LPLLs and DPLLs.

Case study 1: Dynamic performance of the ADPLL using the EXOR PD. Start the simulation program, select the *CONFIG* menu and click at the *Params* item again. In the *PLL Type Selection* dialog box, check the *ADPLL* radio button. In the *ADPLL Circuit Selection* box, choose the *EXOR* phase detector. In the next dialog box, entitled *Parameters of ADPLL*, enter the following values:

$K = 8$
$M = 32$
$N = 16$

Having exited this dialog box, a message box indicates an estimate of the lock range Δf_L. Because the lock range is proportional to the center frequency f_0, the ratio of lock range to center frequency is specified. This value is 0.125 in our example.

Now select the *SIMULATION* menu. Enter the following parameter values:

f-Step radio button	Checked
f_0	100,000 Hz
df	6000 Hz
dphi	Any value
T	400 μs (precalculated)
ZoomFact	1

Because the ratio of lock range to center frequency is 0.125 for this ADPLL, the predicted lock range becomes Δf_L = 12.5 kHz. According to the theory of this type of ADPLL, the time constant of the loop is $\tau = 2\,T_0$ = 20 μs, where $T_0 = 1/f_0$. Duration T of the simulation has been computed in advance. T is set to 20τ in order to see the whole transient. The frequency step chosen in this simulation is about half the lock range. Fig. 4.23 shows the outcome of the simulation. The thin line is the phase error, the thick line is the deviation Δf_2 of the output frequency from the center frequency f_0. A linear system would settle to within 5 percent of its final state in about three time constants, i.e., in 60 μs. The curve for the phase error is quite jittery, but the settling time is seen to be well in that range. The difference frequency curve looks very noisy indeed. Unlike a classical DPLL, the frequency of the output signal does not asymptotically approach the reference frequency but oscillates around that value. This is the most typical property of the ADPLL. By the nature of the ID counter, its output signal can only perform state changes on the edges of the ID clock signal. Hence the instantaneous frequency f_2 of the output signal can take only a number of discrete values. It is easily shown that f_2 can take values of

$$f_2 = f_0 \left(\frac{2N}{2N \pm 1}, \frac{2N}{2N \pm 2}, \cdots \right) \tag{4.18}$$

For positive deviations, f_2 can have the values 100, 103.2, 106.7, . . . kHz. The curve for Δf_2 indicates that the output frequency equals the reference frequency *on average*, however. Because the phase error curve is relatively smooth at about 45°, the ADPLL is firmly locked. Let us increase now the frequency step to say 12 kHz. This simulation is shown in Fig. 4.24. The instantaneous output frequency now oscillates between 110 and 114 kHz approximately, and from the relatively quiet phase error curve we see that the system is still locked. The phase error is near its limit of stability, i.e., slightly less than 90°. If the frequency step is made even larger (13 kHz), the system can no longer maintain lock, Fig. 4.25. The loop pulls the output frequency to about 114 kHz after some reference cycles, but because the phase

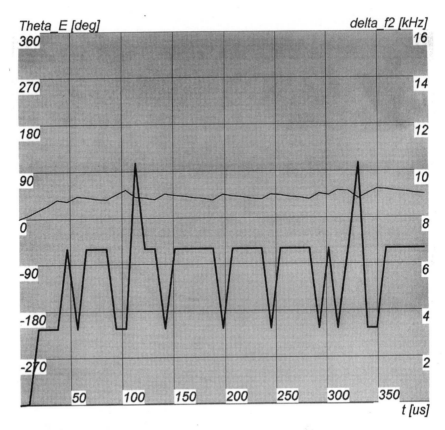

ADPLL Simulation
Tue Jan 30 15:06:33 1996•
PD = EXOR
LF = K-Counter
OSC = I/D Counter
Center frequency = 100000 Hz
Response on frequency step 6000 Hz

Figure 4.23 The simulation program is used to demonstrate the dynamic response of the 74HC/HCT297 type ADPLL. The circuit operates at a center frequency of 100 kHz. The EXOR phase detector is used. The parameters are $M = 32$, $K = 8$, and $N = 16$. The waveforms show the reponse on a frequency step of $\Delta f_1 = 6$ kHz. Note that the ADPLL settles within approximately 50 μs.

error exceeds 90°, the loop locks out momentarily. When the phase error becomes about 180°, the gain of the control system reverses, and the output frequency is corrected in the wrong direction. This experiment shows that the realizable hold range is only slightly less than the predicted one.

Let us look now at how the ADPLL reacts onto phase steps. Applying a phase step of 90° causes the output frequency to be increased in

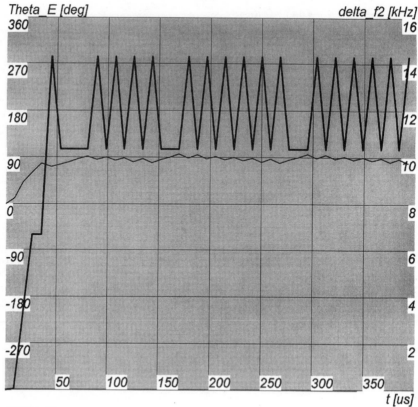

ADPLL Simulation
Tue Jan 30 15:07:03 1996•
PD = EXOR
LF = K-Counter
OSC = I/D Counter
Center frequency = 100000 Hz
Response on frequency step 12000 Hz

Figure 4.24 Like Fig. 4.23, but reference frequency step Δf_1 = 12 kHz. The final phase error is near 90°.

a few cycles after occurrence of the step, but the loop settles to zero phase error very quickly, Fig. 4.26. When the polarity of the phase step is inverted, the output frequency is reduced during a few cycles, until the ADPLL is locked again with zero phase error; this is shown in Fig. 4.27. The steady-state error approaches 0 as predicted by theory, cf. Sec. 2.6.5.

Case study 2: Dynamic performance of the ADPLL using the JK-flipflop PD. Now we stop that simulation and go back to the *CONFIG* menu. We choose again the ADPLL but use the *JK-flipflop* instead of

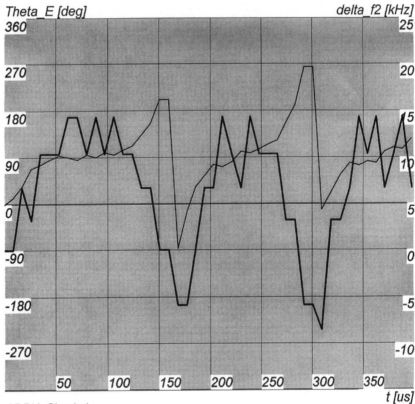

ADPLL Simulation
Tue Jan 30 15:07:32 1996•
PD = EXOR
LF = K-Counter
OSC = I/D Counter
Center frequency = 100000 Hz
Response on frequency step 13000 Hz

Figure 4.25 Like Fig. 4.23, but the reference frequency step $\Delta f_1 = 13$ kHz, i.e., larger than the hold range $\Delta f_H = 12.5$ kHz. The ADPLL is pulled out.

the *EXOR* phase detector. Let us specify the parameters for minimum ripple, i.e., $M = 16, K = 8$, and $N = 8$. This loop has again a theoretical hold range of 12.5 kHz for $f_0 = 100$ kHz. Theory further predicts a time constant of $\tau = 40$ μs. We leave the center frequency at 100 kHz and apply a frequency step of 11 kHz at the reference input. As Fig. 4.28 shows, the loop locks, but the phase error nearly approaches the stability limit of 180°. If the frequency step is increased to 12 kHz, the loop is no longer able to maintain lock. This is shown in Fig. 4.29. As an exercise, the reader is asked to find the actual size of the hold

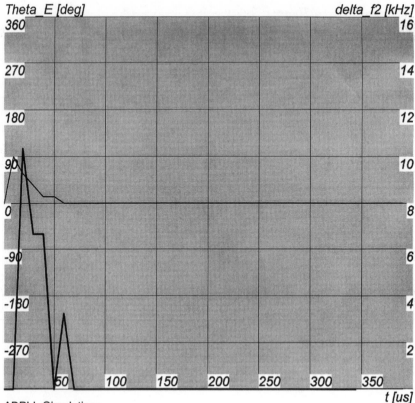

ADPLL Simulation
Tue Jan 30 15:08:01 1996•
PD = EXOR
LF = K-Counter
OSC = I/D Counter
Center frequency = 100000 Hz
Response on phase step 90 deg

Figure 4.26 Same circuit as used in preceding simulations, but the ADPLL is subject to a phase step ($\Delta\Phi = 90°$).

range, i.e., that frequency step for which the loop locks out permanently.

Case study 3: Dynamic performance of the FSK decoder.   Let us investigate now the behavior of the FSK decoder we designed in Sec. 4.5. Going back to the *CONFIG* menu, we select the ADPLL again and specify the *JK-flipflop* PD. In the following dialog, we enter the parameters $M = 16$, $K = 8$, and $N = 4$. The center frequency is $f_0 = 2400$ Hz now. The simulation program indicates a theoretical hold range of 600 Hz. Furthermore, theory predicts a time constant equal to 2 reference cycles, $\tau = 2\,T_0 \approx 0.83$ ms. In the *Simulation* dialog we

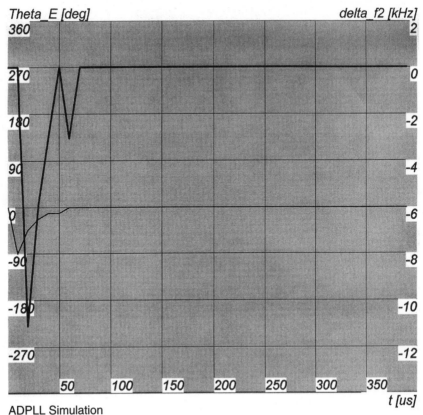

ADPLL Simulation
Tue Jan 30 15:08:30 1996•
PD = EXOR
LF = K-Counter
OSC = I/D Counter
Center frequency = 100000 Hz
Response on phase step -90 deg

Figure 4.27 Same as Fig. 4.25, but $\Delta\Phi = -90°$.

specify a center frequency of 2400 Hz and a frequency step of 300 Hz. Figure 4.30 shows the results of the simulation. Clearly, the ADPLL locks within a few reference cycles. The phase error is around 90°. There is quite a bit of ripple on the output frequency, which is not critical in this application. As Eq. (4.16) indicates, the frequency ripple could be reduced by increasing N. Doubling N, e.g., would reduce the hold range, however, by a factor of 2, which is not acceptable in this application. To maintain the hold range of 600 Hz, M would have to be doubled, too. The reader is encouraged to make his or her own trials on improving the frequency ripple of this ADPLL.

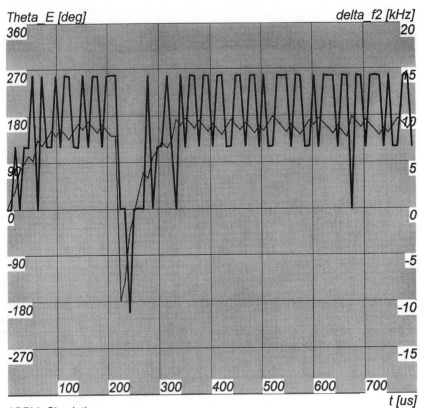

ADPLL Simulation
Tue Jan 30 15:24:52 1996•
PD = JK-Flipflop
LF = K-Counter
OSC = I/D Counter
Center frequency = 100000 Hz
Response on frequency step 11000 Hz

Figure 4.28 In this ADPLL simulation, the JK-flipflop phase detector is used. The parameters are $M = 16$, $K = 8$, and $N = 8$. Theoretical hold range $\Delta f_H = 12.5$ kHz. A frequency step of $\Delta f_1 = 11$ kHz is applied to the reference input. The circuit is just able to maintain lock.

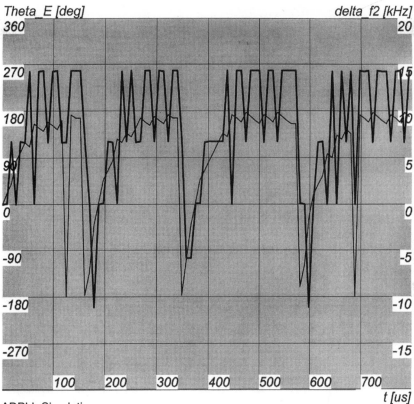

ADPLL Simulation
Tue Jan 30 15:26:19 1996•
PD = JK-Flipflop
LF = K-Counter
OSC = I/D Counter
Center frequency = 100000 Hz
Response on frequency step 12000 Hz

Figure 4.29 The frequency step is increased to 12 kHz. Although this is less than the theoretical hold range (12.5 kHz), the ADPLL is not able to get locked.

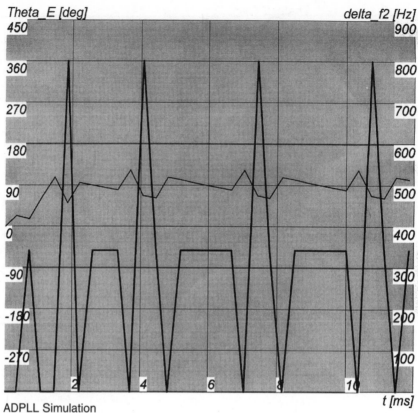

ADPLL Simulation
Tue Jan 30 15:27:48 1996•
PD = EXOR
LF = K-Counter
OSC = I/D Counter
Center frequency = 2400 Hz
Response on frequency step 300 Hz

Figure 4.30 Here the FSK decoder designed in Sec. 4.5 is simulated. Center frequency $f_0 = 2400$ Hz. The response of the ADPLL onto a frequency step of $\Delta f_1 = 300$ Hz is shown.

5

The Software PLL (SPLL)

5.1 The Hardware-Software Trade-off

In the age of microcontrollers and digital signal processors (DSP) it is an obvious idea to implement a PLL system by software. When doing so, the functions of the PLL are performed by a computer program. The designer realizing a software PLL trades electronic components for microseconds of computation time. As the parts count for a hardware PLL increases with the level of sophistication, the number of computer instructions rises with the complexity of the required PLL algorithms.

Of course the SPLL can compete with a hardware solution only if the required algorithms are executing fast enough on the hardware platform which is used to run the program. If a given algorithm performs too slowly on a relatively cheap microcontroller (one of the popular 8051 family, for example) and the designer is forced to resort to more powerful hardware (e.g., a DSP), a price trade-off also comes into play. The high speed and low cost of available PLL ICs makes it difficult for the SPLL to compete with its hardware counterpart. Nevertheless, SPLLs can offer particular advantages, especially when computing power is already available.

When comparing SPLLs with hardware PLLs, we should recognize first that an LPLL or a DPLL actually is an analog computer which continuously performs some arithmetic operations. When a computer algorithm has to take over that job, it must replace these continuous operations by a discrete-time process. From our previous discussion of hardware PLLs we know that every signal of such a system contains a fundamental frequency, which can be equal to its reference frequency f_1 or twice that value. According to the sampling theorem, the algorithm of the SPLL must be executed two or even four times in each cycle of the reference signal. If the reference frequency is a modest 100 kHz, for example, the algorithm must execute 200,000 times per second in the most favorable case, which leaves not more than 5 μs for one pass-through.

Today's microcontrollers easily work with clock frequencies of 200 MHz or more, which says that one machine cycle is 5 ns or less. The clock frequency of some newer 64 bit-μC's even exceeds 500 MHz. For most microcontrollers, however, one instruction needs more—often much more—than one machine cycle to execute. There is a risk, therefore, that the microcontrollers on the lower end of the price scale fail to deliver the required computational throughput. Using DSPs instead brings us a big step forward, because they not only are fast with respect to clock frequency, but also offer Harvard-Plus and pipeline architecture.[18] Harvard architecture means that the DSP has physically separated data and program memories, hence can fetch instructions and data within the same machine cycle. In even more sophisticated DSPs, the machine can fetch one instruction and several data words at the same time. The term "pipeline" implies that the arithmetic and logic units of the machine are fully decoupled, so that the DSP chip is able, for example, to perform one instruction fetch, some operand fetches (data fetches), one or more floating-point additions, one or more floating-point multiplications, one or more instructions decodings, one or more register-to-register operations, and perhaps even more *in one single machine cycle*. This greatly enhances computational throughput but results in higher cost, of course.

In the next section we discuss the steps required to check the feasibility and economy of an SPLL realization.

5.2 Feasibility of an SPLL Design

An SPLL design offers the most degrees of freedom available in any one PLL design, because the SPLL can be tailored to perform similarly to an LPLL or a DPLL or to execute a function which none of these hardware variants is able to do. To check whether a software implementation can economically be justified, we recommend going through the steps described in the following.

Step 1. Definition of the SPLL algorithm. The SPLL design procedure should start with the formal presentation of the algorithm(s) to be performed by the SPLL. Examples of such algorithms will be given in Sec. 5.3. For the moment it is sufficient to write down these algorithms in symbolic form, i.e., by algebraic and/or logic equations. Structograms are ideally suited to define the sequence of the operations, to describe conditional or unconditional program branchings, to describe loops which are repeatedly run through, and the like. Examples of structograms are also given in Sec. 5.3.

Step 2. Definition of the language. Having defined the algorithms, the language which will be used to encode them should be defined, at least tentatively. The programming effort is minimized when a high-level language such as C, FORTRAN, or PASCAL is used. Other frequently used languages are FORTH, BASIC, PL/M, and ADA. If the program is required to finally run on a microcontroller, a language must be chosen for which a compiler is available. Manufacturers of microcontroller or software houses mostly provide compilers for C and PL/M. When the compiled assembly-language program is available, the time required to execute it can be estimated. It should be noted that different assembler instructions may require different execution times.

Not every compiler is able to generate a time-efficient assembly code. If it is necessary to use a DSP, this point is even more important. When the DSP makes use of pipeline techniques,[18,19] the compiler must generate parallel assembly code, i.e., an assembler program where a number of different instructions are executed in any one instruction. In cases where efficient compiler programs are not available, the software designer could even be forced to write the program immediately in assembly code. With parallel-computing DSPs this is not a simple task, however. Some manufacturers of DSP chips offer signal processing libraries written in assembly code, which can be used to perform most elementary signal-processing tasks, e.g., digital filtering and the like.

Whatever language is used, the assembly code must be available to get an estimate of the approximate execution time of the algorithm(s).

Step 3. Estimation of real-time bandwidth. Having estimated the program execution time, the designer must calculate the real-time bandwidth of the SPLL system. If the execution time of the full SPLL algorithm is 50 μs, for example, and two passes are required in one cycle of the reference signal, at least 100 μs of computation time is needed in one reference period. Probably the microcontroller or whatever hardware is used will need some more time for timekeeping, input/output operations, and the like; the real-time bandwidth is likely to fall well below 10 kHz in this example.

Step 4. Real-time testing. To check if the system performs as planned, the designer will have to implement a breadboard and test its system in "real time." Only such a test can make sure that the real system is not even slower than the designer imagines.

5.3 SPLL Examples

Because every known LPLL, DPLL, or ADPLL system can be implemented by software, the number of variants becomes virtually unlimited. We therefore restrict ourselves to a few examples. The required algorithms for the SPLL will be described in great detail, so the reader should be able to adopt the methods to other SPLL realizations. The PLL simulation program delivered with the disk is a good example for SPLLs, because it demonstrates the ability of software to implement a great number of different linear and digital PLL configurations. We should be aware, however, that the simulation program does not represent a real-time system, since it does not work with real signals or execute the algorithms in real time. Nevertheless, it uses a great deal of the algorithms described in the following sections.

5.3.1 An LPLL-like SPLL

We are going to develop an SPLL algorithm which performs similarly to a hardware LPLL. To derive the required SPLL algorithm, we plot a signal flow diagram which shows the arithmetic operations within the loop (Fig. 5.1). The input signal u_1 is supposed to be an arbitrary analog signal, e.g., a sine wave. It is periodically sampled with the frequency $f_s = 1/T$ by an analog-to-digital converter (ADC), where T is the sampling interval. Thus samples are taken at times $t = 0$, T, $2T, \ldots, nT$. $u_1(n)$ is the simplified notation for the input signal sampled at time $t = nT$, i.e., $u_1(n) = u_1(nT)$. All other signals of the SPLL are sampled signals, too, and must be calculated at the sampling in-

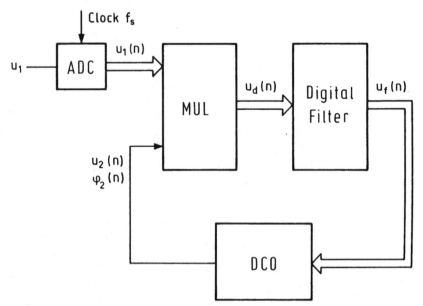

Figure 5.1 Block diagram showing the arithmetic operations to be performed by an SPLL whose performance is similar to the LPLL.

stants $t = 0, T, 2T, \ldots, nT$. Consequently, all function blocks of the signal flow diagram are working synchronously with the ADC clock. In Fig. 5.1 the signals shown by double lines are word signals. The output signal of the DCO, however, is a bit signal and is therefore represented by a single line.

There are three function blocks in the signal flow diagram: a digital multiplier, a digital filter, and a DCO. The multiplier is used as phase detector and corresponds exactly with the already known Nyquist rate PD discussed in Sec. 4.1.1; refer also to Fig. 4.2. Its output signal is denoted $u_d(n)$. The digital filter serves as loop filter, its output signal is $u_f(n)$. Finally, the DCO is supposed to generate a square-wave output signal $u_2(t)$, which is of course known only at the sampling instants. The sampled DCO output signal is denoted $u_2(n)$. As we will see, the DCO is not able to compute $u_2(t)$ directly; this signal must rather be calculated indirectly from the phase $\phi_2(t)$ of the DCO. If a VCO were used instead of the DCO, its instantaneous output angular frequency would be given by

$$\omega_2(t) = \omega_0 + K_0 u_f(t) \tag{5.1}$$

The continuous output signal $u_2(t)$ then would be given by

$$u_2(t) = w[\omega_2(t)t] \tag{5.2}$$

where w denotes the Walsh function; refer to Eq. (2.6b). The total phase $\phi_2(t)$ of the VCO output signal then would be

$$\phi_2(t) = \int \omega_2(t) \, dt = \omega_0 t + K_0 \int u_f \, dt \tag{5.3}$$

Note that we dealt only with the differential phase $\theta_2(t)$ hitherto, which corresponds to the second term only on the right side of Eq. (5.3). Here the *total phase* $\phi_2(t)$ is used to compute the instantaneous value of the DCO output signal $u_2(t)$. If we assign the values $+1$ and -1 to the square-wave signal, it follows from the definition of the Walsh function that $u_2(t)$ is $+1$, when the phase $\phi_2(t)$ is either in the interval $0 \le \phi_2 < \pi$ or in the interval $2\pi \le \phi_2 < 3\pi$, etc. In all other cases, $u_2 = -1$. We are going now to adapt this computation scheme to the time-discrete case we are dealing with. When we know the digital filter output signal $u_f(n)$ at sampling instant $t = nT$ and assume furthermore that it stays constant during the time interval $nT \le t$ $(n + 1)T$, the total phase of the DCO output signal will change by an amount

$$\Delta\phi_2 = [\omega_0 + K_0 u_f(n)]T \tag{5.4a}$$

in that interval. If the phase $\phi_2(n)$ at sampling instant $t = nT$ were known, we would be able to *extrapolate* the total phase $\phi_2(n + 1)$ at sampling instant $t = (n + 1)T$ from

$$\phi_2(n + 1) = \phi_2(n) + [\omega_0 + K_0 u_f(n)]T \qquad (5.4b)$$

This computation is possible because we can initialize the total phase with $\phi_2(0) = 0$ before the SPLL algorithm is started. Hence we can extrapolate $\phi_2(1)$ at time $t = 0 \cdot T$, $\phi_2(2)$ at $t = 1 \cdot T$, etc. Given $\phi_2(n + 1)$, we can also extrapolate the value of $u_2(n + 1)$ at $t = (n + 1)T$,

$$u_2(n + 1) = 1 \qquad \text{if } 2k\pi \le \phi_2(n + 1) < (2k + 1)\pi$$

or

$$u_2(n + 1) = -1 \qquad \text{if } (2k - 1)\pi \le \phi_2(n + 1) < 2k\pi \qquad k = \text{integer}$$

The signals of our SPLL are depicted in Fig. 5.2. The dashed lines represent continuous signals. The sampled signals are plotted as dots. Only the continuous signal $u_1(t)$ really exists; all others are only fictive. The required algorithm is easily derived from these waveforms. At a given sampling instant $t = nT$, the output signal $u_d(n)$ of the multiplier has to be computed by

$$u_d(n) = K_d u_1(n) u_2(n)$$

where K_d is the gain of the phase detector. Given $u_d(n)$, a new sample of $u_f(n)$ must be computed; the corresponding filter algorithm will be given below. Given $u_f(n)$, the value of $\phi_2(n + 1)$ at the next sampling instant is *extrapolated*. This enables us to extrapolate $u_2(n + 1)$ also. This value must be known, because we need $u_2(n)$ in the following sampling instant to compute the next value of $u_d(n)$.

The SPLL algorithm is now represented symbolically in the structogram of Fig. 5.3. When the algorithm is started, initial values are assigned to all relevant variables. The program enters an endless cycle thereafter; i.e., the algorithm within the box is repeatedly executed until the system is halted or switched off. It is assumed that the clock signal (refer to Fig. 5.1) periodically generates interrupts in the microcontroller or whatever hardware is used. Thus interrupt requests show up at time instants $t = T, 2T, \ldots, nT$. As soon as the interrupt is recognized by the hardware, the SPLL algorithm is executed. It starts with the acquisition of a sample of the input signal $u_1(t)$. The next three statements of the structogram correspond with the computation scheme already described. Finally, when all variables of the

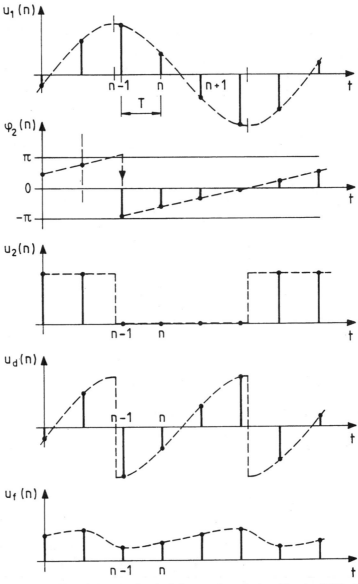

Figure 5.2 Plot of the signals which have to be calculated by the SPLL algorithm.

SPLL have been updated, they must be delayed (or shifted in time). The variable $u_d(n - 1)$ is overwritten by the value $u_d(n)$, which means that the "new" value of $u_d(n)$ computed in this cycle will be the "old" value $u_d(n - 1)$ in the next cycle. The same holds true for all other variables.

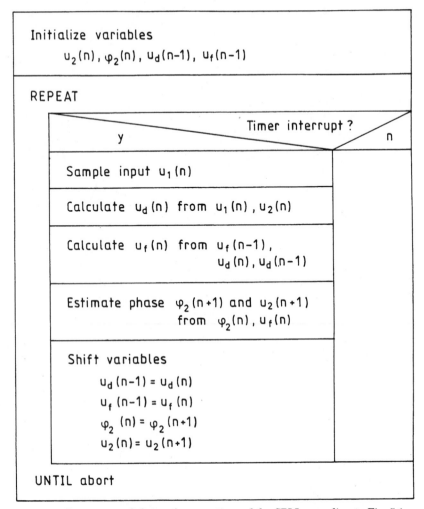

Figure 5.3 Structogram defining the operations of the SPLL according to Fig. 5.1.

Knowing what has to be calculated in every step, we can develop the algorithm in mathematical terms. The full procedure is listed in Fig. 5.4. First all relevant variables are initialized with 0. Depending on the particular application, other values can be appropriate. The operation of the multiplier is trivial. The next statement is the digital filter algorithm.

$$u_f(n) = -a_1 u_f(n-1) + b_0 u_d(n) + b_1 u_d(n-1) \tag{5.5}$$

This is the recursion of a first-order digital filter. As pointed out in Sec. 4.1.2, an analog filter is described by its transfer function

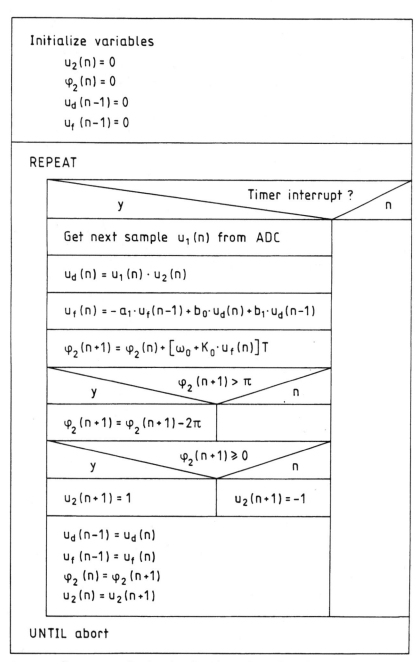

Figure 5.4 Structogram showing the algorithm to be performed by the SPLL in each sampling interval.

$$F(s) = \frac{U_f(s)}{U_d(s)} \tag{5.6}$$

where s is the Laplace operator and $U_f(s)$ and $U_d(s)$ are the Laplace transforms of the continuous signals $u_f(t)$ and $u_d(t)$, respectively. To get a digital filter performing nearly the same function we usually transform $F(s)$ into the z-domain

$$F(z) = \frac{U_f(z)}{U_d(z)} \tag{5.7}$$

where z now is the z-operator.[12,13,14,19] There are a number of transforms which can be used to convert $F(s)$ into $F(z)$. The most often used is called *bilinear z-transform*. It will be covered in some more detail in Appendix C. Before the digital filter is designed, we start with the definition of the (fictive) analog filter. Because we know that the active PI filter offers best PLL performance, we assume that $F(s)$ is the transfer function of the active PI filter; refer also to Eq. (2.3). Using the bilinear z-transform, we get

$$F(z) = \frac{b_0 + b_1 z^{-1}}{1 + a_1 z^{-1}} \tag{5.8}$$

where the filter coefficients are given by

$$a_1 = -1$$

$$b_0 = \frac{T}{2\tau_1}\left[1 + \frac{1}{\tan{(T/2\tau_2)}}\right]$$

$$b_1 = \frac{T}{2\tau_1}\left[1 - \frac{1}{\tan{(T/2\tau_2)}}\right]$$

and T is the sampling interval. Transforming Eq. (5.8) back into time domain, we get the recursion

$$u_f(n) = -a_1 u_f(n) + b_0 u_d(n) + b_1 u_d(n-1) \tag{5.9}$$

which is also listed in the structogram of Fig. 5.4. Using Eq. (5.4b) the total phase of the DCO output signal at the next sampling instant will be

$$\phi_2(n+1) = \phi_2(n) + [\omega_0 + K_0 u_f(n)]T$$

When the algorithm is executed over an extended period of time, the values of $\phi_2(n+1)$ will become very large and could soon exceed the allowable range of a floating number in the processor used. To avoid arithmetic overflow, ϕ_2 is limited to the range $-\pi \leq \phi_2 < \pi$. Whenever the computed value of $\phi_2(n + 1)$ exceeds π, 2π is subtracted to confine it to that range. Now the value of $u_2(n + 1)$ is easily computed by checking the sign of the range-limited total phase. If $\phi_2(n + 1) \geq 0$, $u_2(n + 1) = 1$; otherwise $u_2(n + 1) = -1$. Finally, the calculated values of $u_d(n)$, etc., are delayed by one sampling interval, i.e.,

$$u_d(n - 1) = u_d(n) \qquad \text{etc.}$$

As we stated in Sec. 2.10 when simulating the LPLL on the PC, the sampling rate f_s for this SPLL algorithm must be chosen at least 4 times the reference frequency in order to avoid aliasing of signal spectra.

5.3.2 A DPLL-like SPLL

When the input signal u_1 of a PLL is a binary signal, it is more adequate to implement an SPLL which performs like a DPLL. We develop an algorithm now performing like the DPLL using the phase-frequency detector and a passive lag filter (refer to Fig. 3.26, for example). Though the mathematical and logical operations within such a DPLL seem simpler compared with an LPLL, it turns out that the algorithm for the corresponding SPLL becomes much more complicated.

Let us again represent the required functions by a signal flow diagram (Fig. 5.5). It essentially consists of three functional blocks: a PFD algorithm, a digital filter, and a DCO. The digital filter is required to operate like the passive lag filter in a DPLL, which is once more shown in Fig. 5.6. Before going into details of the last two figures, we consider the signals of this SPLL (Fig. 5.7). The only signal which physically exists—at least at the beginning—is the input signal $u_1(t)$, a square whose frequency can vary within the frequency range of the DCO. The (fictive) output signal $u_2(t)$ of the PLL would be a square wave, too. As we know from Sec. 3.1, the logic state Q of the PFD depends on the positive edges of these two signals (or from the negative edges, whatever definition is made). When the PLL has settled to a steady state, the signals $u_1(t)$ and $u_2(t)$ are nearly in phase. The output Q of the PFD is then in the 0 state most of the time. Should the output frequency of the DCO drift away, the PFD would generate correction pulses; i.e., Q would become $+1$ or -1 for a very short time. The width of the correction pulses is mostly less than $1/1000$ of one period of the reference signal. If we tried to detect the edges of $u_1(t)$ and $u_2(t)$ by sampling these signals, the sampling frequency would have to be at

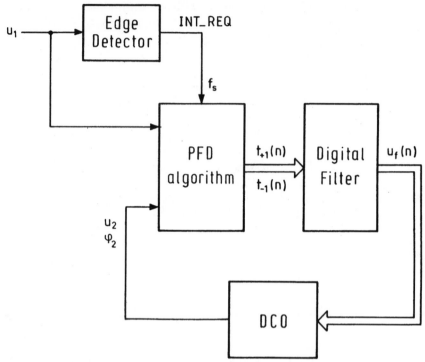

Figure 5.5 Block diagram showing the arithmetic and logic operations to be performed by an SPLL whose performance is similar to the DPLL.

least 1000 times the reference frequency, which is highly unrealistic. Another scheme must be used, therefore, to detect the instants where the state of u_1 and u_2 is changing.

Because we need to know the times where u_1 and u_2 are switching from low to high, we use the (positive and negative) edges of $u_1(t)$ to generate interrupt requests to the computer; refer to the signal INT_REQ in Figs. 5.5 and 5.7. The computer is supposed to have a

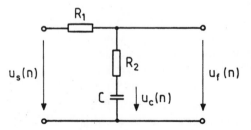

Figure 5.6 Schematic of the passive lag loop filter. This drawing is used to define the variables of the loop filter.

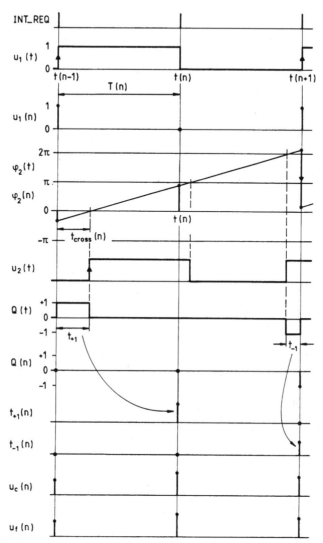

Figure 5.7 Plot of the signals which have to be calculated by the SPLL algorithm.

timer/counter chip such as the Intel 8253^{32} or the AMD $9513.^{33}$ As soon as the interrupt is recognized, a "time stamp" is taken; i.e., the time where the interrupt occurred is stored. The instants where interrupts have been detected are called $t(0), t(1), \ldots, t(n), \ldots$. Three of them are marked on top of Fig. 5.7. Before the SPLL algorithm can be discussed, we have to define a number of signals; refer to Fig. 5.7. $u_1(n)$ is the sampled version of the continuous reference signal $u_1(t)$

immediately after occurrence of the interrupt request. At time $t(n - 1)$, e.g., $u_1(n) = 1$, and at time $t(n)$, $u_1(n) = 0$. $\phi_2(t)$ is the (fictive) continuous phase of the DCO output signal. $\phi_2(n)$ is the sampled version of $\phi_2(t)$. Of course, the samples are also taken at the instants where an interrupt occurred. $u_2(t)$ is the (fictive) continuous output signal of the DCO. It will be calculated from the phase $\phi_2(t)$, as in the example of Sec. 5.3.1. $Q(t)$ is the (fictive) continuous output signal (or state) of the PFD. It can have the values -1, 0, or 1. $Q(n)$ is a sampled version of $Q(t)$ and is defined to be the state of the PFD just *prior to occurrence of the interrupt* at time $t(n)$. For example, $Q(n - 1)$ has the value 0, because $Q(t)$ was in the 0 state before the interrupt at $t = t(n - 1)$ was issued. $T(n)$ is defined to be the time interval between the time of the most recent interrupt $t(n)$ and the time of the preceding interrupt at $t = t(n-1)$; thus $T(n) = t(n) - t(n - 1)$. When $Q(t)$ is in the $+1$ state in a fraction of the $T(n)$ interval, the corresponding duration is stored in the variable $t_{+1}(n)$, as shown by the arrow in Fig. 5.7. When $Q(t)$ is in the -1 state in a fraction of the $T(n)$ interval, however, the corresponding duration is stored in the variable $t_{-1}(n)$; this is indicated by another arrow on Fig. 5.7. Finally, $u_c(n)$ is used to denote the signal on the (fictive) capacitor C in the schematic of Fig. 5.6; $u_f(n)$ is used to denote the sampled output signal of the digital filter in Fig. 5.5. With reference to Fig. 5.6, $u_f(n)$ is nearly identical with $u_c(n)$ but can slightly differ when "current" flows in the (fictive) resistor R_2.

The enumeration of that large set of variables has been quite cumbersome, but the elaboration of the algorithms will be even more fatiguing. The structogram of Fig. 5.8 shows what has to be done on every interrupt service. The signals appearing in the algorithm are shown in Fig. 5.7. The uppermost portion of the SPLL algorithm is trivial and lists the initialization of some variables. As in the previous example, the program then enters an endless loop, where it first waits for the next interrupt. When the interrupt has been detected, the time lapsed since the last interrupt is taken, $T(n) = t(n) - t(n - 1)$. Next, the current value of the reference signal $u_1(t)$ is sampled, $u_1(n) = u_1(t)$. This is necessary because we need to know whether we are in the positive or negative half-cycle of the square wave $u_1(t)$. We assume that the current time t is $t(n)$ right now, which corresponds to the second interrupt request shown in the middle of Fig. 5.7. In contrast to the previous SPLL example, we do not know the value of the phase $\phi_2(t)$ at that time! The reason for this is simple: At time $t = t(n - 1)$ the value of the digital filter output signal $u_f(n - 1)$ could be calculated, and consequently we also knew the instantaneous (angular) frequency $\omega_2(n - 1)$ of the DCO. But since we did not yet know at time $t = t(n - 1)$ how long the duration of the following half-cycle of $u_1(t)$ would be, we could not extrapolate $\phi_2(n)$ but had to postpone that

```
┌─────────────────────────────────────────────────────────────┐
│ Initialize variables                                          │
│      t(n-1), u_c(n-1), φ₂(n-1), Q(n-1), u₁(n-1), u_f(n-1)     │
├──┬──────────────────────────────────────────────────────────┤
│ REPEAT                                                        │
```

Initialize variables
$$t(n-1), u_c(n-1), \varphi_2(n-1), Q(n-1), u_1(n-1), u_f(n-1)$$

REPEAT

WAIT FOR INT_REQ

- Measure current interval $T(n) = t(n) - t(n-1)$
- Sample input signal $u_1(n)$
- Calculate phase $\varphi_2(n)$ from $\varphi_2(n-1), u_f(n-1), T(n)$
- Check if $\varphi_2(t)$ crossed 0 or 2π boundary
  - yes : $t_{cross}(n)$ = time of crossing $- t(n-1)$
  - no : $t_{cross}(n) = 0$
- From previous states $Q(n-1), u_1(n-1), t_{cross}(n)$
  compute intervals where $Q(t)$ was in +1 or -1 state
  - $t_{+1}(n)$ interval $Q(t)$ was +1
  - $t_{-1}(n)$ interval $Q(t)$ was -1
- From $t_{+1}(n), t_{-1}(n), u_c(n-1)$ calculate $u_c(n)$
- From $u_c(n), t_{+1}(n), t_{-1}(n)$ calculate $u_f(n)$

Shift variables
$$u_f(n-1) = u_f(n)$$
$$Q(n-1) = Q(n)$$
$$\varphi_2(n-1) = \varphi_2(n)$$
$$u_1(n-1) = u_1(n)$$
$$u_c(n-1) = u_c(n)$$
$$t(n-1) = t(n)$$

UNTIL abort

Figure 5.8 Structogram defining the arithmetic and logic operations within the SPLL of Fig. 5.5.

until $t = t(n)$. Only now, at $t = t(n)$, are we able to compute $\phi_2(n)$ from

$$\phi_2(n) = \phi_2(n - 1) + [\omega_0 + K_0 u_f(n - 1)]T(n) \qquad (5.10)$$

Note that the phase $\phi_2(n - 1)$ at time $t = t(n - 1)$ was known, because the phase signal is computed recursively and was initialized with $\phi_2(0) = 0$ at $t = 0$. Next we must determine whether or not the (fictive) signal $u_2(t)$ showed up a positive edge in the interval $T(n)$. This is the case when the continuous phase signal $\phi_2(t)$ "crossed" the value 0 or 2π during interval $T(n)$; this is sketched in the waveforms of Fig. 5.7. (The attentive reader will have noted that positive edges also would occur at phase crossing with 4π, 6π, . . . , etc. As will be shown later, we periodically reduce the total phase by 2π whenever it becomes larger than 2π. This is necessary to avoid arithmetic overflow in the computer.) When the phase crossed such a boundary, the corresponding time [i.e., the time interval from $t(n - 1)$ to the crossing] is stored in the variable $t_{\text{cross}}(n)$. When no crossing was detected, $t_{\text{cross}}(n)$ is set 0. The algorithm for the computation of $t_{\text{cross}}(n)$ is indicated in the structogram of Fig. 5.9a. It starts with the "normalization" of the phase signal $\phi_2(t)$, as noted above.

We are ready now to compute the state $Q(t)$ of the PFD during the interval $T(n)$. The signal $Q(t)$ depends on a number of other variables. First of all, the state of $Q(n - 1)$ prior to time $t = t(n - 1)$ must be known. If, as sketched in Fig. 5.7, $Q(n - 1)$ was 0 and $u_1(t)$ made a positive transition at $t = t(n - 1)$, $Q(t)$ goes into the +1 state. When

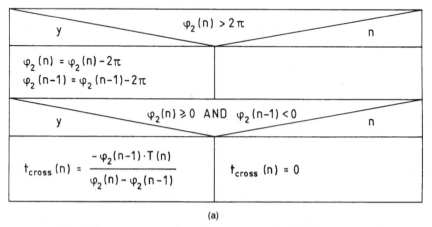

(a)

Figure 5.9 Detailed structograms of the algorithms to be performed by the SPLL of Fig. 5.5; (a) algorithm to determine the variable $t_{\text{cross}}(n)$ [time when output phase $\phi(t)$ crosses boundary of 0 or 2π]; (b) algorithm for the PFD; (c) algorithm for the digital filter.

CASE Q(n-1) OF

+1:
$t_{cross}(n) = 0$
- y: $t_{+1}(n) = T(n)$, $t_{-1}(n) = 0$, $Q(n) = 1$
- n: $t_{+1}(n) = t_{cross}(n)$, $t_{-1}(n) = 0$, $Q(n) = 0$

0:
$u_1(n-1) = 1$
- y:
  $t_{cross}(n) = 0$
  - y: $t_{+1}(n) = T(n)$, $t_{-1}(n) = 0$, $Q(n) = 1$
  - n: $t_{+1}(n) = t_{cross}(n)$, $t_{-1}(n) = 0$, $Q(n) = 0$
- n:
  $t_{cross}(n) = 0$
  - y: $t_{+1}(n) = 0$, $t_{-1}(n) = 0$, $Q(n) = 0$
  - n: $t_{+1}(n) = 0$, $t_{-1}(n) = T(n) - t_{cross}(n)$, $Q(n) = -1$

-1:
$u_1(n-1) = 1$
- y:
  $t_{cross}(n) = 0$
  - y: $t_{+1}(n) = 0$, $t_{-1}(n) = 0$, $Q(n) = 0$
  - n: $t_{+1}(n) = 0$, $t_{-1}(n) = T(n) - t_{cross}(n)$, $Q(n) = -1$
- n: $t_{+1}(n) = 0$, $t_{-1}(n) = T(n)$, $Q(n) = -1$

(b)

$t_{+1}(n) > 0$
- y:
$$u_c(n) = u_c(n-1) + \frac{t_{+1}(n)}{\tau_1 + \tau_2}\left[U_B - u_c(n-1)\right]$$
$$u_f(n) = u_c(n) + \frac{t_{+1}(n)}{T(n)} \cdot \frac{\tau_2}{\tau_1 + \tau_2}\left[U_B - u_c(n)\right]$$
- n:
  $t_{-1}(n) > 0$
  - y:
$$u_c(n) = u_c(n-1)\left[1 - \frac{t_{-1}(n)}{\tau_1 + \tau_2}\right]$$
$$u_f(n) = u_c(n)\left[1 - \frac{\tau_2}{\tau_1 + \tau_2} \cdot \frac{t_{-1}(n)}{T(n)}\right]$$
  - n:
$$u_c(n) = u_c(n-1)$$
$$u_f(n) = u_c(n)$$

(c)

Figure 5.9 (*Continued*)

$u_2(t)$ also makes a positive transition thereafter, $Q(t)$ goes back to the 0 state. If $Q(n - 1)$ had already been in the $+1$ state at $t = t(n - 1)$, however, it could not have changed its state on the positive edge of $u_1(t)$. The behavior of the PFD is therefore case-sensitive; the algorithm in Fig. 5.9b demonstrates that as many as nine different cases are possible. This algorithm determines the values of $t_{+1}(n)$ and $t_{-1}(n)$ and also computes the state of Q at the end of the $T(n)$ interval. This state will be used as initial condition $Q(n - 1)$ in the next interrupt service.

When it turns out that $t_{+1}(n)$ is greater than zero, this means that the "supply voltage U_B" must be applied during interval $t_{+1}(n)$ to the RC filter of Fig. 5.6. When $t_{-1}(n)$ is nonzero, however, the "capacitor" C would have to be discharged to ground during the interval $t_{-1}(n)$. The digital filter algorithm in Fig. 5.9c explains how the voltage $u_c(n)$ on capacitor C must be computed from the previous value $u_c(n - 1)$. If no current flowed into or out of the capacitor C (Fig. 5.6), the output signal $u_f(n)$ would be identical with capacitor voltage $u_c(n)$. In the intervals where current flows, however, $u_f(n)$ can be higher or lower than $u_c(n)$, depending on the polarity of the current. Because $u_f(t)$ is nonconstant in the interval $T(n)$, we define $u_f(n)$ to be the average of $u_f(t)$ in the interval $t(n - 1) \le t < t(n)$. This yields the expression listed in Fig. 5.9c. When deriving the equations in this algorithm, it was assumed that the duration of a $T(n)$ cycle (half a cycle of the reference signal) is much smaller than the filter time constant τ_1 in Fig. 5.6. Under this condition the current flowing into or out from capacitor C remains constant during the charging or discharging intervals. This assumption leads to simpler expressions for $u_c(n)$ and $u_f(n)$.

The algorithm used to compute the filter output $u_f(n)$ differs considerably from conventional digital filter algorithms. In a classical filter algorithm, the sample $u_f(n)$ of the output signal is calculated from a number of delayed samples of the output signal and from a number of delayed samples of its input signal. This scheme does not apply, however, to the current example, because the input signal of this circuit is applied only during a *fraction of the sampling interval*. The input is "floating" in the remaining time. Hence the output signal must be calculated like the output of an analog filter, where the input is applied *continuously*.

All computations of one interrupt service are done now. Because most of the computed samples at $t = t(n)$ will be used as starting values in the next interrupt service, they must be shifted in time. This is indicated in the bottom of the structogram of Fig. 5.8. Finally, the structogram of Fig. 5.10 lists the full algorithm in mathematical statements. To avoid overloading the graph, the algorithms for $t_{\text{cross}}(n)$ for

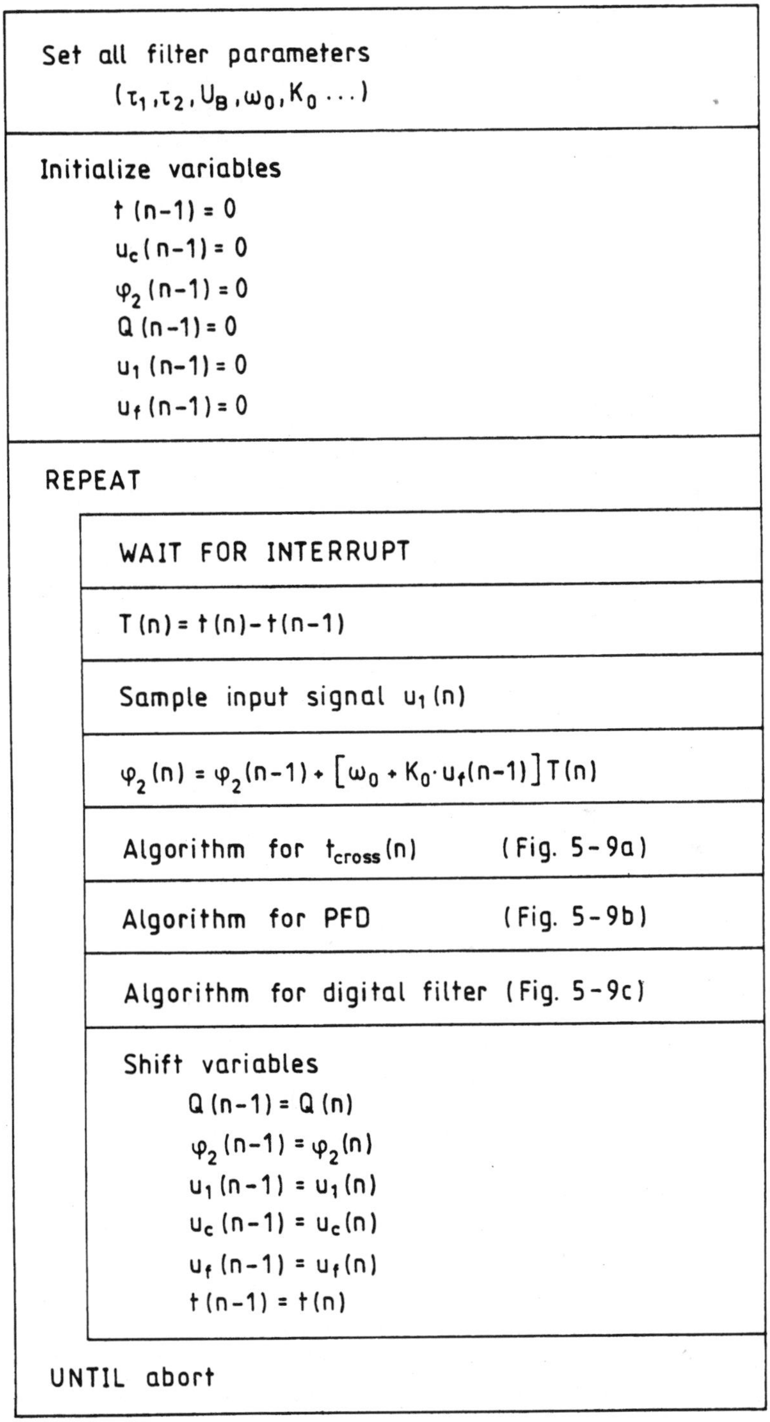

Set all filter parameters
$$(\tau_1, \tau_2, U_B, \omega_0, K_0 \ldots)$$

Initialize variables
$t(n-1) = 0$
$u_c(n-1) = 0$
$\varphi_2(n-1) = 0$
$Q(n-1) = 0$
$u_1(n-1) = 0$
$u_f(n-1) = 0$

REPEAT

WAIT FOR INTERRUPT

$T(n) = t(n) - t(n-1)$

Sample input signal $u_1(n)$

$\varphi_2(n) = \varphi_2(n-1) + \left[\omega_0 + K_0 \cdot u_f(n-1)\right] T(n)$

Algorithm for $t_{cross}(n)$ (Fig. 5-9a)

Algorithm for PFD   (Fig. 5-9b)

Algorithm for digital filter (Fig. 5-9c)

Shift variables
$Q(n-1) = Q(n)$
$\varphi_2(n-1) = \varphi_2(n)$
$u_1(n-1) = u_1(n)$
$u_c(n-1) = u_c(n)$
$u_f(n-1) = u_f(n)$
$t(n-1) = t(n)$

UNTIL abort

Figure 5.10 Structogram showing the complete algorithm of the SPLL of Fig. 5.5.

the PFD and for the digital filter are shown separately (Fig. $5.9a$ to c).

We note that the only signal which physically exists hitherto is the reference signal $u_1(t)$. A realistic SPLL system should also have one or more real output signals. When the SPLL is used as FSK decoder, for example, the demodulated signal is represented by $u_f(n)$, which is directly computed in every interrupt service. It would be quite simple to apply the u_f to an output port of the microcontroller whenever it has been computed. The situation would become more troublesome if the SPLL were requested to deliver the continuous DCO output signal $u_2(t)$. As we see from the waveforms in Fig. 5.7, the SPLL algorithm determines at time $t = t(n)$ when this signal made *its last transition,* i.e., it only knows what happened in the past. To output a real-time signal, the SPLL algorithm could extrapolate at time $t(n)$ how much time should lapse until the next transient of u_2. The corresponding time delay could then be loaded into a timer, which causes another interrupt request when timed out. The corresponding interrupt routine would finally set a bit of an output port with the current state of the u_2 signal. This simple example demonstrates that it can become quite cumbersome if only a simple hardware device has to be replaced by software.

5.3.3 A Note on ADPLL-like SPLLs

There is no question that all hardware ADPLLs can easily be implemented by software. Every hardware ADPLL operates under the control of one or several clock signals. On each clock pulse, a number of arithmetic and/or logic operations are performed. In the corresponding SPLL algorithm, these clock signals would have to be replaced by interrupt requests, and the interrupt service routines would have to perform the operations triggered by the clock in the hardware ADPLL. It is a relatively simple task to implement the algorithm of the popular 74HC/HCT297 IC (discussed in Sec. 4.3) by software. The author has built in such an algorithm in the simulation program distributed with the disk. Assume that the 74HC/HCT297 circuit is used for clock signal recovery in a modem operating at a baud rate of 9600 bauds. This circuit would then operate with a center frequency of $f_0 = 9600$ Hz. The K modulus of the K counter cannot be smaller than $K = 8$, as pointed out in Sec. 4.3. To get minimum ripple, M (the multiplier of the K clock) would be chosen $M = 4K = 32$. Moreover, the assumption $M = 2N$ is made whenever possible, where N is the scale factor of the external $\div N$ counter. Consequently, the K counter and the ID counter would both operate at a frequency of $Mf_0 = 307,200$ Hz. If the ADPLL algorithm must be implemented by software, the corresponding interrupt service routine must execute more than 300,000 times in a second, and one single pass of the routine should take less than 3 μs, including the operations required to service the interrupt request. By the present state of the art, this is out of the reach of simpler microcontrollers such as the popular 8051 family.[34] Unfortunately, most hardware ADPLLs use clock signals whose frequency is a multiple of the center frequency; hence their realization by software stays restricted to low-frequency applications. Of course, the upper frequency limit can be extended if the more powerful DSPs are used as hardware platforms.

6

The PLL in Communications

6.1 Types of Communications

In this chapter we will discuss applications of the PLL in the domain of data communication. There are many different kinds of communications, however, we will first have a look on the most important variants. First of all, analog or digital data can be transmitted over a data link. Historically, the PLL was developed to be used in the analog domain: the inventor Henri de Bellescize[22] designed a vacuum tube based synchronous demodulator for an AM receiver. The first important application of the PLL was in the recovery of the color subcarrier in television receivers around 1950[2].

6.1.1 From analog to digital

In the last years digital communications have become increasingly important, even in the classical analog domains such as telephone, radio and television. Because synchronization is an extremely important task whenever digital data is transmitted, the PLL and related circuits find widespread applications.

Analog and digital signals can be transmitted either as baseband signals or by modulation of a carrier; in the latter case we speak about bandpass modulation. If analog signals are communicated in the baseband, there is no need for synchronization, hence this is not an issue for the PLL. PLL's and related circuits come into play, however, when analog signals are modulating a high-frequency carrier. For analog signals, the classical modulation schemes still are *amplitude modulation* (AM), *frequency modulation* (FM), and *phase modulation* (PM).

Historically, *AM* is the oldest modulation scheme. To modulate an analog signal (e.g. voice, music, measurements such as temperature or humidity) onto a carrier, an analog multiplier can be used. The multiplication can also be performed by a digital multiplier; this operation can be executed as well by software on a suitable platform, e.g. on a microcontroller. If the multiplication is done digitally, the modulated carrier signal must be converted into analog form by a DAC, of course. Demodulation of AM signals can be carried out by a LPLL circuit similar to that shown in Fig. 2.24. When a linear PLL locks onto an AM modulated carrier whose frequency is identical to the center frequency of the PLL, there is a phase difference of 90° between the carrier (u_1) and the VCO output signal (u_2). The 90° phase shifter in Fig. 2.24 delivers an output signal u_1' which is in phase with the carrier, hence it can be used to synchronously demodulate the signal by a four quadrant multiplier. A low pass filter eliminates the unwanted high-frequency component at about twice the carrier frequency. Of course, the Schmitt trigger in Fig. 2.24 would have to be removed from this circuit when it is applied as an AM demodulator. The bandwidth of the low pass filter must be matched to the bandwidth of the information signal. If the information signal has a bandwidth of 4.5 kHz (as in AM radio), the low pass filter should have a cutoff frequency of about 4.5 kHz.

Frequency modulation (FM) is easily realized by PLL's and related circuits, too. A VCO is a frequency modulator by itself (*cf.* Fig. 2.30). Moreover, an ordinary linear or digital PLL is able to demodulate FM signals without additional circuitry. A circuit as shown in Fig. 2.29 (FSK demodulator) can serve as an FM decoder as well. The demodulated signal can be taken from the output of the loop filter (u_f).

FM modulators and demodulators are easily converted to PM circuits with only a few additional components. As we already know, a VCO (as shown in Fig. 2.30) acts as an FM modulator. The instantaneous radian frequency ω_2 is

$$\omega_2(t) = \omega_0 + K_0 u_f(t)$$

and hence is proportional to the information signal u_f. The phase of the VCO output is the integral of ω_2 over time

$$\varphi_2(t) = \omega_0 t + K_0 \int u_f(t) dt$$

which can be written as

$$\varphi_2(t) = \omega_0 t + \vartheta_2(t)$$

where $\theta_2(t)$ represents the phase that carries the information. (The relationship between phase and radian frequency was explained in

Sec. 2.2.) In PM, $\theta_2(t)$ is required to be proportional to the information signal u_f *itself* and *not to its integral*. PM is therefore realized by inserting a differentiator between the information signal source and the control input (u_f) of the VCO. In the same way, a PLL used as an FM receiver is converted to a PM decoder by integrating the u_f signal. The integrated u_f signal then represents the information signal. We should be aware, however, that an integrator is driven into saturation if an offset voltage exists at its input (which is the normal situation). Saturation is avoided when an "ac integrator" is used in place of an ordinary integrator, i.e. a circuit that integrates higher frequencies only but has finite gain at dc. The transfer function $F(s)$ of an ac integrator can be given by

$$F(s) = \frac{1}{1 + sT_i}$$

where T_i is the integrator time constant. Roughly speaking, such a circuit would integrate signals whose radian frequency is above $1/T_i$, but would have gain 1 for lower frequencies. (Note: such an "ac integrator" is nothing more than a simple first-order low-pass filter, which can be realized as a passive or active RC network.)

It is quite clear that the bandwidth of such data transmissions cannot extend down to dc, but will be restricted to some lower band edge (maybe 300 Hz), as is the case with voice signals.

As mentioned digital communications are of much greater interest today, so we will concentrate on digital applications. Baseband transmission of digital signals has already been discussed in Sec. 3.5.4, where we also considered various coding schemes such as NRZ, RZ, biphase, and delay modulation formats. We also became aware of the problem of recovering the clock frequency from the information signal and we considered a number of circuits that extract clock information from the data. In broadband communications, bandpass modulation is applied almost exclusively, and we will now consider the techniques of bandpass communication in more detail.

6.2 Digital Communications by Bandpass Modulation

Let us start with a review of the most important modulation schemes for digital communications.

6.2.1 Amplitude shift keying

Amplitude shift keying (ASK) was one of the earliest methods of digital modulation used in radio telegraphy around the year 1900. In its sim-

plest form, a high-frequency carrier is turned on and off by a binary
data signal, which is also called "on-off keying". (In the first era of
communication technology, the transmission of Morse symbols effectively used a pseudo-ternary code: a "dot" was represented by a short
carrier burst, a "dash" by a long carrier burst, and the pauses by
"nothing".) On-off keying is very simple, but has severe drawbacks.
Switching the carrier on and off creates steep transients, which broad-
ens the spectrum of the signal in an undesired way. Though quite
primitive, ASK still finds applications where low bit rates are sufficient and low bandwidth is not of primary concern. In a typical ASK
setup, a high-frequency carrier is switched by an NRZ encoded binary
data stream. The receiver is built from an ordinary linear or digital
PLL whose center frequency is tuned to the carrier frequency. The PLL
is equipped with an in-lock detector (as shown in Fig. 2.24). Whenever
the carrier is on, the PLL locks, and the output of the in-lock detector
switches into the 1 state. When the carrier is off, the PLL unlocks,
and the output of the in-lock detector goes to 0. Such receiver circuits
are commonly referred to as *tone decoders*. A number of tone decoder
IC's are available, such as the XR2213 (Exar) or the LM567 (National):
Table 7.1.

6.2.2 Phase shift keying

Phase shift keying (PSK) is the most widely used digital modulation
today. Some of its descendants (QPSK, OQPSK, QAM) are more bandwidth efficient and will be discussed later. In classical PSK (also referred to as BPSK [binary PSK] or PSK_2), the polarity of a carrier is
controlled by a binary data signal. Such a binary signal is shown in
Fig. 6.1a, the modulated carrier in Fig. 6.1b. When the data signal is
a logical 1, the carrier phase is 0 by definition. For a data signal of 0,
the carrier phase becomes π (180°). The definition can be inverted if
desired. As usual in the theory of alternating currents, amplitude and
phase of the modulated carrier can be represented as a vector (phasor)
in the complex plane, Fig. 6.1c. The length of the phasor represents
its amplitude, the angle with the horizontal axis its phase. For BPSK,
the phase can only take two values 0 and π.

As we will see in Sec. 6.4, there is a distinct relationship between
the symbol rate R (number of binary symbols transmitted in one second) and the bandwidth W (in Hz) required to transmit the modulated
signal. It will show up that for BPSK, the (two-sided) bandwidth W
becomes approximately equal to R. Digital data are often carried by
phone lines. Old analog telephone cables are known for its poor bandwidth, so if we planned to increase the symbol rate of an existing link
by a factor of 10, we cannot hope to reach that goal simply by increasing the carrier frequency by a factor of 10 and modulate it with the

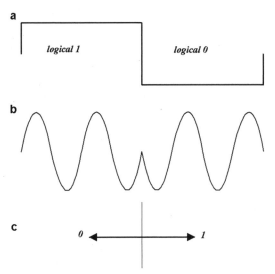

Figure 6.1 Binary phase shift keying (BPSK). (*a*) Binary signal. (*b*) Modulated carrier. (*c*) Vector plot of the modulated carrier signal.

faster data signal. Therefore we must try to increase the channel capacity (the throughput of binary symbols) without increasing the channel bandwidth. This seems contradictory at the first glance, but we will see shortly that it is possible.

6.2.3 Quadrature phase shift keying

QPSK (*quadrature phase shift keying*, also called quaternary PSK or PSK_4) brings us a step further. How does it work? In BPSK, we created just one carrier and switched its polarity by a binary data stream. In QPSK, we generate *two* carriers that are offset in phase by 90°. One of the carriers is called the *in-phase* component; it is assigned to a phase of 0 and is mathematically represented by a *cosine wave*. The other carrier is called *quadrature carrier* and is assigned to a phase of 90°. The latter is mathematically represented by a *sine wave*. (The ensemble of both carriers is often called a *complex signal*. This sounds a little bit curious, because a signal never can take on complex values—try to think of a complex temperature or humidity! The term complex only makes sense if we consider the in-phase component as the real part of a complex signal, and the quadrature component as the imaginary part thereof.)

We are now going to switch the polarity of the in-phase carrier by one binary signal, s_1. Moreover we switch the polarity of the quadrature carrier by *another* binary signal, s_2. Hence, the phase of the in-phase carrier can take on the values 0 or π (180°), but the phase of the quadrature carrier can take on the values $\pi/2$ (90°) or $3\pi/2$ (270°). If we add the two carriers, the phase of the resulting signal can have values 45°, 135°, 225°, or 315°. Adding the two modulated carriers increases the symbol rate by a factor of two, but does not require additional bandwidth. Because the two carriers are "orthogonal" (perpendicular on each other), the two data signals s_1 and s_2 can be decoded by synchronous demodulation at the receiver, and therefore they do not interfere. The symbol rate of the QPSK signal is now half the symbol rate of the original data stream, as is easily seen from Fig. 6.2. If we call the symbol rate of the QPSK signal, we can state that QPSK transmits 2 bits in one symbol period.

Figure 6.2 explains the principle of QPSK. The binary data stream as delivered by the data source is shown in Fig. 6.2a. This sequence is partitioned into two data streams now: the even bits (b_0, b_2 etc.) form the binary sequence that modulates the in-phase carrier (Fig. 6.2b), while the odd bits (b_1, b_3 etc.) modulate the quadrature carrier (Fig. 6.2c). Fig. 6.3a represents the four possible phases of the sum of in-phase and quadrature signals, and Fig. 6.3c shows the corresponding phasor plot.

As can be seen from Fig. 6.2, both in-phase and quadrature data streams change their values at half the original symbol rate, i.e. on every second symbol delivered by the data source. In other words, the two data streams are *aligned*. In OQPSK (*offset* quadrature phase shift keying), the second data stream (Fig. 6.2c) is delayed by one symbol period of the source signal. In-phase and quadrature data stream no longer change their states simultaneously, but are "staggered" in time. This does not have any impact on channel capacity or bandwidth, but rather offers some practical benefits. When the data streams are lowpass filtered (as will be explained in Sec. 6.4), the data signals are no longer square waves, but will become "smoothed". Whenever a data signal changes polarity then, the modulated carrier

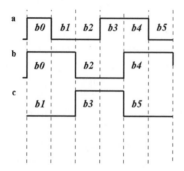

Figure 6.2 The principle of QPSK. (a) Binary signal to be transmitted. (b) Data stream corresponding to the even bits in a. (c) Data stream corresponding to the odd bits in a.

a

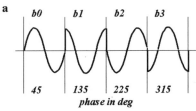

b0   *b1*   *b2*   *b3*

45   *135*   *225*   *315*

phase in deg

b

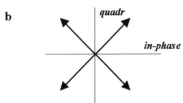

quadr

in-phase

Figure 6.3 Phase relationship of BPSK signals: (*a*) The modulated carrier can take on 4 different phases (cf. text); (*b*) Phasor diagram of modulated carrier.

temporarily fades away, hence we temporarily loose the signal which the receiver tries to lock onto. This drawback becomes less severe if not both of the data streams can change state at the same time. This is the motivation to use OQPSK.

6.2.4 QAM (m-ary phase shift keying)

Having increased the channel capacity by a factor of two without sacrificing bandwidth, we are going now to expand this principle to even greater benefit. The amplitudes of the two carriers for QPSK have always been the same, so we can normalize these amplitudes to 1. What about using more than one amplitude value, say 1 or 2? When doing so, the modulated in-phase carrier could take on a phase of 0 with an amplitude of 1, or a phase of 0 with an amplitude of 2, or a phase of 180° with an amplitude of 1, or a phase of 180° with an amplitude of 2. The same would hold true for the quadrature signal. If the receiver is not only able to determine the phases of the two signal components but also would be in the position to discriminate different signal amplitudes, this would increase the channel throughput by another factor of 2. In other words, if we compare the channel capacity with ordinary BPSK, this modulation scheme would transmit 4 bits in one symbol period. This modulation scheme is called QAM (quadrature amplitude modulation). Actually QAM is a combination of PSK with ASK, because we alter both phase and amplitude. Of course we must not restrict the amplitudes to only 2 values—more values can be used. A phasor diagram of QAM can look like Fig. 6.4. Here two amplitude values are used. The end points of the phasor have identical distance on the horizontal and on the vertical axis. Note that the in-

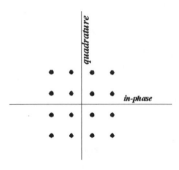

Figure 6.4 Phasor diagram of QAM.

phase component can take on relative amplitudes of 0.5 and 1.5 and not 1 and 2, respectively. With this choice of amplitudes, all phasors become equidistant on both axes. Using the scheme shown in Fig. 6.4, the combined carrier can take on 16 states, hence 4 bits per symbol are transmitted. The code shown in Fig. 6.4 is commonly called QAM_{16}. Note that QAM_4 is identical with QPSK. Theoretically, the set of amplitudes can be made as large as desired, but noise superimposed to the transmitted signal sets limits. To discriminate between different amplitude levels, the receiver must define *decision limits*. With reference to Fig. 6.4, the receiver would split the phasor plane into a number of squares of equal size, and the marked dots would sit in the center of each square. Noise on the signal would cause the phasor to deviate from its nominal position. As long as the end point stays within the corresponding decision region, no error would result. Errors occur, however, when the noise causes the phasor to migrate into another decision region. The actual bit error rate of a particular code can be computed using statistics and Shannon's information theory[20]. Today, QAM is the most widely used technique. In modern modems QAM_{64}, QAM_{128}, and QAM_{256} come into play. As we will see in Sec. 6.6, QAM will be used very efficiently in DVB (Digital Video Broadcasting).

6.2.5 Frequency shift keying

FSK (frequency shift keying) is another modulation scheme that is frequently used in modems. An FSK signal can be transmitted in the base band, or it can be used to modulate a carrier. Let us discuss baseband communication first. In its simplest form (FSK_2), two different frequencies f_1 and f_2 are defined, where f_1 represents a binary 0, and f_2 represents a binary 1 or vice versa (Fig. 6.5a). The two frequencies can be chosen arbitrarily, e.g. f_1 = 1650 Hz, f_2 = 1850 Hz.

a

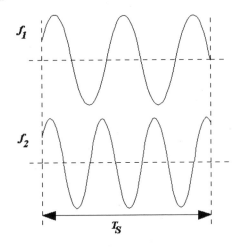

b

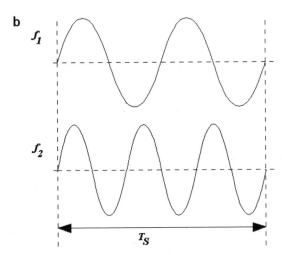

Figure 6.5 Definition of binary FSK: (*a*) Nonorthogonal FSK: arbitrary values for frequencies f_1 and f_2 are chosen; (*b*) Orthogonal FSK: there is a tight mathematical relationship between the frequencies f_1 and f_2, cf. text.

Moreover it is not required that the symbol interval T_S is an integer multiple of one period of the f_1 or of the f_2 signal. Normally a linear or digital PLL serves as an FSK decoder. The difference between the two frequencies must simply be chosen such that the PLL can savely determine whether it "sees" frequency f_1 or f_2. This type of FSK is called non-orthogonal FSK. The other choice is *orthogonal FSK*, Fig. (6.5b). Here the frequencies f_1 and f_2 are chosen as to fulfill the relationship

$$f_i = k_i \cdot \frac{1}{2T_s} \qquad (6.1)$$

where $i = 1, 2$ and k_i is a positive integer greater 0. Both frequencies then are integer multiples of half the symbol rate $1/(2T_s)$. In the example of Fig. 6.5b, $k_1 = 4$ and $k_2 = 6$. If we denote the signal waveforms for logical 0 and logical 1 by $s_1(t)$ and $s_2(t)$, respectively, we have

$$\int_0^{T_s} s_i(t)s_j(t)dt = 1, \quad i = j$$

$$= 0, \quad i \neq j \qquad (6.2)$$

which means that the signals $s_1(t)$ and $s_2(t)$ are orthogonal. This property can be beneficial when the received symbols are buried in noise. A coherent FSK decoder then would store replicas of the symbol waveforms $s_1(t)$ and $s_2(t)$ and correlate the incoming symbol with both of these. Without noise, the correlation with one of the replicas would yield 1, the correlation with the other 0. Whenever the correlation with $s_1(t)$ leads to the result 0, the receiver would decide for logical 1. With added noise, the correlation coefficients depart from their ideal values 0 or 1, hence the receiver would decide for 0 if the correlation with $s_1(t)$ becomes greater than the correlation with $s_2(t)$ or for 1 if the reverse is true. Such a decision is a *maximum likelihood decision* in the statistical sense.

We only considered binary FSK hitherto, but FSK can be extended to *m*-ary FSK when more than 2 frequencies are chosen. For example, if $m = 4$, 4 different frequencies f_1 to f_4 would be used. When doing so, 2 bits per symbol can be transmitted. There are a number of simple modems that use FSK_2 or FSK_4 and are working immediately with baseband signals. Of course, the baseband signal can be modulated onto a high-frequency carrier. When doing so, many FSK modulated carriers (with different carrier frequencies of course) can share one single link.

In case of nonorthogonal FSK, switching abruptly from one frequency to another can cause sharp transients, which extends the bandwidth of the signal in an undesired manner. This drawback is avoided in a special version of FSK called *MSK* (*minimum shift keying*). With MSK, every symbol waveform starts and ends with a zero crossing of the same direction, e.g. from negative to positive values (as can be seen from Fig. 6.5b). Adjacent symbol waveforms then never execute phase steps. This reduces the bandwidth required to transmit the FSK signal.

6.3   The Role of Synchronization in Digital Communications

Before discussing the various modulation techniques applied in digital communication, some notes on the synchronization problem appear appropriate. Whenever digital data are transmitted by bandpass modulation, synchronization on different signal levels will be required. Assume for the moment that binary phase shift keying (BPSK) is used to modulate the phase of a high-frequency carrier. At the receiver, the data signal has to be recovered by demodulation. In most cases the demodulation is performed synchronously, i.e. the receiver generates a replica of the carrier and multiplies the incoming signal with that reconstructed carrier. This shifts the spectrum of the received signal by a frequency offset which is equal to the carrier frequency, and the data signal is obtained by simple low pass filtering. Because the replica of the carrier must be in phase with the latter, *phase synchronization* is required. This can be considered the first level of synchronization and is usually realized by PLL circuits. For reasons to be explained in the next sections, locking onto a modulated carrier is not always trivial, so extended versions of PLL's are required in many cases (e.g. the Costas Loop).

After demodulation, the data stream in the receiver is almost always a filtered version of the original data signal. While the original data signal represented a square wave signal, the transients of the filtered data signal become smoothed. In many situations, the receiver performs a correlation of the received data with template functions that are stored in the receiver. This is done to reduce the effects of noise which has been added to the signal on the transmission link. This correlation is usually performed during a time interval which must be identical with the symbol duration, hence the receiver must know when an incoming symbol starts and when it is over. A second level of synchronization must be performed then, which is referred to as *symbol synchronization*. The circuits that perform symbol synchronization are still other special versions of PLL's; we will discuss some of them in following sections, including the one on the *Early-Late-Gate* synchronizer.

In some communication systems, an even higher level of synchronization is required. This is usually called *frame synchronization*. This kind of synchronization comes into play when the information is organized in blocks. Such a block may consist of a block header, followed by the actual data and eventually by a trailer that contains check code (e.g. a cyclic redundancy check, CRC). Information is organized in blocks when the communication channel is time-shared by a number of transmitters. In this case, the receiver must be able to decide when a block starts. Frame synchronization will not be discussed here[20].

6.4 Digital Communications Using BPSK

6.4.1 Transmitter considerations

Filtering the data: yes or no? Communicating binary signals by binary phase shift keying looks quite trivial, but imposes serious problems if we consider bandwidth requirements. As shown in Fig. 6.1, the polarity of a high-frequency carrier is switched by the binary information sequence. If the data are not filtered, the carrier is modulated by a square wave. It is not strictly a square wave, but rather a binary random sequence which can change its amplitude at the start of each symbol period. What the data signal has in common with a square wave are its very steep transients. Normally the NRZ format is used to modulate the carrier (Fig. 3.27). If we transmitted an infinite sequence of binary 0's or 1's, the binary random sequence would degenerate to a dc level, hence the bandwidth of the data signal would be zero. Because we actually send an arbitrary sequence of 0's and 1's,

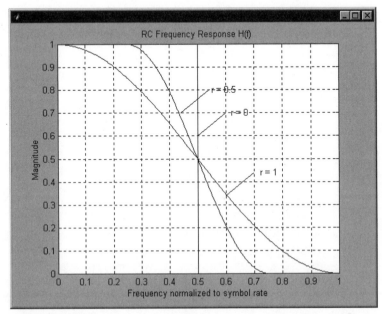

Figure 6.6 Amplitude response of brickwall filter ($r = 0$) and raised cosine filters ($r > 0$).

the spectrum of the data signal will contain some higher frequencies. It is easy to recognize that the frequency of the data sequence becomes maximum when we transmit a sequence of alternating 1's and 0's. In this case, the data signal would be a square wave signal whose frequency is half the symbol rate.

As Nyquist realized in 1928[40] a receiver will be able to detect the data signal if only the fundamental component of this square wave signal is transmitted. Consequently, we would low pass filter the data signal at the transmitter. The most appropriate filter would be a "brickwall filter", i.e. an ideal low pass filter having gain 1 at frequencies below half the symbol rate and gain 0 elsewhere. The impulse response of the brickwall filter is

$$h(t) = \frac{\sin(\pi t / T)}{\pi t / T} \tag{6.3}$$

which is referred to as a *sinc* function. T is the symbol period. The duration of the impulse response is infinite, and it starts at $t = -\infty$; the filter delay is also infinite. The impulse response is plotted in Fig. 6.7 by the curve labelled $r = 0$ (the meaning of r will be explained in the following). Of course such a filter cannot be realized. An approxi-

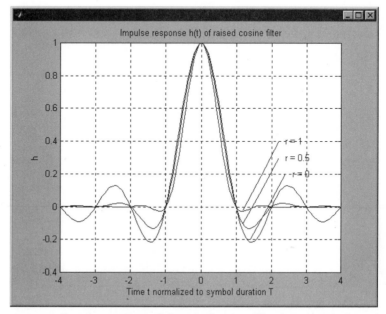

Figure 6.7 Impulse response of the raised cosine filter for various values of (relative) excess bandwidth r.

mation to the brickwall filter can be implemented by a FIR digital filter (FIR = finite impulse response, cf. Appendix C) to any desired level of accuracy. However, because the impulse response decays slowly to 0, this leads to excessively long FIR filters, i.e. to filters with a large number of taps. Nevertheless, we will consider the impulse response in more detail because it exhibits a very useful property. Every FIR filter causes the signal that is passed to be delayed, where the delay τ is given by

$$\tau = \frac{L-1}{2} T_F \qquad (6.4)$$

where L is the length of the FIR filter and T_F is the sampling interval of the filter. In Fig. 6.7, the delay of the filter has been neglected, hence its maximum (+1) occurs at time $t = 0$. Normally such a filter would be sampled at a frequency f_F which is an integer multiple of the symbol rate $R = 1/T$, where T is the symbol period. If we used a FIR filter of length 33, it would have a delay of 16 T_F; if we assume furthermore that the filter sampling frequency f_F is four times the symbol rate R, the filter would delay the signal by 4 symbol periods. For this FIR filter, the impulse response $h(t)$ would no longer be symmetrical about $t = 0$, but rather about $t = 4T$. For simplicity, the delay has been omitted in Fig. 6.7, i.e. we assume that the filter is a so-called *zero phase filter*, which causes no delay. Note that the impulse response of the brickwall filter becomes 0 at $t = T, 2T, 3T$ etc. Now we remember that the impulse response is nothing more than the response of the filter to a delta function with amplitude 1, applied at $t = 0$ to its imput. Sending a logical 1 therefore corresponds to applying a positive delta function, and sending a logical 0 corresponds to applying a negative delta function (amplitude -1). Next we consider the case where a number of symbols it passed through the filter in succession. If two succeeding 1's are supplied, the first logical 1 will be represented by a delta function with amplitude +1 applied at $t = 0$, the second logical 1 by another delta function with amplitude +1 applied at $t = T$. This situation has been plotted in Fig. 6.8. The upper trace shows the two delta functions, the lower traces the corresponding impulse responses (solid curves). The zeros of the impulse responses are marked by circles. The superposition of the two impulse responses is shown by the dotted curve (thick). In order to recover the data, the receiver would have to sample the filtered signal exactly at time $t = 0, T, 2T$ etc. As can be seen from Fig. 6.8, the amplitude of the combined signal at $t = 0$ (marked by a square) depends uniquely on the impulse applied at $t = 0$; the impulse response caused by the second delta function is zero at this time. The same holds for the combined signal at $t = T$.

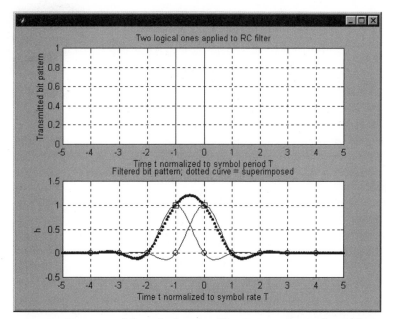

Figure 6.8 Response of the raised cosine filter to a sequence of two binary zeros applied to its input. Upper trace: a sequence of two delta functions corresponding to two binary 1's transmitted in succession. Lower trace: the solid curve (thin) whose maximum is at $t/T = -1$ is the response of the filter to the delta function applied at $t/T = -1$, the solid curve (thin) whose maximum is at $t/T = 0$ is the response to the delta function applied at $t/T = 0$, and the thick (dotted) curve is the superposition of these two responses.

Its amplitude at $t = T$ (marked by a square) uniquely depends on the amplitude of the second delta function and does not contain any contribution from the first delta function. In other words, the impulse responses caused by various delta functions *do not interfere,* i.e. when using the brickwall filter there will be no *inter-symbol interference* (ISI). To recover the data signal without error, the applied filter must be chosen such that no ISI can occur. (Note: It is easy to demonstrate how ISI can be created: assume that we filter the data signal by a Butterworth lowpass filter e.g. Its impulse response will be a damped oscillation. It also passes through zero, but these zeros are not at $t = 0, T, 2T$ etc., hence the impulse responses coming from delta functions applied to the input will interfere, or in other words, an output sample taken at $t = kT$ (k = positive integer) is the sum of the contributions of many symbols. This cannot be tolerated, of course.)

How to get rid of ISI? As we have seen, a useful approximation to a brickwall filter would lead to excessive filter length. Nyquist has shown a valuable alternative: if the transition region of the filter is widened, its impulse response decays faster towards zero. Moreover, if the amplitude response $H(f)$ of the filter is symmetrical about half the symbol rate frequency $(R/2)$, the location of the zeros of its impulse response remains unchanged. Filters having this property are commonly referred to as "*Nyquist filters*". One possible realization of the Nyquist filter is the *raised cosine filter* (RCF). Its transition region (the region between passband and stopband) is shaped by a "raised" cosine function, i.e. by a cosine wave which stands "on a piedestal". The width of the transition band is determined by the parameter r which is called excess bandwidth. If the frequency corresponding to half the symbol rate is denoted W_0 ($W_0 = 1/2T$), the transition band starts at $f = (1 - r)\,W_0$ and ends at $(1 + r)\,W_0$. The amplitude response $H(s)$ of the RCF has been plotted in Fig. 6.6 for three values of r, $r = 0$, 0.5, and 1. (The case $r = 0$ applies for the brickwall filter.) Note that the frequency response is symmetrical about half the symbol rate. Mathematically the frequency response of the RCF is given by

$$H(s) = \begin{cases} 1 & \text{for } |f| < \dfrac{1-r}{2T} \\[2ex] \cos^2 \dfrac{\pi}{4} \dfrac{2T|f| + r - 1}{r} & \text{for } \dfrac{1-r}{2T} \le |f| \le \dfrac{1+r}{2T} \\[2ex] 0 & \text{for } |f| > \dfrac{1+r}{2T} \end{cases} \quad (6.5)$$

Figure 6.7 shows the impulse response of the RCF for $r = 0$, 0.5, and 1. The larger r is chosen, the faster the decay becomes. To economize bandwidth, however, large values of r must be avoided. In practice, values of r in the range 0.15 to 0.35 are customary. For completeness, we also give the expression for the impulse response $h(t)$ of the RCF[41]:

$$h(t) = \frac{\sin(\pi t/T)\cdot\cos(\pi rt/T)}{(\pi t/T)\cdot\left[1 - \left(\dfrac{2rt}{T}\right)^2\right]} \quad (6.6)$$

Designing an FIR raised cosine filter is an easy task: Eq. (6.6) can be used to compute the filter coefficients. Care must be taken, however, since there exist values of t where both nominator and denominator become 0. This happens if the expression in the square brackets of the nominator becomes 0, explicitly for $t/T = 1/(2r)$. For the same value of t, the cosine term also becomes 0. This singularity is removed by replacing both nominator and denominator by their derivatives; mathematically this is called L'Hôpital's rule. The computation is made even easier if Matlab's Signal Processing Toolbox is available[25]. It contains a function called FIRRCOS, that calculates the coefficients of the RCF.

So far the raised cosine filter seems to offer the optimum solution for low pass filtering because it completely suppresses ISI. As we will see in Sec. 6.4.2, the receiver also needs a low pass filter for different reasons. It will show up that we cannot use another raised cosine filter at the receiver: when doing so, the data signal would have to pass through two cascaded RCF's: one in the transmitter and one in the receiver. The overall frequency response then would be the RCF transfer function *squared!* The resulting frequency response would no longer be symmetrical about half the symbol rate, and ISI would occur. We will see in the next section that so called *root raised cosine filter* (RRCF) will fix that problem.

6.4.2 Receiver considerations

A number of tasks has to be performed in the receiver to reconstruct the symbol stream. The most important of these will be discussed here. Because the data modulate the carrier, the symbol stream must be demodulated by a convenient procedure. Basically we can differentiate between coherent (synchronous) and noncoherent demodulation. Coherent demodulation is more efficient, especially if noise has been added to the signal when it propagated across the data link. When coherent detection has to be applied, the receiver must know the phase of the carrier exactly and by what means that the receiver must reconstruct a replica of the latter. As we will see in the following, this is more difficult than it appears at a first glance. If noncoherent detection is to be realized, the receiver must not know the phase of the carrier. It rather takes the phase of the currently detected symbol as a reference for the detection of the next symbol. If the phase does not change, the receiver knows that the next symbol is the same as the previous. Otherwise it decides that the symbol changed its value from 0 to 1 or from 1 to 0. For correct operation, some initial synchronization becomes necessary, i.e. the receiver must know at the start of a transmission whether the currently received symbol is a 0 or a 1. Noncoherent detection has been extensively analyzed in[20]. Due to its higher noise immunity, coherent detection is the preferred method in most receiver circuits today[42], so we will concentrate on this technique.

Phase synchronization within the receiver. When working with BPSK, the carrier is modulated by a (filtered or unfiltered) symbol sequence. The carrier has the frequency f_s, and its spectrum consists of just one line at f_s, but the symbol sequence has a broader spectrum which

ranges approximately from 0 to half the symbol rate, when it is filtered. When it is not filtered, the spectrum can be much broader. Assume for the moment that the data signal consists of only one frequency f_d. With BPSK, the data signal and carrier are multiplied,
hence their spectrum will contain lines at the sum frequency $f_s + f_d$
and at the difference frequency $f_s - f_d$.

Normally the data signal is an arbitrary sequence of logical 1's and
0's, where a 1 is represented by a positive voltage (+1) and a 0 by a
negative voltage (−1). On a long run there will be approximately the
same number of 1's and 0's, hence this signal will have no dc component or only a very small one. Consequently, the spectrum of the data
signal has almost no power at $f = 0$. This implies that the modulated
carrier does not contain appreciable power at the carrier frequency;
the power is concentrated at other frequencies. This simply means
that we cannot use a simple PLL to reconstruct a replica of the carrier!

Quite an arsenal of solutions has been invented to overcome that
problem[1,20].

The squaring loop.   One of the earliest circuits has been the *squaring
loop* (Fig. 6.9). The radian frequency of the carrier is ω_1 here, and the
carrier is multiplied by the symbol signal $m(t)$. In the simplest case,
$m(t)$ is a binary random sequence taking on the values +1 or −1,
respectively. The modulated carrier signal $s(t)$ is therefore represented
by

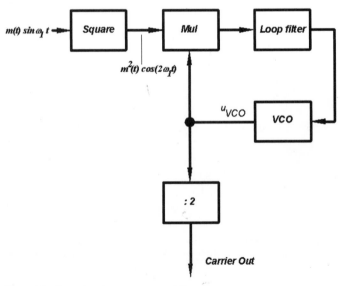

Figure 6.9   Squaring loop.

$$s(t) = m(t)\sin(\omega_1 t)$$

The output of the (analog) squaring device is then

$$s^2(t) = m^2(t)\,\sin^2(\omega_1 t)$$

Making use of the multiplication theorems of trigonometry, this consists of a dc term (which can be discarded here) and a high-frequency term in the form $\cos(2\omega_1 t)$, i.e. of a term having twice the frequency of the carrier. A conventional PLL is used then to lock onto that frequency. A replica of the carrier is obtained by scaling down that signal by a factor of 2.

The Costas loop. Because squaring devices were hard to implement by analog circuitry, alternatives have been examined. A very successful solution is the *Costas loop*[1,20]. It originally was realized as an analog circuit, but is implemented mostly by digital techniques today. An example is the digital Costas loop HSP50210 manufactured by Harris[42]. The basic Costas loop is shown in Fig. 6.10. The explanation becomes easier if we assume that the circuit has locked. The input signal can be represented by $m(t)\sin(\omega_1 t + \theta_1)$, and the reconstructed carrier (the output signal of the VCO) by $\sin(\omega_1 t + \theta_2)$. In the locked state, the phases θ_1 and θ_2 will be nearly the same. Let us see how the Costas loop manages to lock onto the carrier frequency. Two closed loops are recognized, one in the upper branch (I branch) and one in the lower branch (Q branch). Assume for the moment that only the

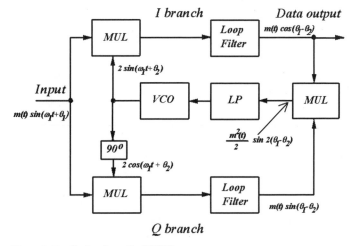

Figure 6.10 Costas loop for BPSK.

lower loop does exist, and discard the multiplier at center right, i.e. connect the output of the loop filter in the lower branch directly to the input of the center low pass filter. This forms a rather conventional PLL. The multiplier at left in the Q branch acts as a phase detector. This PLL circuit would adjust the frequency of the VCO in such a way that it is phase-locked to the carrier. The output of the VCO would have nearly the same phase as the input signal ($\theta_1 \approx \theta_2$), and the two input signals of the lower multiplier would be out of phase by 90°, as usual with linear PLL's. This arrangement would work as long as the modulating signal $m(t)$ does not change polarity. When $m(t)$ changes polarity, however, the output of the multiplier is inverted as well, which means that the multiplier no longer sees a phase error $\theta_e = \theta_1 - \theta_2$, but rather a phase error of $\theta_e = \theta_1 - \theta_2 + \pi$. Consequently, the loop would lock in antiphase with the carrier now, which is not the desired action. Rather, we are looking for a circuit which does not change polarity when $m(t)$ is inverted. To overcome the problem, the I branch is added, together with the additional multiplier at center right. As indicated in Fig. 6.10, the output signal of the loop filter in the I branch is given by $m(t) \cos(\theta_1 - \theta_2)$. If phase is nearly correct for tracking (i.e. $\theta_1 - \theta_2$ is small), the cosine term in the I branch is nearly 1, thus the output of the upper loop filter is the demodulated signal $m(t)$. Because the polarity of the output signal changes with $m(t)$, it is used to cancel the change in polarity at the output of the lower loop filter. Multiplying both loop filter output signals (I and Q branch) creates a signal of the form

$$\frac{m^2(t)}{2} \sin 2(\theta_1 - \theta_2)$$

hence a signal that is proportional to the phase error θ_e when the loop is locked. When lock is achieved, the output of the VCO is in phase with the carrier, so the multiplier in the I branch synchroneously demodulates the data signal. The names of the branches are self-explanatory: the reconstructed carrier fed to the I branch is the in-phase component, and the reconstructed carrier fed to the Q branch is the quadrature component.

Symbol detection by correlation filters: the matched filter. As we recognize from Fig. 6.10, the received data signal passes through the loop filter in the I branch of the Costas loop. Assume for the moment that the data signal has not been low pass filtered at the transmitter, so it originally had a square wave shape. The transients of the data signal are now flattened by the loop filter within the receiver. To decide whether an incoming symbol is a logical 1 or 0, the receiver must sample the demodulated signal at the right time. But what is the "right time"? Taking a sample at the beginning of the symbol would certainly be wrong, since the filtered symbol must first be allowed to settle to its final amplitude (e.g. $+1$ for logical 1, or -1 for logical 0). Hence it would be more appropriate to sample that signal at the end of the symbol. The receiver then would decide that the symbol is 1 if the final amplitude is positive, or a 0 if it is negative. But what if there is noise on the signal? If the current incoming symbol is a 1 and a noise spike appears just at the end of the symbol causing the signal to take on a negative value, the receiver would falsely decide for a 0. Had it taken a sample just prior to that noise spike, the result would probably have been correct. This leads us to the idea that a decision for a symbol could be more meaningful if the receiver would not only check one single sample of the received symbol, but would rather form some statistic (e.g. an average or a correlation) from all samples of the symbol waveform within the symbol period. It has been shown that the best estimate — in the statistical sense – is obtained from a *correlation filter*, which is also referred to as a *matched filter*[20].

What type of correlation filter do we need?   There are many variants of the correlation filter. First of all, the correlation can be performed with the undemodulated or with the demodulated symbol waveform. Correlating the undemodulated waveform is simpler to explain, hence we start with this topic. Figure 6.11 demonstrates the operating principle. It is assumed that the data signal has not been filtered at the transmitter, and that a logical 1 is represented by one full cycle of a sine wave (solid curve at top left) having initial phase 0. A logical 0 is also represented by one cycle of the sine wave, but has initial phase π (dashed curve). Of course, the symbols could have longer duration, so one symbol could extend over a number of sine wave cycles. The symbol waveform is sampled at the input of the correlation receiver; here it was assumed that the sampling rate is four times the symbol rate. We first consider the case where the incoming symbol is a 1 (solid curve). The samples are digitized by an ADC which is not shown in the diagram. The digitized samples are shifted into a FIR filter having length 4. To be more general, assume that the 4 samples of the symbol have the values $a\ b\ c\ d$; these values form a vector. The coefficients of the upper FIR filter have the same values by definition, but in reversed order, so they take on the values $d\ c\ b\ a$. When the samples are shifted into the FIR filter, after 4 shifts the first sample, a, will be at the last tap of the filter (coefficient a). The second sample will be at the second last tap (coefficient b) etc. At this instant, the upper filter outputs an output sample having the value $a^2 + b^2 + c^2 + d^2$. If there is no noise on the received symbol, this equals the autocor-

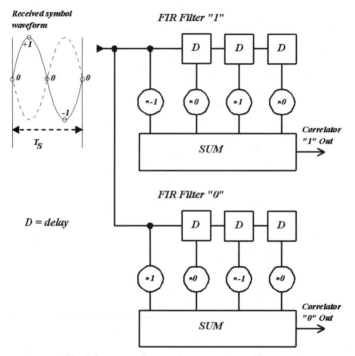

Figure 6.11 Block diagram of a correlation receiver. The incoming symbol is correlated with template functions that are stored in the receiver.

relation of the symbol sequence (with time delay $\tau = 0$). We now consider what happens with the lower FIR filter. Its coefficients defined by $[-d \ -c \ -b \ -a]$ correspond to the undistorted values of the symbol waveform for a logical 0. In our example, the lower correlator outputs the value $-a^2 \ -b^2 \ -c^2 \ -d^2$ at the end of the symbol period, which is the cross correlation of the logical 1 waveform with the logical 0 waveform (with time delay $\tau = 0$). Numerically the upper correlator delivers an output of $+2$ when a 1 is received, and the lower correlator delivers -2. If a logical 0 is received, the reverse happens: the upper correlator outputs -2, and the lower outputs $+2$. To decide whether a symbol is 0 or 1, the receiver compares the two correlator outputs at the end of each symbol. Whenever the upper correlator output is more positive than the lower, the receiver decides that the symbol is a 1 and vice versa. The benefit of the correlation filter becomes obvious when we assume that a logical 1 was received, but the second sample (nominal value 1) was corrupted by noise and has a value of 0. The upper correlator then would output $+1$, and the lower -1, which means that this error does not lead to a wrong decision. An error occurs only if the noise level is so high that it reverses the polarities of the correlator output signals.

In this example, the waveform for logical 0 was chosen identical with the waveform for logical 1, but with inverted polarity. In such simple cases the receiver does not need two correlators. The upper correlator in Fig. 6.11 would be sufficient: decisions would be made by looking only at the sign of the correlator output. The filters shown in the diagram are called *matched filters* because their coefficients exactly *match* the samples of the undistorted symbol waveform(s), but are arranged in reversed order. It can be shown that the decision made by a correlation receiver is a *maximum likelihood* decision in the statistical sense[20], since the receiver compares the incoming symbol waveform with all existing theoretical symbol waveforms and checks which of the stored templates has the greatest similarity (likelihood) with the received symbol.

What about raised cosine filters?   As stated, the described correlation method shown works best with carriers that are modulated with a square wave data signal. To reduce the bandwidth, modern digital communications mostly work with filtered data, as discussed in Sec. 6.4.1. When considering the working principle of the Costas loop (refer again to Fig. 6.10), we became aware that the output of the loop filter in the I branch provides the recovered data signal. When filtering the symbol stream in the transmitter, we must design the filter such that no ISI can occur. The raised cosine filter has been considered an optimum solution for this problem. But now in the Costas loop the data signal passes through additional low pass filter. As was demonstrated by the correlation receiver in the example of Fig. 6.11, the qualitiy of symbol decision in the receiver would be optimum when using the same RCF in place of the I branch of the Costas loop. Unfortunately this does not work in the desired way, because now the signal goes through two cascaded raised cosine filters. The frequency response of the combined filter then would no longer be symmetrical about half the symbol rate, and ISI would be generated. The frequency response of the low pass filter in the receiver should be chosen such that the combined filter represents an RCF in order to realize minimum ISI. Theoretically this goal is achieved only if the loop filter is an ideal low pass filter. In the best case, we could try to approximate a brickwall filter, but this is suboptimal and results in a FIR filter with a very great number of taps. Moreover, the brickwall filter would not be matched to the symbol waveform and therefore it would have poor noise suppressing capabilities.

There is a more elegant solution: the *root raised cosine filter* (*RRCF*). The frequency response of the root raised cosine filter is defined as the square root of the frequency response of the raised cosine filter. Thus the RRCF has a frequency response of the kind

$$H(s) = \begin{cases} 1 & \text{for } |f| < \dfrac{1-r}{2T} \\[2ex] \cos\dfrac{\pi}{4} \dfrac{2T|f| + r - 1}{r} & \text{for } \dfrac{1-r}{2T} \le |f| \le \dfrac{1+r}{2T} \\[2ex] 0 & \text{for } |f| > \dfrac{1+r}{2T} \end{cases} \qquad (6.7)$$

The impulse response $h(t)$ of the RRCF is given by[41]

$$h(t) = \frac{\sin\left[\dfrac{\pi t}{T}(1-r)\right] + \dfrac{4rt}{T}\cos\left[\dfrac{\pi t}{T}(1+r)\right]}{\dfrac{\pi t}{T}\left[1 - \left(\dfrac{4rt}{T}\right)^2\right]} \qquad (6.8)$$

Similar to the impulse response of the RCF, the impulse response given by Eq. (6.8) also has singularities. For some values of time t, $h(t)$ becomes a division 0/0. By analogy, this problem is mastered by applying L'Hôpital's rule and replacing nominator and denominator by its derivatives.

The frequency response of the RRCF is no longer symmetrical about half the symbol rate, hence it leads to ISI which, of course, is not desired. But if the receiver also contains an identical RRCF, the overall frequency response is identical to the RCF, hence after the demodulator, ISI becomes zero again. Two second goal—a maximum likelihood demodulator—is achieved simultaneously: the impulse response of the RRCF in the receiver closely matches the symbol waveform, which is nothing else than the impulse response of the RRCF in the transmitter. Consequently, the RRCF in the receiver is a *matched filter*. The operating principle of the matched filter is sketched by Fig. 6.12. This schematic can be considered part of the block diagram of the Costas loop in Fig. 6.10. The modulated carrier is first fed to the input of a multiplier, which serves as a demodulator. The multiplier can be an analog circuit, though in many cases a digital multiplier is used today. The matched filter is given by the RRCF. This filter is usually clocked with a frequency f_C which is an integer multiple of the symbol rate f_s. (In Fig. 6.13, $f_c = 1/T_c$ and $f_s = 1/T_s$.) The output of the matched filter (correlator) is read out at the symbol rate f_s. It must be emphasized that the data signal at the output of the matched filter is no longer given by pulses or square waves, but by the superposition of oscillating waveforms as was shown by Fig. 6.8. To reconstruct the

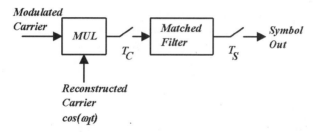

Figure 6.12 Principle of matched filter that correlates the demodulated symbol waveform with a template function stored in the receiver.

data correctly, it is very important that the receiver samples the matched filter ouput at the correct times, i.e. precisely at $t = 0$, T, $2T$. This implies that the receiver must know when a symbol starts and when it is over. This must be accomplished by appropriate symbol synchronization circuits which will be discussed at the end of this section.

The "integrate and dump" circuit. When the data signal was not low pass filtered at the transmitter, the demodulated symbol stream represents nearly a square wave signal. The input to the matched filter then would be almost constant during the symbol period ($+1$ or -1).

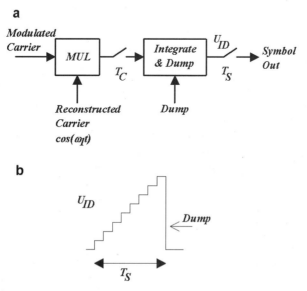

Figure 6.13 Integrate and Dump circuit: (a) Block diagram; (b) Output of the integrator for a square wave input.

Under this condition, the correlator can be built from an integrator, as shown in Fig. 6.13a. At the end of the preceding symbol, the integrator is reset (dumped) to zero. When the next symbol is a logical 1 and is shifted into the integrator, its output builds up a staircase (Fig. 6.13b). The staircase function reaches its maximum at the end of the symbol. If the incoming symbol were a logical 0, the staircase would run negative. The symbol is determined at the end of the symbol period. Because the integrator is dumped after every symbol period, this configuration is also called *integrate and dump circuit*. Integrated digital Costas loops such as the Harris HAS50210[42] provide both RRCF and integrate and dump circuits. They can be configured individually according to a given application and symbol format.

Symbol synchronization.   To conclude, we will deal with the problems related to symbol synchronization. When performing phase synchronization, the receiver reconstructs a replica of the carrier which is locked in frequency and phase to the incoming carrier. When synchronizing with symbol timing, the receiver must recreate the symbol clock in order to know exactly when the end of a symbol period has been reached. We remember that the receiver has to sample the output of the correlation filter at the instant where the incoming symbol is terminated and the next symbol period is starting. There are many methods to achieve symbol synchronization; we will discuss only the most often used.

"Mid-symbol" sampling.   We first discuss a method called mid-symbol synchronization. It is mainly used when the data are low pass filtered, e.g. by the root raised cosine filter. The correlator output is then sampled at the end of each symbol period (which corresponds to the instants $t = -2T, -T, 0, T, 2T$ etc. in Fig. 6.8). If a sequence of 1's is received, the end-symbol amplitudes will all be around $+1$. If a sequence of 0's is transmitted, all succeding end-symbol amplitudes will be around -1. Whenever a 1 is following a 0 or vice versa, however, the data signal crosses zero; this is shown in Fig. 6.14. When the symbol clock is properly adjusted, the zero crossing will be in the center between two succeding decision points (labelled "end-symbol 1" and "end-symbol 0" in the figure). The expected mid-symbol time is now calculated by taking the average of the two end-symbol times. The theoretical location of the zero-crossing is labelled "mid-symbol (expected)". If the actual zero crossing of the symbol signal deviates from this precalculated value (cf. the point labelled "mid-symbol (actual)"), the symbol timing is erroneous and must be corrected. The timing error (the difference between actual and expected location of mid-symbol) is used to adjust the frequency of a local symbol oscillator.

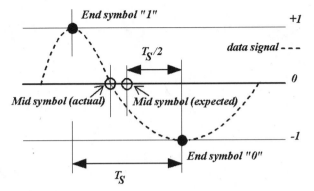

Figure 6.14 Symbol synchronization using mid-symbol values (cf. text).

The mid-symbol synchronization makes use of zero crossings. No timing information is available when a logical 1 is followed by another logical 1. The same holds true for a sequence of 0's.

The Early-Late Gate. Another symbol synchronization technique that is widely used is the *Early-Late Gate*, (Fig. 6.15). This method works best when the data are square wave signals. In this case the symbol signal $s(t)$ is constant and positive during a symbol period if the incoming symbol is a 1, or constant and negative if the symbol is a 0. The "early gate" is a gated integrator summing up the symbol during the first half of the symbol period. The "late gate" is another integrator which is gated on during the second half of the symbol period. When symbol timing is correct, the outputs of both integrators are identical,

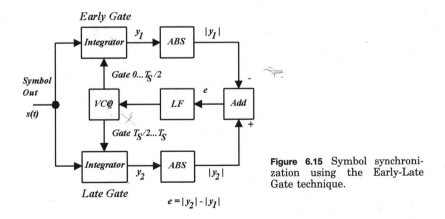

Figure 6.15 Symbol synchronization using the Early-Late Gate technique.

as shown in Fig. 6.16a (cf. the shaded areas). The output signals y_1 and y_2 are equal then, and the error signal e in Fig. 6.15 is zero. When there is an offset in symbol timing, the outputs of the integrators become unequal (Fig. 6.16b), and the error signal becomes positive or negative. Its polarity depends on the sign of the timing error. The waveforms in Fig. 6.16 are shown for an incoming symbol representing logical 1. If the symbol is a 0, the polarities of the integrator outputs get inverted. Because the sign of the error signal must not change when the symbol switches from 1 to 0, absolute value circuits follow the integrators (Fig. 6.15).

Additional symbol synchronization circuits have been discussed in the references[1,20].

6.5   Digital Communications Using QPSK

6.5.1   Transmitter considerations

QPSK has been defined in Sec. 6.2, refer also to Fig. 6.3. To implement QPSK, two carriers have to be generated in the transmitter which are offset in phase by 90°. As explained by Fig. 6.2, the symbol stream is partitioned into two data streams; one of those is used to modulate the in-phase carrier, while the other modulates the quadrature carrier. An example of a QPSK modulator is shown in Fig. 6.17. This is a block

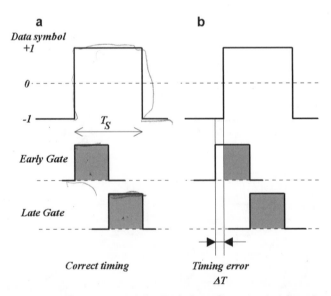

Figure 6.16 Gate timing of the Early-Late Gate circuit (cf. text): (a) Waveforms for correct timing. (b) Waveforms for timing error ΔT.

Block Diagram

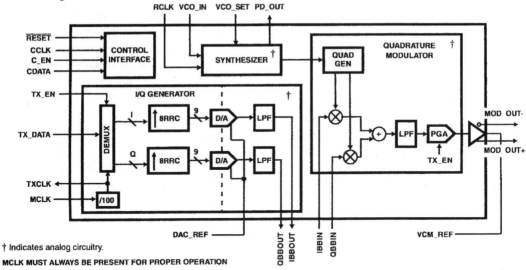

† Indicates analog circuitry.

MCLK MUST ALWAYS BE PRESENT FOR PROPER OPERATION

Figure 6.17 Block diagram of burst QPSK modulator Harris HSP50307. Used with the permission of the Harris Corporation.

diagram of the Harris HSP50307 Burst QPSK Modulator[43]. Data are fed at a rate of 256 kbit/s to the TX_DATA input. The partitioning into I and Q streams is made by the demultiplexer. The bit rate after partitioning is reduced to 128 kbit/s. I and Q data are filtered by an RRCF. This filter has order 64 (length 65) and is clocked at a frequency of 1.024 MHz, i.e. the filter input signal is oversampled by a factor of 8. Oversampling is accomplished by zero padding: 7 zeros are inserted between any two succeeding samples of the I and Q data stream. The excess bandwidth of the RRCF is chosen $r = 0.5$. The filtered data are applied to the inputs of two DAC's. The unwanted high-frequency signals caused by D-to-A conversion are removed by two analog low pass filters. These filter outputs (IBBOUT and QBBOUT) modulate the two carriers, as shown in the right half of the block diagram. The carrier frequency can be programmed within a range of 8 to 15 MHz in steps of 32 kHz.

This integrated circuit is a mix of analog and digital circuitry. All operations that can be made digitally are performed by digital circuits.

6.5.2 Receiver considerations

The receiver for BPSK had to regenerate a carrier that was in phase with the carrier; this was done by the Costas loop in Fig. 6.10. With

QPSK, we no longer have to deal with one single carrier, but receive a signal which is the addition of two carriers modulated by two different data signals $m_1(t)$ and $m_2(t)$, respectively. If we assume that $m_1(t)$ modulates the in-phase carrier and $m_2(t)$ the quadrature carrier, the combined output signal u_{QPSK} of the transmitter can be written as

$$u_{QPSK}(t) = m_1(t) \cos(\omega_1 t + \theta_1) - m_2(t) \sin(\omega_1 t + \theta_1)$$

where ω_1 is the radian frequency of the carrier and θ_1 is the phase. The Costas loop was extended to lock onto a modulated QPSK signal[1,20]. This circuit is shown in Fig 6.18. It is quite similar to the original arrangement (Fig. 6.10). The two low pass filters (loop filters) in the I and Q branch are now followed by limiters. By virtue of the limiters the output signals become independent of the levels of the modulating signals, so the circuit can also be used to demodulate QAM signals (cf. Sec. 6.6). When the loop has locked, the VCO will generate a signal of the form

$$u_{VCO}(t) = \sin(\omega_1 + \theta_2)$$

where θ_2 is the phase of the output signal. Following the analysis of Gardner[1], the output v_d of the adder (at center right in Fig. 6.18) is given by

$$v_d(t) = [m_1(t) \sin\theta_e + m_2(t) \cos\theta_e] \operatorname{sgn}[m_1(t) \cos\theta_e - m_2(t) \sin\theta_e]$$
$$- [m_1(t) \cos\theta_e - m_2(t) \sin\theta_e] \operatorname{sgn}[m_1(t) \cos\theta_e + m_2(t) \sin\theta_e]$$

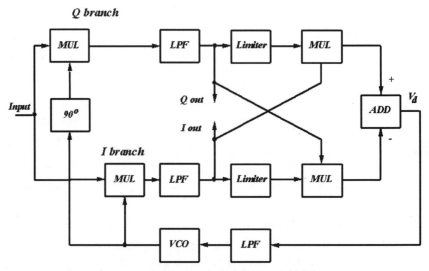

Figure 6.18 Costas loop extended for the demodulation of QPSK.

where sgn (signum) specifies the action of the limiter and $\theta_e = \theta_1 - \theta_2$ is the phase error. For rectangular data signals, the average dc output v_d of the adder becomes

$$v_d(t) = \begin{cases} \sin\theta_e & if \quad -45° < \theta_e < 45° \\[1em] -\cos\theta_e & if \quad 45° \;< \theta_e < 135° \\[1em] -\sin\theta_i & if \quad 135° < \theta_e < 225° \\[1em] \cos\theta_e & if \quad 225° < \theta_e < 315° \end{cases}$$

There are four values for θ_e where the loop can lock stably: $\theta_e = 0°$, 90°, 180°, or 270°, respectively. If it locks at $\theta_e = 0°$, $v_d(t) = \sin \theta_e$, which can be approached by θ_e for small phase error. The Costas loop then operates like a conventional PLL. If the loop locks at $\theta_e = 90°$, $v_d(t)$ becomes $-\cos \theta_e$, which is near zero. When the phase error deviates only slightly from 90°, $v_d(t)$ is proportional to that deviation, and the circuit again operates like a simple PLL. Analog considerations can be performed with the remaining values of phase error where the Costas loop can lock.

A realization of the Costas loop for QPSK reception is found in the circuit HSP50210 manufactured by Harris[42]. Figure 6.19 shows a block diagram of this device. Actually the full receiver is made up from cascading two integrated circuits, the HSP50110 (Digital Quadrature Tuner) and the HSP50210 to be described here. Demodulation of the I and Q signals takes place at the complex multiply circuit at the left. "Complex multiplication" simply means that the incoming signal is multiplied with a cosine wave in one arm and with a sine wave in the other arm of the demodulator. The complex multiplier output delivers the unfiltered baseband data. These are fed to two root raised cosine filters. The sine and cosine waveforms that are required for demodulation are taken from a table look up ROM, shown at bottom left. These signals are generated digitally by interpolating values of sine and cosine functions stored in a ROM. Also shown are two integrate and dump circuits which can be used when the data signals are rectangular. More detailed information is available from application notes[45,46].

6.6 Digital Communications Using QAM

As was demonstrated in Sec. 6.2, QAM can be considered an extension to QPSK. While all symbols had the same amplitude with QPSK, the amplitude of both in-phase and quadrature components now can take on a number of different values. The circuits for transmission and

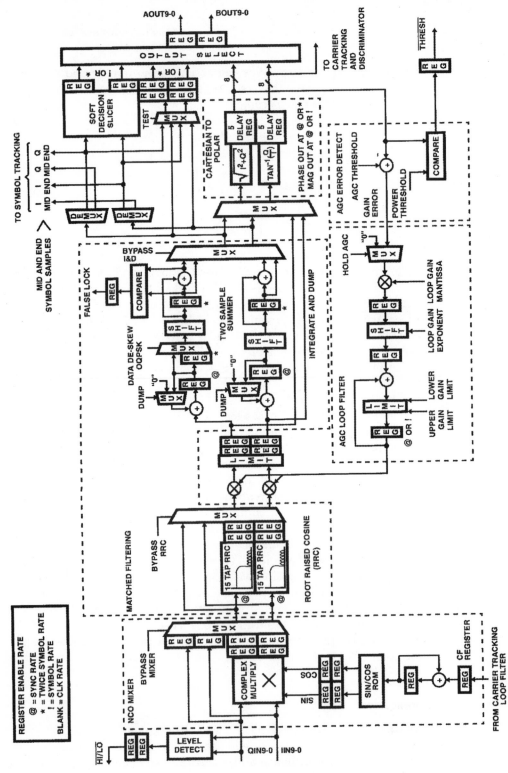

Figure 6.19 Block diagram of the Digital Costas Loop Harris HSP50210. Used with the permission of the Harris Corporation.

reception of QAM signals are similar to those used with QPSK, but the receiver must be equipped with additional level discriminators. The major benefit of QAM is increased symbol throughput without increasing the bandwidth. A nice application of QAM is found in DVB (Digital Video Broadcasting). In Europe, a proposal for digital video broadcasting was presented under the name DVB-C[47]. In analog video broadcasting, channel spacing is standardized to 8 MHz. It was one of the goals of this standard to use the same bandwidth with digital TV broadcasting. This can only be realized by extensive use of source and channel coding[20]. Source coding is a means to reduce the symbol rate at the data source. An analog TV signal has a bandwidth of 5 MHz approximately. When transmitting the signal digitally, we therefore would have to sample it at 10 MHz. Using 8 bits per pixel, the luminance signal alone needs a channel capacity of around 80 Mbit/s. When the color signal is added, we easily end up with 100 Mbit/s. To transmit the signal using BPSK, the bandwidth would become around 50 MHz which is far from realistic. By source coding some redundancy is removed from the signal. Application of the MPEG-2 standards reduces the bit rate to about 38 Mbit/s. Because MPEG-2 adds little error protection, the redundancy of the compressed signal is slightly increased by making use of Reed-Solomon codes[20]. RS(188,204) is recommended in the proposal. This represents a Reed-Solomon code that adds 16 check bytes for each block of 188 bytes: instead of transmitting 188 bytes, we transmit $188 + 16 = 204$, hence the code RS(188,204). With this amount of added redundancy, the bit rate goes up to 41.4 Mbit/s. Using QAM_{64}, 6 bits per symbol can be transmitted (refer to Fig. 6.4). Consequently the symbol rate becomes 41.4 Mbit/s/6 = 6.9 Mbaud. When transmitting the signal in the baseband, a one-sided bandwidth of 6.9/2 = 3.45 MHz would be required. Because the proposal recommends the RRCF (with $r = 0.15$) for filtering the data, the bandwidth is slightly increased to $(1 + r) \cdot 3.45 = 3.97$ MHz. When modulating a high-frequency carrier with that signal, its two-sided bandwidth becomes $2 \cdot 3.97 = 7.94$ MHz, which is below the required value of 8 MHz.

6.7 Digital Communications Using FSK

6.7.1 Simple FSK decoders: easy to implement, but not effective

FSK is still one of the most widely used techniques in the communication of digital signals. Binary FSK has been extensively used in modems for moderate speeds. In such applications, a linear, digital or all-digital PLL can immediately serve as a FSK demodulator, as demonstrated by the examples in Fig. 2.29 (LPLL) and Fig. 4.21 (ADPLL).

Such circuits are very easily implemented, but have the disadvantage that the maximum symbol rate stays far below the limits predicted by theory. Before discussing alternatives, let us recall what should be achieved theoretically, and why the simpler circuits cannot reach these goals.

As stated in Sec. 6.2, there are two basic types of FSK decoding—coherent and noncoherent. Eq. (6.1) says that for coherent FSK each frequency representing a symbol value must be an integer multiple of half the symbol rate, where the symbol rate is given by $1/T_s$. The difference between any two frequencies used in FSK must therefore be at least half the symbol rate. If a binary coherent FSK system with a symbol rate of 2400 bit/s is envisaged, we could use the frequencies $f_1 = 1200$ Hz and $f_2 = 2400$ Hz to represent the binary values 0 and 1, respectively. A binary 0 would then be given by just one half cycle of a 1200 Hz oscillation, and a binary 1 would be given by one full cycle at 2400 Hz. If a LPLL, DPLL or ADPLL were used as FSK decoder (cf. Fig. 4.21 e.g.), it would have to lock onto a new frequency whenever an incoming symbol differs from the preceding. Although the ADPLL is the fastest circuit from those mentioned, it would certainly not be able to settle to another input frequency within one half cycle only. The decoder circuit shown in Fig. 4.21 sets the D flipflop on every positive edge of the input signal u_1, therefore it needs at least one cycle to react onto a frequency step. To accomplish the required symbol rate, the frequencies f_1 and f_2 would have to be chosen much higher. This would result in much larger bandwidth, of course.

As will be demonstrated later, a coherent detector for FSK must know the phase of the FSK signal. This requirement is dropped if the receiver uses noncoherent detection. For noncoherent detection, the minimum frequency spacing is larger by a factor of 2, i.e. the minimum spacing becomes equal to the symbol rate $1/T_s$. This can be explained by looking at the spectrum of the tones. We assume that FSK$_2$ is applied and two tones having frequencies f_1 and f_2 are used to represent binary 0 and 1, respectively. Tone 1 is called $s_1(t)$ and is represented by a burst of a cosine wave of frequency f_1 and duration T_s, hence can be written as

$$s_1(t) = (\cos 2\pi f_1 t)\, rect(t/T_s) \tag{6.9}$$

where $rect(t/T_s)$ denotes a rectangular pulse of duration T_s and is defined by

$$rect(t/T_s) = 1 \ for \ -T/2 \le t \le T/2$$

$$0 \ for \ |t| > T/2$$

The spectrum of tone 1 is a sinc function

$$S_1(f) = T_s \frac{\sin(\pi T_s(f - f_1)}{\pi T_s(f - f_1))} \qquad (6.10)$$

The spectrum is symmmetrical about the frequency $f = f_1$, as shown in Fig. 6.20 (solid curve). It has zeros at frequencies $f = f_1 \pm k/T_s$, where k is an integer > 0. In order that two tones in binary FSK not interfere with each other during detection, the peak of the spectrum of tone 1 must coincide with one of the zero crossings of the spectrum of tone 2. This requirement is fulfilled if the frequency of tone 2 is displaced by exactly $1/T_s$. The spectrum of tone 2 is shown by the dotted curve in Fig. 6.20. Its peak is at a location where the spectrum of tone 1 is exactly 0. Hence the two tones can be detected without interference. This proves that the frequency spacing must be equal to the symbol rate $1/T_s$ and is therefore larger by a factor 2 than the spacing required for coherent detection. We are going now to discuss some of the most familiar coherent and noncoherent FSK decoder schemes.

6.7.2 Coherent FSK detection

A coherent binary FSK decoder is shown in Fig. 6.21. The input is given by the demodulated FSK signal. This demodulator is not shown

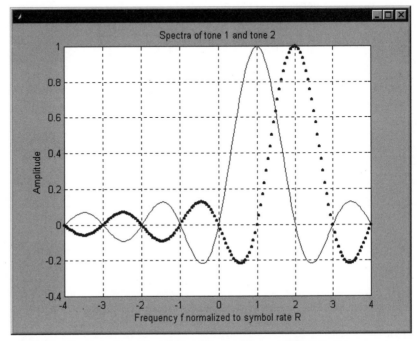

Figure 6.20 Spectra of tone 1 and tone 2 in noncoherent FSK detection.

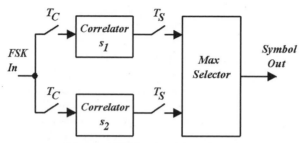

Figure 6.21 Coherent FSK decoder (cf. text).

in the block diagram. The incoming symbol is correlated by two correlators. These are usually implemented by FIR filters; their coefficients are identical with the samples of the corresponding symbol waveforms, cf. Fig. 6.5b. The FIR filters operate at the clock frequency $f_C = 1/T_C$, which is mostly an integer multiple of the symbol rate. The correlator outputs are sampled at the symbol rate $f_S = 1/T_s$ at the end of each symbol. If $s_1(t)$ is the waveform for a logical 0 and the incoming symbol is a logical 0, the upper correlator delivers a large positive output at the end of the symbol; the result is actually the autocorrelation of the signal $s_1(t)$ at a delay of 0. The lower correlator will then output a signal which is close to 0; actually it is the cross correlation between $s_1(t)$ and $s_2(t)$ which is 0 by definition. To decide whether an incoming symbol is a 0 or a 1, the detector compares the outputs of both correlators. If output of the upper correlator is larger than the output of the lower, the decoder decides that the symbol was a 0, and vice versa. In m-ary FSK, more than two frequencies are used to represent m-ary symbols, e.g. four. For arbitrary m, m correlation filters must be provided in the FSK decoder.

6.7.3 Noncoherent FSK detection and quadrature FSK decoders

A noncoherent binary FSK decoder is shown in Fig. 6.22. With noncoherent detection, the receiver does not know the phase of the symbol waveforms. It therefore generates *quadrature signals* for every existing symbol waveform. This decoder is therefore referred to as *quadrature FSK* decoder. Assume that the FSK signal can have the two radian frequencies ω_1 and ω_2. Consequently the decoder provides four waveforms, i.e. $\cos \omega_1 t$, $\sin \omega_1 t$, $\cos \omega_2 t$, and $\sin \omega_2 t$. It is still assumed that radian frequency ω_1 represents a logical 0, and radian frequency ω_2 represents a logical 0. If the incoming symbol is a 0, its waveform can be written as

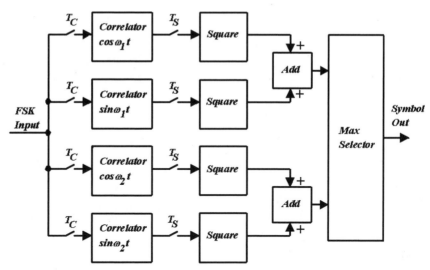

Figure 6.22 Noncoherent (quadrature) FSK decoder (cf. text).

$$s_1(t) = \cos(\omega_1 t + \varphi)$$

where φ is an arbitrary phase. If φ were close to 0, there would be a strong correlation of the FSK signal with cos $\omega_1 t$, so the uppermost correlator would deliver a large positive output at symbol end, but the second correlator output would be near 0, because there is almost no correlation with the sine wave sin $\omega_1 t$. If φ were close to $\pi/2$, the reverse would happen: the uppermost correlator output would be near 0, but the second correlator would deliver a large correlation. For arbitrary phase φ, there would be a correlation with both cosine and sine waves. By squaring and adding the outputs of the upper two correlators, the combined correlator output becomes independent of phase φ. In fact the output signal of the adder is proportional to the energy of the symbol waveform. The lower two correlators provide correlation of the incoming symbol with the stored replicas for a logical 1. Again, a maximum selector checks which of the summed correlator outputs is greater at symbol end. If it happens that the upper correlator had a larger output, the decoder decides that the received symbol was a logical 0. This circuit can also be extended for m-ary FSK. It then must provide a pair of correlators (cosine and sine waves) for each state the symbol can take on.

State of the Art of Commercial PLL Integrated Circuits

The designer of a PLL system has a broad choice of ICs built from different semiconductor technologies. The first available ICs were mainly manufactured in "bipolar" technology, i.e. they used a process known from bipolar operational amplifiers. Shortly after, TTL circuits appeared, and somewhat later CMOS devices became available. Faster circuits were produced in ECL technology. In the late 1980s, some GaAs PLLs were added to the already big family of silicon devices. The GaAs devices extended the frequency range to the GHz region. The GaAs technology could not realize the expected breakthrough, however; in the last years, silicon devices came on the market which even outperformed the former GaAs types in speed. In the last few years production of these GaAs devices has been discontinued.

There are a great number of ICs which contain a complete PLL system. In addition, there exist ICs which only include one or more phase comparators, a single VCO or just an analog multiplier. Furthermore many PLL frequency synthesizer systems are available in one single package. Beside the usual PLL components, the synthesizer ICs mostly include fixed or variable ratio down scalers for the reference input and one or more programmable down scalers for the VCO output signal. The first of these down scalers is used to scale down the frequency of a quartz oscillator which typically operates at 5 or 10 MHz. Most of these synthesizer ICs include the circuitry of the reference oscillator.

In the last few years, a great number of PLL-based semiconductor devices has been introduced that contain the full circuitry for radio and television receivers or for mobile phones. Because the frequency range of many such circuits exceeds the capabilities of standard CMOS devices, the number of high-speed prescalers is increasing very fast.

Many of the offered ICs contain some "extras" such as an additional op amp, an additional "in-lock detector" and the like. Table 7.1 lists the presently available ICs containing a full PLL, a full frequency synthesizer or parts of a PLL system.

The PLL ICs are sorted by device family, which is listed in the first column. The second column describes the function of the IC. When looking for an ADPLL circuit e.g., we recognize that only one type is on the market, the 74..297. It is available in two device technologies however: in CMOS and in LS-TTL. Furthermore, the general-purpose DPLL chips are all based on the old 4046 chip, which is a member of the 4000 CMOS family originally introduced by RCA. Two descendants have been derived from this circuit: the 74HC/HCT7046 and the 74HC/HCT4046. The 74HC/HCT7046 is functionally equivalent to the former 4046, but the 74HC/HCT4046 is different, because it contains three different phase detectors instead of two. The choice is largest among the LPLL chips. Besides the PLL systems, the market offers a wide selection of related circuits such as phase detectors, VCOs, prescalers, frequency synthesizers and the like.

Searching for PLL devices has become much simpler because of the Internet. Practically every available integrated circuit is found in *IC Master*, a data book published yearly *by Hearst Business Communications*, Garden City, New York. On the World Wide Web, the IC master is found on http://www.icmaster.com. Upon registration, every user can browse through IC master free of charge. Integrated circuits can be searched by part number, by device technology, or by function. In addition, there are many links to application notes, e.g. on PLL circuits. Of course most manufacturers of semiconductor devices maintain their own web sites.

TABLE 7.1   Commercially Available PLL ICs and Related Components, Fall 1998

Note: Asterisks (**) preceding and/or following a part number mean that this device is registered with different prefixes and/or suffixes. Example: **4046** is delivered as CD4046B, MC14046B, CD4046BC, etc.

Device Technology	Function	Device	Source	Features
CMOS	ADPLL	74HC297	Philips Harris	General purpose ADPLL including two phase detectors: EXOR and JK-flipflop. Operation to approx. 60 MHz (K counter) and 40 MHz (ID counter).

TABLE 7.1 Commercially Available PLL ICs and Related Components, Fall 1998 (*Continued*)

Device Technology	Function	Device	Source	Features
CMOS		74HCT297	Philips	as 74HC297
		74AC297	Harris	
		74ACT297	Harris	
	DPLL	**74HC4046*	AVG Semi	General purpose DPLL containing three phase detectors: EXOR, JK-flipflop and PFD. Frequency range to 19 MHz.
			Harris	
			Motorola	
			National	
			Philips	
			Rochester	
		74HCT4046	Harris	as 74HC4046
			Philips	
		74HCT7046A	Philips	functionally equivalent to **4046** (listed below)
			Harris	
CMOS 4000	DPLL	**4046**	Harris	General purpose DPLL containing two phase detectors: EXOR and PFD.
			Motorola	
			National	
			R&E Intl	
			SGS	
			Thomson	
CMOS	PLL Synthesizer	IMI145146**	IMI	4-bit data bus input. Programmable reference divider ($\div 3$–4096), oscillator, PFD, in-lock detector, programmable $\div N$ counter, programmable $\div A$ counter (for external 2-modulus prescaler)
TTL	ADPLL	74LS297	TI	functionally equivalent to 74HC/HCT297 (listed above)
Linear	PLL	AD802-**	AD	Cordless telephone + data retiming
		AD805	AD	
		CLC016	Comlinear	
	PLL	TB2109	Toshiba	Cordless telephone, with prescaler
		TB31209	Toshiba	
	Frequency Synthesizer	MC145106	Motorola	Chip contains reference oscillator, $\div 2$ reference divider and $\div 2$ reference output, another reference divider (scaling factor 2^9 or 2^{10}), phase detector and 9-bit programmable divide-by-N counter. Programming by parallel data input (9 + 1 bits).
		SM5133D	NPC	
		SM5133E	NPC	
		SM5142A	NPC	
		SM5157	NPC	
		SM5165	NPC	
		SM5166	NPC	
		U2781B	Temic	
	Frequency Synthesizer,	SP5510	GEC Plessey	for TV/VCR tuning systems

TABLE 7.1   Commercially Available PLL ICs and Related Components, Fall 1998 (*Continued*)

Device Technology	Function	Device	Source	Features
Linear	Bidirectional, 1.3 GHz	SP5510S	GEC Plessey	
		SP5512	GEC Plessey	
		SP5611	GEC Plessey	
	Frequency Synthesizer, Bidirectional, 2.6 GHz	SP5055	GEC Plessey	for Satellite TV
	Frequency Synthesizer, Bidirectional, 2.7 GHz	SP5655	GEC Plessey	for TV tuning systems
	Frequency Synthesizer, CMOS	TDD1742T	Philips	
	Frequency Synthesizer	SP2002	GEC Plessey	direct, up to 400 Mz, square, triangular, or sine wave output
	Frequency Synthesizer	TDA7326	SGS-Thomson	for AM/FM radio
	Frequency Synthesizer	NJU6104	NJR	for DTS
	Frequency Synthesizer	UMA1019M	Philips	for mobile telephones
	Frequency Synthesizer	UMA1016T	Philips	for radio communications equipment
		UMA1017M	Philips	
		UMA1019AM	Philips	
	Frequency Synthesizer	BU2611**	ROHM	for tuners
		BU2614**	ROHM	
		BU2615**	ROHM	
		BU2616	ROHM	
		BU2618	ROHM	
		BU2619	ROHM	
		BU2622	ROHM	
		BU2624	ROHM	
	Frequency Synthesizer	M54937	Mitsubishi	for TV/VCR tuners, serial input
		M54938	Mitsubishi	
		M54939	Mitsubishi	
		M56768	Mitsubishi	
		M56769	Mitsubishi	
		M56770	Mitsubishi	
		M56771	Mitsubishi	
		M56772	Mitsubishi	
		M56773	Mitsubishi	
		M56776	Mitsubishi	
		M64092	Mitsubishi	
		M64093	Mitsubishi	
		M64892	Mitsubishi	
		M64894	Mitsubishi	
	Frequency Synthesizer	TRF2040	TI	Fractional-N/Integer-N operation, Triple-channel
	Frequency Synthesizer	SA7015	Philips	Fractional-N (1 GHz)
		SA7025	Philips	
	Frequency Synthesizer	SA8015	Philips	Fractional-N (2 GHz)
		SA8025A	Philips	
	Frequency Synthesizer	NJ88C33	GEC Plessey	I²C bus programmable
	Frequency Synthesizer	SP5658**	GEC Plessey	Low-phase noise, 2.7 GHz
		SP5659	GEC Plessey	

TABLE 7.1   Commercially Available PLL ICs and Related Components, Fall 1998 (*Continued*)

Device Technology	Function	Device	Source	Features
Linear	Frequency Synthesizer	SM5160**	NPC	programmable
	Frequency Synthesizer	SY89429V	Synergy	programmable, 31.25–510 MHz
	Frequency Synthesizer	MC12181	Motorola	programmable, 125–1000 MHz
	Frequency Synthesizer	MB87094	Fujitsu	serial-input
		MC145156-2	Motorola	Programmable reference divider (scaling factor 8/64/128/256/ 640/1000/1024/2048), programmable $\div$N-counter (3–1023) and programmable $\div$A-counter (0...127). Digital programming of A and N by serial input signal (10 + 7 bits). Uses external 2-modulus prescaler.
		MC145157-2	Motorola	Chip contains reference oscillator, 2 phase detectors, in-lock detector, 14-bit programmable reference divider and 14 bit programmable divide-by-N counter. Programming by serial data input.
		MC145158-2	Motorola	Programmable reference divider (scaling factor 3–16384), oscillator, PFD, in-lock detector, 10-bit programmable $\div$N-counter, programmable $\div$A-counter, uses external 2-modulus prescaler. Programming by serial signal.
		PLL0305**	NPC	
		PLL2001	NPC	
		SM5158	NPC	
	Frequency Synthesizer	MC145159-1	Motorola	Serial-input, with analog phase detector. Same specs as MC145158-2, but phase detector has differing output signal specs. Refer to data sheet.
	Frequency Synthesizer	UMA1014	Philips	universal, for mobile telephones
	Frequency Synthesizer	SP8861	GEC Plessey	with phase detector, 1.3 GHz
	Frequency Synthesizer	SP8853**	GEC Plessey	with phase detector, 1.3/1.5 GHz
	Frequency Synthesizer	SP8858	GEC Plessey	with phase detector, 1.5 GHz
	Frequency Synthesizer	SP8855D	GEC Plessey	with phase detector, 1.7 GHz
	Frequency Synthesizer	SP8855E	GEC Plessey	with phase detector, 2.7 GHz
	Frequency Synthesizer	MB87093A	Fujitsu	with power saving function
		MB87095A	Fujitsu	
		MB87096A	Fujitsu	
	Frequency Synthesizer	MC145170	Motorola	with serial interface
		MC145170-1	Motorola	
		MC145170-2	Motorola	
	Frequency Synthesizer	UMA1005T	Philips	Dual
		ST7162	SGS-Thomson	

TABLE 7.1   Commercially Available PLL ICs and Related Components, Fall 1998 (*Continued*)

Device Technology	Function	Device	Source	Features
Linear	Frequency Synthesizer	MC145162	Motorola	Dual, for cordless phones, 60 MHz
		MC145162-1	Motorola	
		MC145165	Motorola	
		MC145166	Motorola	
		MC145167	Motorola	
		MC145168	Motorola	
		MC145169	Motorola	
	Frequency Synthesizer	UMA1015M	Philips	for mobile telephones
		UMA1018M	Philips	
		UMA1020**	Philips	
		UMA1021M	Philips	
	Frequency Synthesizer	BU2630**	ROHM	Dual-PLL
	Frequency Synthesizer	W43C94A	IC Works	dual, serially programmable
	Frequency Synthesizer	MC145149	Motorola	dual, with serial interface
	Frequency Synthesizer	WB1315	IC Works	dual, with 2.5 GHz prescalers
	Frequency Synthesizer	M64074	Mitsubishi	Dual 1 GHz
	Frequency Synthesizer	LMX2335**	National	Dual, 1.1 GHz/1.1 GHz
	Frequency Synthesizer	MC12302	Motorola	Dual, 1.1 GHz/500 MHz
	Frequency Synthesizer	LMX2332**	National	Dual, 1.2 GHz/510 MHz
	Frequency Synthesizer	LMX2336**	National	Dual, 2 GHz/1.1 GHz
	Frequency Synthesizer	LMX2331**	National	Dual, 2 GHz/510 MHz
	Frequency Synthesizer	MC12306	Motorola	Dual, 2.0 GHz/500 MHz
	Frequency Synthesizer	MC12310	Motorola	Dual, 2.5 GHz/500 MHz
	Frequency Synthesizer	LMX2330**	National	Dual, 2.5 GHz/510 MHz
	Frequency Synthesizer	HFA3524	Harris	Dual, 2.5 GHz/600 MHz
	Frequency Synthesizer	S4503	AMCC	Dual (10–300 MHz)
	Frequency Synthesizer	NJ88C50	GEC Plessey	Dual 125 MHz (AMPS, ETACS, DECT)
	Frequency Synthesizer	M54958	Mitsubishi	Dual 400 MHz
	Frequency Synthesizer	WB1337	IC Works	Dual, 550 MHz/550 MHz
		LMX2337	National	
	Frequency Synthesizer	MC145173	Motorola	Dual-Band with ADC/Frequency Counter
	Frequency Synthesizer	SY89424	Synergy	1 GHz
	Frequency Synthesizer	SP5026	GEC Plessey	1 GHz (for TV/VCR tuning systems)
	Frequency Synthesizer	MC145192	Motorola	1.1 GHz
		LMX1501A	National	
		LMX9301	National	
	Frequency Synthesizer	MC145190	Motorola	1.1 GHz, with phase/frequency detector, prescaler
	Frequency Synthesizer	MC145191	Motorola	1.1 GHz, with 5 V phase detector
	Frequency Synthesizer	TSA5511	Philips	1.3 GHz, for TV tuning systems
		TSA5512	Philips	
		TSA5514	Philips	
		TSA5515T	Philips	
	Frequency Synthesizer	SP5024	GEC Plessey	1.3 GHz, for TV/VCR tuning systems
		TSA5520	Philips	
		TSA5521	Philips	

TABLE 7.1   Commercially Available PLL ICs and Related Components, Fall 1998 (*Continued*)

Device Technology	Function	Device	Source	Features
Linear		TSA5526**	Philips	
		TSA5527**	Philips	
	Frequency Synthesizer	TSA5522	Philips	1.4 GHz, for TV/VCR tuning systems
	Frequency Synthesizer	SP8852D	GEC Plessey	1.7 GHz
		SP8854D	GEC Plessey	
	Frequency Synthesizer	Q3236	Qualcomm	2 GHz
	Frequency Synthesizer	CXA1787	Sony	2 GHz, for mobile telephones
	Frequency Synthesizer	SA8025	Philips	2 GHz, fractional-N
	Frequency Synthesizer	MC145200	Motorola	2 GHz, with 8–9.5 V phase detector
	Frequency Synthesizer	SP5070	GEC Plessey	2.4 GHz, for satellite receivers
	Frequency Synthesizer	SP5054	GEC Plessey	2.6 GHz, for satellite TV tuning systems
	Frequency Synthesizer	SP8852E	GEC Plessey	2.7 GHz
		SP8854E	GEC Plessey	
		M64896	Mitsubishi	
	Frequency Synthesizer	SP5657	GEC Plessey	2.7 GHz, for satellite TV
	Frequency Synthesizer	SP5668	GEC Plessey	2.7 GHz, for TV/VCR tuning systems
	Frequency Synthesizer	SP5654	GEC Plessey	2.7 GHz, 3-wire bus controlled
	Frequency Synthesizer	SP5502	GEC Plessey	1.3 GHz, 4-bit address
	Frequency Synthesizer	SP5511**	GEC Plessey	1.3 GHz, for TV/VCR tuning systems
	Frequency Synthesizer	MC145145-2	Motorola	4-bit data bus input. Chip contains reference oscillator, 12-bit programmable reference divider, 2 phase detectors, in-lock detector, 14-bit programmable divide-by-N counter and data latches for parameter storage.
		MC145146-2	Motorola	4-bit data bus input. Programmable reference divider ($\div 3$–4096), oscillator, PFD, in-lock detector, programmable $\div$N counter, programmable $\div$A counter (for external 2-modulus prescaler).
	Frequency Synthesizer	SY89439	Synergy	50–1020 MHz
	Frequency Synthesizer	SC11346	Int. Semi	80 MHz
	Frequency Synthesizer	AD809	AD	155.52 MHz
	Frequency Synthesizer	LMX2301	National	160 MHz, for RF communications
	Frequency Synthesizer	MD115	MicroNet	200 MHz–1 GHz
		MD315	MicroNet	
	Frequency Synthesizer	SY89439V	Synergy	400–1020 MHz
	Frequency Synthesizer	SY89429**	Synergy	400–800 MHz
	Frequency Synthesizer	M56760	Mitsubishi	500 MHz
	Frequency Synthesizer	LMX2305	National	550 MHz, for RF communications
	Frequency Synthesizer	M64080	Mitsubishi	410 MHz 2 System
	FSK Modulator/ Demodulator	RC2211AM	Raytheon	

TABLE 7.1 Commercially Available PLL ICs and Related Components, Fall 1998 (*Continued*)

Device Technology	Function	Device	Source	Features
Linear	Intercarrier frequency modulator	4002Y-A	NCM	internal PLL and comparator
	Mixer	MC12002	Motorola	double balanced analog mixer
	Phase comparator and programmable counters	MC14568B	Motorola	Contains a PFD with in-lock detector and 2 programmable counters. First counter has scaling factor of 4/16/64/100, second counter is a binary 4-bit divide-by-N counter.
	Phase detector (CMOS)	IMI4345	IMI	Phase-frequency detector with built-in lock detector. Operating frequency >30 MHz (typical).
	Phase-frequency detector	**12140	Motorola	
	Detector	12040	Motorola	
		12540	Motorola	
	Phase-locked loop	MB1501H	Fujitsu	with 1.1 GHz 2-modulus prescaler ÷ 64/65 or ÷ 128/129
	PIF processor/PLL	μPC1820	NEC	
	PLL (linear)	XR215	Exar	0–35 MHz frequency range.
		SL652	GEC Plessey	LPLL having 4 timing current inputs and 2 binary control inputs for switching of timing currents. Main application is FM, FSK, PSK modulator/demodulator etc. Frequency range 0–500 kHz.
		NE565	Ideal Semi	Frequency range 0–500 kHz. VCO has parallel square and triangular wave outputs.
		SE565	Ideal Semi	Frequency range 0–500 kHz. VCO has parallel square and triangular wave outputs.
		MN6152	Panasonic	
		MN6153	Panasonic	
		MN6155	Panasonic	
		NE564	Philips	Post-detection processor and limiting circuit at the input of the PD integrated on chip. Frequency range to 50 MHz.
		LC7150	Sanyo	
		LC7152	Sanyo	
	PLL and prescaler	MB15A16	Fujitsu	1 GHz
		MB15B03	Fujitsu	
		MB15B11	Fujitsu	
		MB15B13	Fujitsu	
		MB1503	Fujitsu	
	PLL and prescaler	MB15E05	Fujitsu	2 GHz

TABLE 7.1 Commercially Available PLL ICs and Related Components, Fall 1998 (*Continued*)

Device Technology	Function	Device	Source	Features
Linear	PLL and prescaler	MB15E06	Fujitsu	2.5 GHz
	PLL clock driver	MC88LV930	Motorola	
		MC88LV950	Motorola	
		MC88LV970	Motorola	
	PLL clock driver	MPC932	Motorola	programmable
	PLL clock driver	MC88PL117	Motorola	Power PC601, Pentium applications
	PLL clock recovery and data retiming	AD800	AD	
	PLL clock recovery and data retiming	AD800-45	AD	45 MHz
	PLL clock recovery and data retiming	AD800-52	AD	52 MHz
	PLL clock recovery and data retiming	AD800-155	AD	155 MHz
	PLL clock recovery and data retiming	AD803	AD	30 Mbps
	PLL controller for DC motor drive	UC1637	Unitrode	
		UC2637		
		UC3637		
	PLL Controller	LC7230	Sanyo	LCD driver for radio receiver (CMOS)
	PLL (linear)	LMC568	National	demodulates FM/FSK signals
	PLL for radio receiver	LC7218	Sanyo	CMOS
	PLL for radio receiver	LC7001	Sanyo	NMOS
	PLL frequency controller	UC1633	Unitrode	PLL for motor speed control. Contains sense amplifier for speed feedback signal (from hall sensor or other speed detector).
		UC1634	Unitrode	Similar to UC1633, but is intended to drive two phase brushless motors.
		UC1635	Unitrode	
		UC2633	Unitrode	
		UC2634	Unitrode	
		UC2635	Unitrode	
		UC3633	Unitrode	
		UC3634	Unitrode	
		UC3635	Unitrode	
	PLL frequency synth.	MB1503/13	Fujitsu	
		PMB2307R	Siemens	
		TBB200	Siemens	
	PLL frequency synth.	IMISC434	IMI	Clock chip
		IMISC464	IMI	
		IMISC465	IMI	
		IMISC466	IMI	
		IMISC468	IMI	
		IMISC470	IMI	
		IMISC471	IMI	

TABLE 7.1 Commercially Available PLL ICs and Related Components, Fall 1998 (*Continued*)

Device Technology	Function	Device	Source	Features
Linear	PPL frequency synth.	IMI2XFSUT157	IMI	Two independent serial input PLL frequency synthesizers equivalent to two IMI145157's on one chip. Input frequency >60 MHz (typical)
	PLL frequency synth.	TDA8735	Philips	for satellite receivers
	PLL frequency synth.	SM5160	NPC	high-speed
	PLL frequency synth.	ILC1080	Impala Linear	high-speed, 40–120 MHz
		ILC1088	Impala Linear	
	PLL frequency synth.	IMI4347	IMI	with phase detector
	PLL frequency synth.	PMB2306	Siemens	serial control
	PLL frequency synth.	MC12179	Motorola	serial input
		MC12202	Motorola	
		MC12206	Motorola	
		MC12210	Motorola	
		MC122179	Motorola	
	PLL frequency synth.	SM5158	NPC	serial input, high-speed
	PLL frequency synth.	GM6530	LG Semicon	with serial interface
	PLL frequency synth.	IMI145145	IMI	4-bit data bus input. Chip contains reference oscillator, 12-bit programmable reference divider, 2 phase detectors, in-lock detector, 14-bit programmable divide-by-N counter and data latches for parameter storage.
	PLL frequency synth.	SM5153	NPC	fixed-ratio frequency deviding
		SM5162LF1S	NPC	
	PLL/prescaler	MB1504**	Fujitsu	0.5 GHz
		MB1505	Fujitsu	
		MB1507	Fujitsu	
		MB1508	Fujitsu	
		MB1509	Fujitsu	
		MB1519	Fujitsu	
	PLL/prescaler	MB15A01	Fujitsu	1 GHz
		MB15A02	Fujitsu	
		MB15B01	Fujitsu	
		MB1501	Fujitsu	
		MB1502**	Fujitsu	
		MB1510	Fujitsu	
		MB1511	Fujitsu	
		MB1512	Fujitsu	
		MB1513	Fujitsu	
		MB1516A	Fujitsu	
	PLL/prescaler	WB1332	IC Works	1.2 GHz

TABLE 7.1   Commercially Available PLL ICs and Related Components, Fall 1998 (*Continued*)

Device Technology	Function	Device	Source	Features
	PLL/prescaler	MB15A19 MB1514	Fujitsu	1.5 GHz
	PLL/prescaler	MB1506 MB1517**	Fujitsu Fujitsu	2 GHz
	PLL/prescaler	MB1515 MB1518	Fujitsu Fujitsu	2.5 GHz
	PLL/programmable divider	MB87086A	Fujitsu	binary 10-bit counter
	PLL/programmable divider	MB87087	Fujitsu	binary 14-bit counter
	PLL (NMOS)	LM7005	Sanyo	Ranges to UHF receiver
	PLL frequency synth.	IMI145155	IMI	Programmable reference divider (scaling factor 16/512/1024/2048/3668/4096/6144/8192), ÷N-counter with scaling factor of 3–16383. Scaling factor N digitally controlled by serial input signal (14 bits).
		IMI145157	IMI	Chip contains reference oscillator, 2 phase detectors, in-lock detector, 14-bit programmable reference divider and 14 bit programmable divide-by-N counter. Programming by serial data input.
		SAA1057	Philips	
	PLL frequency synth.	IMI145106A	IMI	parallel input
		IMI145151	IMI	Parallel input (14 bit). Programmable reference divider (÷8/128/256/512/1024/2048/2410/8196) and ÷N divider with scaling factor 3–16383.
		MC145151-2	Motorola	Parallel input (14 bit). Programmable reference divider (÷8/128/256/512/1024/2048/2410/8196) and ÷N divider with scaling factor 3–16383.
		MC145152-2	Motorola	Programmable reference divider (scaling factors 8/64/128/256/512/1024/1160/2048), oscillator, PFD, in-lock detector, 10 bit programmable ÷N-counter, 6 bit programmable ÷A-counter for external 2-modulus prescaler. Fully parallel programming.
	PLL frequency synth.	HC0320	Hughes	programmable divider, to 1021 channels, adder, phase comparator

TABLE 7.1   Commercially Available PLL ICs and Related Components, Fall 1998 (*Continued*)

Device Technology	Function	Device	Source	Features
Linear	PLL frequency synth.	IMI145152	IMI	Programmable reference divider (scaling factors 8/64/128/256/512/1024/1160/2048), oscillator, PFD, in-lock detector, 10 bit programmable ÷N-counter, 6 bit programmable ÷A-counter for external 2-modulus prescaler. Fully parallel programming.
	PLL frequency synth.	IMI145156**	IMI	Programmable reference divider (scaling factor 8/64/128/256/640/1000/1024/2048), programmable ÷N-counter (3–1023) and programmable ÷A-counter (0–127). Digital programming of A and N by serial input signal (10 + 7 bits). Uses external 2-modulus prescaler.
		IMI145158**	IMI	Programmable reference divider (scaling factor 3–16384), oscillator, PFD, in-lock detector, 10-bit programmable ÷N-counter, programmable ÷A-counter, uses external 2-modulus prescaler. Programming by serial data input.
		MC145155-2	Motorola	Programmable reference divider (scaling factor 16/512/1024/2048/3668/4096/6144/8192), ÷N-counter with scaling factor of 3–16383. Scaling factor N digitally controlled by serial input signal (14 bits).
	PLL frequency synth.	MB87001A	Fujitsu	serial-input system block, 13 MHz, 5 V, 3 mA
	PLL frequency synth.	IMIFSUT157**	IMI	serial input, 2 channel
	PLL frequency synth.	IMI145146**	IMI	4-bit data input. Programmable reference divider (÷3–4096), oscillator, PFD, in-lock detector, programmable ÷N counter, programmable ÷A counter (for external 2-modulus prescaler)
	PLL tuning circuit	MC44818	Motorola	1.3 GHz, for TV/VCR
		MC44824	Motorola	
		MC44825	Motorola	
		MC44826	Motorola	
		MC44827**	Motorola	
		MC44828	Motorola	
		MC44829	Motorola	

TABLE 7.1   Commercially Available PLL ICs and Related Components, Fall 1998 (Continued)

Device Technology	Function	Device	Source	Features
Linear	PLL tuning circuit	MC44802	Motorola	1.3 GHz prescaler
	PLL tuning circuit	MC44864	Motorola	1.3 GHz prescaler, D/A converters
	PLL, universal	KS8805B	Samsung	
	PLL, Video Genlock	CH9073	Chrontel	pixel clock generator
	Dot	AV9173	IntCirSys	
		AV9173-01	IntCirSys	
	PLL	74HCT9046A	Philips	with bandgap controlled VCO
	PLL with prescaler	MC145220	Motorola	1.1 GHz, $\div$ 64/65 or $\div$ 32/33, dual
	PLL with prescaler	TEA8805	SGS-Thomson	1.3 GHz, $\div$ 8
	PLL with prescaler	MC145201	Motorola	2 GHz, $\div$ 64/65
		MC145202	Motorola	
	PLL with prescaler	MB1501L	Fujitsu	1.1 GHz, $\div$ 64/65 or $\div$ 128/129
	PLL, dual	SM5132**	NPC	for cordless phones, 60 MHz
		SM5134**	NPC	
	PLL, dual	WB1310	IC Works	with 1.2 GHz prescalers (serial input)
	PLL, dual	WB1331	IC Works	with 2 GHz/510 MHz prescalers (serial input)
	PLL, dual	WB1330	IC Works	with 2.7 GHz/510 MHz prescalers (serial input)
	PLL, dual	WB1305	IC Works	with 510 MHz prescalers (serial input)
		WB1333	IC Works	
	PLL, dual	U2782	Temic	1.1 GHz/1.1 GHz
	PLL, dual	SY89420	Synergy	1.12 GHz
		SY89423	Synergy	
	PLL, dual	U2783B	Temic	1250 MHz/400 MHz
	PLL, dual	U2784B	Temic	2200 MHz/200 MHz
	PLL with prescaler	MB87014A	Fujitsu	$\div$ 64/65 or $\div$ 128/129
	PLL	SY89421	Synergy	1.12 GHz
	PLL (linear)	568A	Philips	150 MHz
	PWM Controller	MB3785	Fujitsu	
	PWM/PFC Controller	ML4802	MicroLinear	
	PLL frequency synth.	MB87091	Fujitsu	serial input, power saving function
	PLL system block	MB87006A	Fujitsu	serial input, 10 MHz, 3–5 V, 3.5 mA
	PLL system block	MB87076	Fujitsu	serial input, 15 MHz, 3–5 V, 3 mA
	PLL system block	MB87086	Fujitsu	serial input, 95 MHz, 5 V, 10 mA
	Subcarrier PLL	MC44144	Motorola	for video applications
	Synthesizer	NJ8811	GEC Plessey	mobile radio, 2 device set
		NJ8812	GEC Plessey	
		NJ8901	GEC Plessey	
		NJ8906	GEC Plessey	
	Tone decoder	LM567**	National	Operation to 500 kHz. Has additional quadrature phase detector which can be used as in-lock detector.
		KA567C	Samsung	

TABLE 7.1   Commercially Available PLL ICs and Related Components, Fall 1998 (*Continued*)

Device Technology	Function	Device	Source	Features
Linear	VCO	MC1648	Motorola	
		AN8585	Panasonic	
		AN8586	Panasonic	
		AN8587	Panasonic	
		LC7444	Sanyo	
		SN74LS624	TI	(LS-TTL)
		SN74LS628	TI	(LS-TTL)
	VCO and phase comp.	LM565C	National	
	VCO buffer amplifier	KGF1145	OKI	
		KGF1146	OKI	
		KGF1191		
	VCO buffer amplifier	RF2501	RF Micro	for ISM cellular
		RF2502	RF Micro	
	VCO function generator	LM566**	National	Frequency range 0–1 MHz. Simultaneous square and triangular wave output, no sine.
		NE566	Philips	Like LM566
		SE566	Philips	Like LM566
	VCO function generator	XR2207**	Exar	Frequency range 0–1 MHz. Simultaneous square and triangular wave output, no sine. Four timing resistors can be switched in and out by two binary inputs.
	VCO, low power	MC12148	Motorola	
		MC12149	Motorola	
	PLL (linear)	SL650	GEC Plessey	Chip contains additional op amp. Frequency range 0–500 kHz.
	PLL (linear)	SL651	GEC Plessey	Like SL650, but without op amp.
	VCO waveform generator	XR2206**	Exar	
	VCO waveform generator	ICL8038	Harris	Sine, square, triangular, sawtooth, and pulse waveform output.
	VCO, dual	SN74LS124	TI	(LS-TTL)
		SN74LS625	TI	(LS-TTL)
		SN74LS629	TI	(LS-TTL)
		SN74S124	TI	(S-TTL)
	Video dot clock generator	ICS1394	IntCirSys	
	Video PLL system	LM1291	National	for continuous-sync monitors
		LM1292	National	
	Voltage controlled multivibrator	MC1658	Motorola	
	Voltage controlled multivibrator	MC12101	Motorola	130 MHz
	Voltage controlled multivibrator	MC12100	Motorola	200 MHz
	Voltage controlled oscillator buffer	MC12147	Motorola	low power

TABLE 7.1   Commercially Available PLL ICs and Related Components, Fall 1998 (*Continued*)

Device Technology	Function	Device	Source	Features
Linear	Prescaler	MB15C02	Fujitsu	Single-chip
		RC1505	Rochester	
	Prescaler	MB511	Fujitsu	$\div$ 1, 2, or 8, 1 GHz
	Prescaler	MC12083	Motorola	$\div$ 2, 1.1 GHz
	Prescaler	μPB558	NEC	$\div$ 2, 260 MHz
	Prescaler	MC12090	Motorola	$\div$ 2, 740 MHz
	Prescaler	MC12095	Motorola	$\div$ 2/4, 2.5 GHz
	Prescaler	MC12093	Motorola	$\div$ 2/4/8, 1.1 GHz
	Prescaler	U842BS	Temic	$\div$ 2, 2.4 GHz
	Prescaler	SP8802	GEC Plessey	$\div$ 2, 3.3 GHz
	Prescaler	SP8902	GEC Plessey	$\div$ 2, 5/5.5 GHz
	Prescaler	μPB565	NEC	$\div$ 2 (500 MHz), $\div$ 4/8/64 (1 GHz)
	Prescaler	SP8720	GEC Plessey	$\div$ 3/4
	Prescaler	SP8740	GEC Plessey	$\div$ 4/8
	Prescaler	SP8804	GEC Plessey	$\div$ 4, 3.3 GHz
	Prescaler	SP8904	GEC Plessey	$\div$ 4, 5/5.5 GHz
	Prescaler, 2-modulus	MC12009	Motorola	$\div$ 5/6, 480 MHz
	Prescaler, 2-modulus	MC12509	Motorola	$\div$ 5/6, 600 MHz
	Prescaler, 2-modulus	SP8741	GEC Plessey	$\div$ 6/7
	Prescaler, 2-modulus	SP8400	GEC Plessey	$\div$ 8/9
		SP8743A	GEC Plessey	
	Prescaler, 2-modulus	MC12026**	Motorola	$\div$ 8/9, $\div$16/17, 1.1 GHz
	Prescaler, 2-modulus	MC12011	Motorola	$\div$ 8/9, 550 MHz
	Prescaler, 2-modulus	12511	Motorola	$\div$ 8/9, 600 MHz
	Prescaler	μPB567	NEC	$\div$ 8, 1 GHz
	Prescaler	SP8808	GEC Plessey	$\div$ 8, 3.3 GHz
	Prescaler	SP8908	GEC Plessey	$\div$ 8, 5/5.5 GHz
	Prescaler, 2-modulus	SP8647	GEC Plessey	$\div$ 10/11
		SP8685	GEC Plessey	
		SP8690	GEC Plessey	
	Prescaler, 2-modulus	MC12013	Motorola	$\div$ 10/11, 550 MHz
	Prescaler, 2-modulus	MC12513	Motorola	$\div$ 10/11, 600 MHz
	Prescaler, 2-modulus	SP8401	GEC Plessey	$\div$ 10/11, 300 MHz
	Prescaler, 2-modulus	μPB554	NEC	$\div$ 10/11 (50 MHz), $\div$ 20/22, $\div$ 40/44 (150 MHz)
	Prescaler	MC12080	Motorola	$\div$ 10/20/40/80, 1.1 GHz
	Prescaler	SP8830**	GEC Plessey	$\div$ 10, 1.5 GHz
	Prescaler	SP8910	GEC Plessey	$\div$ 10, 5/5.5 GHz
	Prescaler, 2-modulus	SP8782**	GEC Plessey	$\div$ 16/17, $\div$ 32/33, 1 GHz
	Prescaler, 2-modulus	μPB553A	NEC	$\div$ 16/17, 150 MHz
		μPB556	NEC	
	Prescaler	SP8801	GEC Plessey	$\div$ 16, 3.3 GHz
	Prescaler	SP8831	GEC Plessey	$\div$ 16, 3.5 GHz
	Prescaler	SP8916	GEC Plessey	$\div$ 16, 5/5.5 GHz
	Prescaler, 2-modulus	MC120189	Motorola	$\div$ 20/21, 225 MHz
		MC12019	Motorola	
	Prescaler, 2-modulus	MC12028**	Motorola	$\div$ 32/33, $\div$ 64/65, 1.1 GHz
	Prescaler, 2-modulus	SP8714	GEC Plessey	$\div$ 32/33, $\div$ 64/65, 2.1 GHz
	Prescaler, 2-modulus	MC12033**	Motorola	$\div$ 32/33, $\div$ 64/65, 2.0 GHz
		MC12034**	Motorola	
	Prescaler, 2-modulus	MB504**	Fujitsu	$\div$ 32/33, $\div$ 64/65, 520 MHz

TABLE 7.1   Commercially Available PLL ICs and Related Components, Fall 1998 (*Continued*)

Device Technology	Function	Device	Source	Features
Linear	Prescaler, 2-modulus	MC12015	Motorola	÷ 32/33, 225 MHz
		12515	Motorola	
	Prescaler, 2-modulus	μPB569	NEC	÷ 32/33, ÷ 64/65, 550 MHz
	Prescaler	SP8803	GEC Plessey	÷ 32, 3.3 GHz
	Prescaler	SP8833	GEC Plessey	÷ 32, 3.5 GHz
	Prescaler, 2-modulus	SP8793**	GEC Plessey	÷ 40/41
	Prescaler, 2-modulus	MC12016	Motorola	÷ 40/41, 225 MHz
	Prescaler	MC12079	Motorola	÷ 44/128/256, 2.8 GHz
	Prescaler	MC12073	Motorola	÷ 64, 1.1 GHz
	Prescaler	MC12023	Motorola	÷ 64, 225 MHz
	Prescaler	μPB565C	NEC	÷ 64/8/4, 1 GHz
	Prescaler, 2-modulus	MC12022**	Motorola	÷ 64/65, ÷ 128/129, 1.1 GHz
		MC12036**	Motorola	
		MC12038A	Motorola	
		MC12052A	Motorola	
		MC12053A	Motorola	
	Prescaler, 2-modulus	MC12031**	Motorola	÷ 64/65, ÷ 128/129, 2.0 GHz
		MC12032**	Motorola	
		MC12054A	Motorola	
	Prescaler, 2-modulus	μPB597	NEC	÷ 64/65, ÷ 128/129, 1 GHz
	Prescaler, 2-modulus	PMB2312	Siemens	÷ 64/65, ÷ 128/129, 1.1 GHz
	Prescaler, 2-modulus	MC12052B	Motorola	÷ 64/65, ÷ 128/129, 1.1 GHz
	Prescaler, 2-modulus	MB501**	Fujitsu	÷ 64/65, ÷ 128/129, 1.1 GHz
		SP8715	GEC Plessey	
	Prescaler, 2-modulus	SA701	Philips	÷ 64/65, ÷ 128/129, 1.2 GHz
	Prescaler, 2-modulus	MB501UL	Fujitsu	÷ 64/65, ÷ 128/129, 1.1 GHz
		MB509	Fujitsu	
	Prescaler, 2-modulus	KGL2132	OKI	÷ 64/65, 2.0 GHz
	Prescaler, 2-modulus	MC12022**	Motorola	÷ 64/65, ÷ 128/129, 1.1 GHz
	Prescaler, 2-modulus	MC12017	Motorola	÷ 64/65, 225 MHz
	Prescaler, 2-modulus	MC12025	Motorola	÷ 64/65, 520 MHz
	Prescaler	SP8713	GEC Plessey	÷ 64/65/72, 1.1 GHz
		SA702	Philips	
	Prescaler	μPB562	NEC	÷ 64/65/128/136, 1 GHz
	Prescaler	MC12089	Motorola	÷ 64/128, 2.8 GHz
	Prescaler	U893B	Temic	÷ 64/128/256, 1.3 GHz
	Prescaler	DN8522	Panasonic	÷ 64/128/256, 1.7 GHz
	Prescaler	MB506	Fujitsu	÷ 64/128/256, 2.4 GHz
	Prescaler	AN8523	Panasonic	÷ 64/128/256, 2.7 GHz
	Prescaler	CA3163	Harris	÷ 64/256
		CA3179	Harris	
	Prescaler	μPB564	NEC	÷ 64/128/256, 1.3 GHz
	Prescaler	MC12066	Motorola	÷ 64/256, 1.3 GHz
	Prescaler, 2-modulus	SP8792**	GEC Plessey	÷ 80/81
	Prescaler, 2-modulus	MC12058	Motorola	÷ 126/128, ÷ 254/256, 1.1 GHz
	Prescaler, 2-modulus	MB507	Fujitsu	÷ 128/129, ÷ 256/257, 1.6 GHz
	Prescaler, 2-modulus	MB551	Fujitsu	÷ 128/129, with VCO
	Prescaler, 2-modulus	MC12018	Motorola	÷ 128/129, 520 MHz
	Prescaler, 2-modulus	KGL2135	OKI	÷ 128/129/130, 1.7 GHz
	Prescaler, 2-modulus	KGL2115	OKI	÷ 128/129, 1.3 GHz
	Prescaler, 2-modulus	MB508	Fujitsu	÷ 128/130, ÷ 256/258, ÷ 512/514, 2.3 GHz

TABLE 7.1   Commercially Available PLL ICs and Related Components, Fall 1998 (*Continued*)

Device Technology	Function	Device	Source	Features
Linear	Prescaler, 2-modulus	μPB595	NEC	$\div$ 128/136, 1 GHz
		DN8506	Panasonic	
	Prescaler, 2-modulus	M54487	Mitsubishi	$\div$ 128/136, 1 GHz
	Prescaler, 2-modulus	M54477	Mitsubishi	$\div$ 128/136, with ECL output
	Prescaler, 2-modulus	MB510	Fujitsu	$\div$ 128/144, $\div$ 256/272, 2.7 GHz
	Prescaler, 2-modulus	MB517	Fujitsu	$\div$ 128/144, $\div$ 256/272, 2.0 GHz
	Prescaler	MB505-16	Fujitsu	$\div$ 128/256, 1.6 GHz
	Prescaler	SP8402	GEC Plessey	$\div$ 2^1 to 2^8
	Prescaler, 2-modulus	SP8906	GEC Plessey	$\div$ 239/240, $\div$ 255/256
	Prescaler	SP4652	GEC Plessey	$\div$ 256, 1 GHz
	Prescaler	MC12074	Motorola	$\div$ 256, 1.1 GHz
	Prescaler	SP4742	GEC Plessey	$\div$ 256, 1.3 GHz
		MC12076	Motorola	
		MC12078	Motorola	
	Prescaler	M54478	Mitsubishi	$\div$ 256, 1 GHz, with ECL output
	Prescaler	M54473	Mitsubishi	$\div$ 256, 1.25 GHz, with TTL output
	Prescaler, 2-modulus	SP8901	GEC Plessey	$\div$ 478/480, $\div$ 510/512
	Prescaler	SP4780	GEC Plessey	$\div$ 4096/8192, 1.3 GHz
	Prescaler	MC12098	Motorola	$\div$ 8192, 2.5 GHz
	Prescaler, programmable	AN6881	Panasonic	
	Prescaler	MC12075	Motorola	$\div$ 64, 1.3 GHz

Measuring PLL Parameters

This chapter deals with the measurement of PLL parameters. When an ADPLL is used, nothing has to be measured, because all parameters of the circuit are multipliers for clock frequencies and divider ratios of counters. In case of the DPLL, the parameters are normally well specified in the data sheets. The phase detector gain, for example, depends uniquely on the supply voltage in most cases. For a DPLL built from CMOS technology the levels of the phase detector output signal come close to the supply rails, so the phase detector gain K_d is approximately U_B/π for the EXOR, $U_B/2\pi$ for the JK-flipflop, and $U_B/4\pi$ for the PFD. Moreover, the VCO characteristics are well specified on the data sheets of the 4046 and the 7046 ICs (refer to Table 7.1), so there is sufficient information to calculate the values of the external components of the DPLL system. The situation is different in the case of LPLLs, where some parameters are poorly specified for some products. Many LPLL ICs have a large supply voltage range, and some of the PLL parameters can vary with it. Furthermore, the phase detector gain depends on the level of the reference signal.

It is therefore advantageous if the users are able to measure these parameters themselves. It will be shown in this section that the relevant parameters such as K_0, K_d, ω_n, ζ, and many others are easily measured with the standard equipment available even in a hobbyist's lab, i.e., an oscilloscope and a waveform generator. For the following measurements a multifunction integrated circuit of type XR-S200 (EXAR) has been arbitrarily selected as the DUT.

8.1 Measurement of Center Frequency f_0

Only the VCO portion of the DUT is used for this measurement (Fig. 8.1). The two pins of the symmetrical VCO input are grounded. Con-

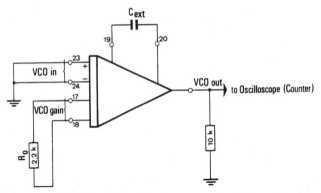

Figure 8.1 Test circuit for the measurement of the center frequency ω_0 and the VCO gain K_0.

sequently the VCO will oscillate at the center frequency f_0. The center frequency is now easily measured with an oscilloscope (Fig. 8.2a). If the values $C_{ext} = 82$ nF and $R_0 = 2.2$ kΩ are chosen for the external components, the VCO oscillates at a center frequency f_0 of 6.54 kHz.

8.2   Measurement of VCO Gain K_0

The same test circuit (Fig. 8.1) can be used to measure the VCO gain K_0. One of the VCO inputs (VCO in) stays grounded; a variable dc voltage is applied to the other. This signal corresponds to the loop filter output signal u_f. By definition the VCO gain K_0 is equal to the variation of VCO angular frequency $\Delta\omega_0$ related to a variation of the u_f signal by $\Delta u_f = 1$ V. As shown in Fig. 8.2b, the VCO frequency falls to 5.78 kHz for $u_f = 1$ V, which corresponds to a variation of 0.76 kHz. The sign of the frequency variation is irrelevant here; it would have been positive if the VCO input pins had been connected with inverted polarity. For the VCO gain K_0 we now obtain

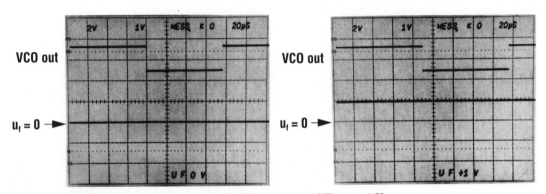

Figure 8.2   (a) Measurement of ω_0, $u_f = 0$ V. (b) Measurement of K_0, $u_f = 1$ V.

$$K_0 = \frac{\Delta\omega_0}{\Delta u_f} = \frac{2\pi \cdot 0.76 \times 10^3}{1} = 4.78 \times 10^3 \ \mathrm{s^{-1}V^{-1}}$$

The test shows furthermore that K_0 would be larger by a factor of 10 if ω_0 had been chosen larger by a factor of 10. Hence, for this device, K_0 varies in proportion to ω_0. When $R_0 = 2.2 \ \mathrm{k\Omega}$ is chosen, K_0 is given by the general equation

$$K_0 = 0.73 f_0 \qquad \mathrm{s^{-1}V^{-1}} \tag{8.1}$$

where f_0 is in Hertz. For example, for $f_0 = 1 \ \mathrm{kHz}$, we find $K_0 = 730$ $\mathrm{s^{-1}V^{-1}}$.

8.3 Measurement of Phase-Detector Gain K_d

The test circuit of Fig. 8.3 is used for the measurement of K_d. The PD of the XR-S200 is shown as an operational multiplier. In the configuration shown, the PD and the VCO are used. The VCO output signal is coupled capacitively to one of the PD inputs. A variable dc level is applied to the other input.

The measurement procedure is explained by the waveforms in Fig. 8.4. In Fig. 8.4a the signal applied to phase-detector input 1 (PD in

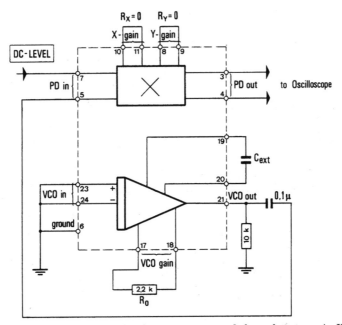

Figure 8.3 Test circuit for the measurement of phase-detector gain K_d.

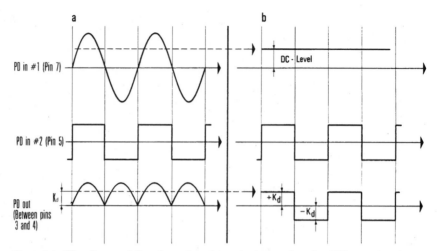

Figure 8.4 Waveforms of the phase-detector output signal u_d for different signals applied to input 1. (a) for a sine-wave signal. (b) For a dc level.

#1) is a sine wave, and the signal applied to phase-detector input 2 (PD in #2) is a square wave (usually the VCO output signal). It is furthermore assumed that both signals are in phase. Consequently the phase error θ_e is 90°. The phase detector output signal (PD out) is a full-wave rectified sine signal; its average value is given by

$$\overline{u_d} = K_d \sin 90° = K_d$$

i.e., the measured output signal is identical with K_d.

If a dc voltage is applied to PD in #1, whose level is equal to the average value of the previously applied sine signal, the output signal of the phase detector becomes a square wave having a peak amplitude K_d. In most data sheets K_d is specified as a function of the reference signal level; a sine signal is usually chosen for the reference signal, and the signal level is usually given as an rms value. In Fig. 8.4b we see that the rms value of the sine signal is 1.11 times its average value. [For a sine signal of peak amplitude 1, the rms value is $1/\sqrt{2}$, the average value (linear average) is $2/\pi$.] Consequently, to measure K_d for a reference signal level of 1 V rms, we have to apply a dc level of $1/1.11 \approx 0.9$ V to PD in #1 and then simply measure the peak value of the output signal.

The phase-detector gain of the DUT in Fig. 8.3 was measured at four different levels of the reference signal: at 10, 30, 40, and 100 mV rms. The measured signals are displayed in Fig. 8.5. The four oscillograms show the results for $u_1 = 10, 20, 30,$ and 40 mV rms, respectively. As Fig. 8.3 demonstrates, this PD has a symmetrical output.

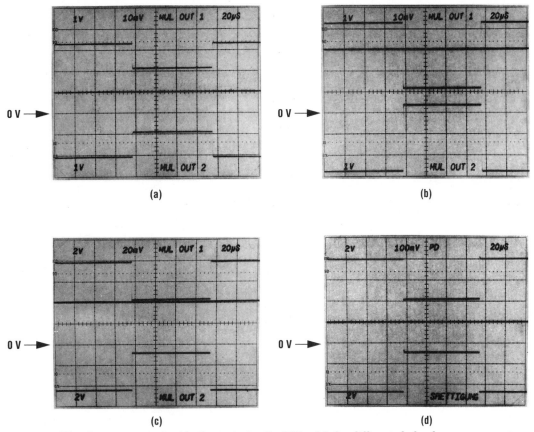

Figure 8.5 Waveforms measured with the test circuit of Fig. 8.3 for different dc levels applied to input 1. (*a*) 10 mV. (*b*) 30 mV. (*c*) 40 mV. (*d*) 100 mV.

Three signals are displayed in each oscillogram: the top trace shows multiplier output 1 (MUL OUT 1), the center trace the reference signal (dc), and the bottom trace multiplier output 2 (MUL OUT 2).

Because the output of the PD is a symmetrical signal, the phase-detector gain is *twice* the measured peak amplitude of the square wave. This means that the PD output signal (shown in the bottom trace of Fig. 8.4) must be measured differentially *between pins 3 and 4*. The phase-detector gain is then the amplitude of the square wave measured from the centerline, as indicated in Fig. 8.4.

The measured K_d values are plotted against the reference signal level in Fig. 8.6. It is clearly observed that the PD becomes saturated at signal levels above 400 mV rms.

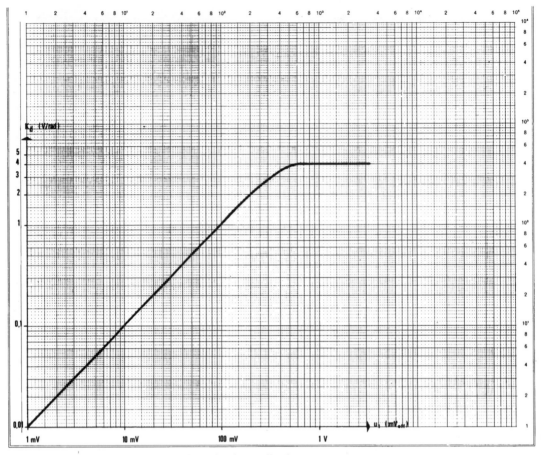

Figure 8.6 Plot of K_d against input voltage level in millivolts rms.

8.4 Measurement of Hold Range $\Delta\omega_H$ and Pull-in Range $\Delta\omega_P$

To measure these parameters a full PLL circuit must be built. This is easily done by adding a loop filter to the test circuit of Fig. 8.3 and by closing the loop. The new test circuit of Fig. 8.7 is then obtained. A passive loop filter (Fig. 2.2) is chosen for simplicity; it consists of two on-chip resistors, $R_1 = 6$ kΩ each, and the external capacitor C. A signal generator is applied to the reference input (pin 7) of the DUT. Both reference signal u_1 and VCO output signal u_2 are displayed vs. time on an oscilloscope. The oscilloscope is triggered on u_2.

The frequency of the signal generator is now varied manually until lock-in is observed (Fig. 8.8). In the case of Fig. 8.8a the reference frequency $|f_1|$ is far away from the center frequency f_0 and the system is unlocked. Figure 8.8b shows the situation where f_1, has come closer

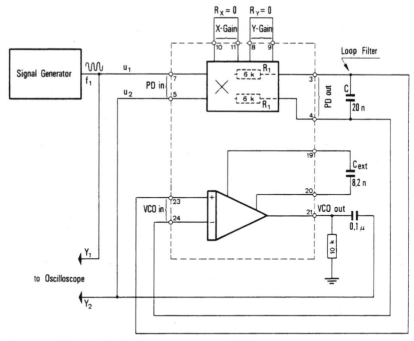

Figure 8.7 Test circuit for the measurement of hold range $\Delta\omega_H$ and pull-in range $\Delta\omega_P$.

to the center frequency. The PLL is trying to pull in the VCO. Frequency modulation of u_2 is easily observable. If the reference frequency approaches f_0 slightly more, the PLL suddenly locks (Fig. 8.8c).

Determination of the hold range Δf_H and the pull-in range Δf_P is very simple. The hold range is measured by slowly varying the reference frequency f_1 and monitoring the upper and lower values of f_1 where the system unlocks.

The pull-in range Δf_P is determined in a similar way. To get Δf_P, we first set the reference frequency f_1 to approximately the center frequency f_0 and then increase f_1 slowly until the loop locks out. To see where the loop pulls in again, we must reduce the reference frequency *slowly* and monitor the value of f_1 where the loop pulls in. The difference between the value of f_1 and the center frequency f_0 is the pull-in range Δf_P.

8.5 Measurement of Natural Frequency ω_n, Damping Factor ζ, and Lock Range $\Delta\omega_L$

The PLL circuit of Fig. 8.7 is used for the following measurements. To measure the natural frequency ω_n and the damping factor ζ of a PLL, we apply a disturbance to the PLL which forces the system to settle

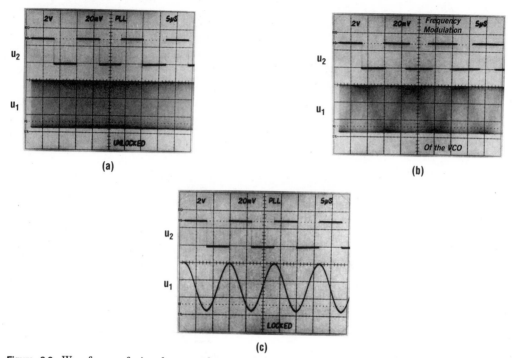

Figure 8.8 Waveforms of signals u_1 and u_2 in the test circuit of Fig. 8.7. (*a*) PLL unlocked. (*b*) PLL near locking. (*c*) PLL locked.

at a different stable state. This is most easily done by modulating the reference frequency with a square-wave signal. The corresponding test circuit is shown in Fig. 8.9. The PLL under test operates at a center frequency of approximately 70 kHz. The frequency of the signal generator is modulated by a square-wave generator. Of course, the mod-

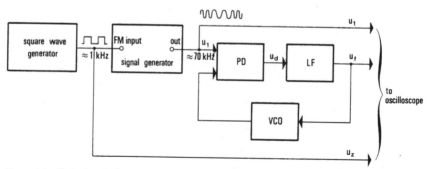

Figure 8.9 Test circuit for the measurement of natural frequency ω_n and damping factor ζ.

ulating frequency must be chosen much smaller than the center frequency, for example, 1 kHz.

If the frequency of the signal generator is abruptly changed, a phase error θ_e results. The output signal u_f of the loop filter can be considered a measure of the average phase error. Thus the transient response of the PLL can be easily analyzed by recording u_f on an oscilloscope. This is shown by the waveforms in Fig. 8.10a, which displays the signals u_1 (reference signal, top trace), u_f (center trace), and the 1-kHz square wave (bottom trace). The oscilloscope is triggered on the 1-kHz square wave. Because the reference signal has a much higher frequency than the 1-kHz square wave and is by no means synchronized with the latter, it is displayed as a bar only.

The u_f signal performs a damped oscillation on every transient of the 1-kHz square wave and settles at a stable level thereafter. Now ζ and ω_n can be calculated from the waveform of u_f. Figure 8.10b is an enlarged view of this signal; ζ can be calculated from the ratio of the amplitudes of two subsequent half-waves A_1 and A_2. (Any pair of subsequent half-waves may be chosen.) The damping factor is given by

$$\zeta = \frac{\ln\,(A_1/A_2)}{\{\pi^2 + [\ln\,(A_1/A_2)]^2\}^{1/2}}$$

The natural frequency ω_n is calculated from the period T of one oscillation in Fig. 8.10b according to

$$\omega_n = \frac{2\pi}{T\sqrt{1 - \zeta^2}}$$

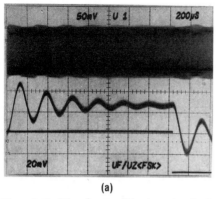
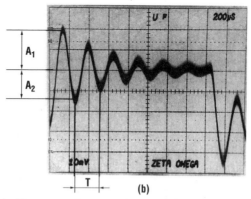

Figure 8.10 Waveforms of the test circuit shown in Fig. 8.9. (a) Top trace—input signal u_1; center trace—output signal u_f of the loop filter; bottom trace—modulating signal, 1-kHz square wave. (b) Enlarged view of the u_f waveform shown in (a), center trace.

Let us now evaluate numerically the waveform of Fig. 8.10b. For A_1 and A_2 we read approximately 1.9 and 1.5 divisions, respectively. Consequently we obtain

$$\zeta \approx 0.08$$

For T we find $T \approx 240\ \mu$s. Hence ω_n is

$$\omega_n \approx 26.0 \times 10^3\ \text{s}^{-1}$$

or

$$f_n \approx 4.1\ \text{kHz}$$

To complete this measurement let us calculate the values of ω_n and ζ using the theory of the linear PLL [Eq. (2.15a)] and check whether our measurements agree with the predicted results. Using Eq. (8.1), we obtain for the VCO gain

$$K_0 = 0.73 \cdot 70 \times 10^3 = 51.1 \times 10^3\ \text{s}^{-1}\text{V}^{-1}$$

According to Fig. 8.10a the amplitude of the reference signal is about 120 mV peak-to-peak, which corresponds to an rms value of 43 mV. For this signal level we read a phase-detector gain $K_d \approx 3.7$ V from Fig. 8.6. The two time constants τ_1 and τ_2 can be derived from the test circuit in Fig. 8.7,

$$\tau_1 = 12\ \text{k}\Omega \cdot 20\ \text{nF} = 240\ \mu\text{s}$$

$$\tau_2 = 0$$

Using Eq. (2.15a) we finally get

$$\omega_0 = \left(\frac{K_0 K_d}{\tau_1}\right)^{1/2} = \left(\frac{51.1 \times 10^3 \cdot 3.7}{240 \times 10^{-6}}\right)^{1/2} = 28 \times 10^3\ \text{s}^{-1}$$

$$\zeta = \frac{1}{2}\ \omega_n \left(\tau_2 + \frac{1}{K_0 K_d}\right) = \frac{28 \times 10^3}{2 \cdot 51.1 \times 10^3 \cdot 3.7} = 0.07$$

This agrees well with the experimental measurements.

Note that this experimental method of measuring ω_n and ζ is applicable only for $\zeta < 1$. This requirement is met, however, in most cases. If ζ were greater than 1, the transient response would become aperiodic, and it would become impossible to define the values A_1 and A_2 in Fig. 8.10a. In the example of Fig. 8.10a the damping factor ζ has purposely been chosen too small in order to get a marked oscillatory transient. An underdamped system is often obtained when a loop filter without a zero is chosen. To increase ζ, τ_2 should be chosen > 0.

Measurement of the lock range $\Delta\omega_L$ is possible, though not so easy. It can be done using the setup shown in Fig. 8.9 which was built to measure the natural frequency and damping factor. The signal generator is still frequency-modulated by the square wave generator as shown. Assume that the radian center frequency of the PLL under test is ω_0. The higher of the two frequencies (ω_{high}) generated by the signal generator must now be chosen higher than the pull-out frequency, i.e. $\omega_{\text{high}} > \omega_0 + \Delta\omega_p$. The lower frequency ($\omega_{\text{low}}$) has to be chosen initially to be as high as the higher ($\omega_{\text{low}} = \omega_{\text{high}}$). In this situation the amplitude of the square wave generator will be zero. The test circuit must now be tuned such that ω_{high} stays always the same, but that ω_{low} decreases when the square wave amplitude is made larger. The square wave amplitude is now continually increased while watching the u_f signal with the scope as shown in Fig. 8.10a. During the interval where the frequency of the signal generator is high, u_f will be a "high-frequency" signal, which is only visible as a "smeared" trace. When the lower frequency reaches a value around $\omega_{\text{low}} \approx \omega_0 + \Delta\omega_L$, the PLL will lock quickly, i.e. the u_f signal will settle to a stable value within at most one cycle (oscillation). The lock range $\Delta\omega_L$ is now given by the difference $\omega_{\text{low}} - \omega_0$. For larger values of ω_{low}, a pull-in process will be observed: the u_f signal also settles to some finite value, but shows up more than one cycle.

8.6 Measurement of the Phase-Transfer Function $H(j\omega)$ and the 3-dB Bandwidth ω_{3dB}

There are different methods of measuring $H(j\omega)$; some of these use PM techniques,[35,49] others use frequency modulation.

The PM technique has the advantage that the maximum phase error θ_e can be limited to small values (e.g., $\theta_l < \pi/4$) so the PD never operates in its nonlinear region. Unfortunately, most signal generators used in the lab can be frequency-modulated but not phase-modulated.

To enable the measurement of $H(j\omega)$, we use FM techniques therefore. The test setup is shown in Fig. 8.11. The frequency of the signal generator is tuned to the center frequency of the PLL. A sine wave is

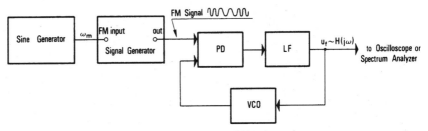

Figure 8.11 Test circuit for the measurement of $H(j\omega)$.

applied now to the FM input. Consequently, the output signal of the generator is given by

$$\omega_1 = \omega_0 + \Delta\omega \sin \omega_m t \qquad (8.2)$$

where ω_m is the modulating frequency and $\Delta\omega$ is the peak frequency deviation. The size of $\Delta\omega$ will be determined later. Because the phase $\theta_1(t)$ of the reference signal is equal to the integral of the frequency deviation $\Delta\omega \sin \omega_m t$ over time, we have

$$\theta_1(t) = -\frac{\Delta\omega}{\omega_m} \cos \omega_m t \qquad (8.3)$$

The amplitude of the phase signal is therefore

$$|\Theta(j\omega_m)| = \frac{\Delta\omega}{\omega_m} \qquad (8.4)$$

From the definition of the phase-transfer function, the amplitude $\Theta_2(j\omega_m)$ of the phase signal $\theta_2(t)$ is given by

$$|\Theta_2(j\omega_m)| = H(j\omega_m)\, \Theta_1(j\omega_m) \qquad (8.5)$$

According to Eq. (2.11), the gain of the VCO at frequency $\omega = \omega_m$ is

$$\frac{\Theta_2(j\omega_m)}{U_f(j\omega_m)} = \frac{K_0}{j\omega_m} \qquad (8.6)$$

For the amplitude of the u_f signal we obtain therefore

$$U_f(j\omega_m) = \frac{j\omega_m\, H(j\omega_m)\, \Theta_2(j\omega_m)}{K_0} \qquad (8.7)$$

Inserting Eq. (8.4) into Eq. (8.7) we finally get

$$|H(j\omega_m)| = \frac{U_f(j\omega_m)\, K_0}{\Delta\omega} \qquad (8.8)$$

This indicates that the phase-transfer function is very easily obtained just by measuring the amplitude of the u_f signal at different modulating frequencies ω_m. What is still unknown is the peak frequency deviation $\Delta\omega$. As Eq. (8.4) says, the peak phase deviation of the reference signal is proportional to $\Delta\omega$. As long as the peak phase deviation is small enough, the PD operates within its linear range and the u_f signal is an undistorted sine wave.

To check whether the initially chosen peak frequency deviation $\Delta\omega$ is adequate, we monitor the u_f waveform on an oscilloscope and manually sweep the modulating frequency f_m from 0 to, say, 10 kHz. If f_m comes close to the natural frequency f_n, a resonance peak is observed on the scope. If the u_f waveform stays undistorted at its peak amplitude, the peak frequency deviation Δf is adequate (refer to Fig. 8.12a).

If too large a peak frequency deviation has been selected, the u_f waveforms become distorted, as shown in Fig. 8.12b. This indicates that $\hat{\theta}_e$ has become so large that the phase detector now operates near saturation. If such a distorted waveform is observed, the peak frequency deviation simply must be reduced.

The phase-transfer function can now be measured by plotting the amplitude of the u_f signal against the modulating frequency ω_m. The absolute level of u_f is unimportant, because $H(j\omega) = 1$ for small frequencies ω_m. This measurement can be done manually, but can also be automated by means of a spectrum analyzer.

In this case the modulating frequency ω_m must be generated by a sweep generator. If the phase-transfer function $H(j\omega)$ has to be recorded within a range of modulating frequencies of, say, 0 to 10 kHz, a sweep range of 10 kHz has to be specified for both the spectrum analyzer and the sweep generator. However, the sweep rate has to be chosen differently for each instrument.

The sweep rate of the spectrum analyzer depends on the specified spectral resolution. In this example, a spectral resolution of 300 Hz has been chosen. The spectrum analyzer then scans the frequency spectrum from 0 to 10 kHz once every second (Fig. 8.13). To avoid intermodulation between spectrum analyzer and sweep generator, the

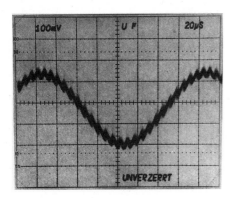
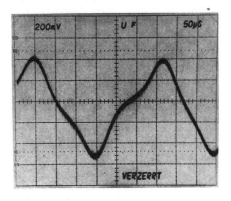

(a)   (b)

Figure 8.12 Output signal of the loop filter u_f measured by the test circuit of Fig. 8.11. (a) The peak frequency deviation of the FM modulation is chosen low enough to enable linear operation of the PD. (b) The peak frequency deviation is too large. The PD is operating in its nonlinear region.

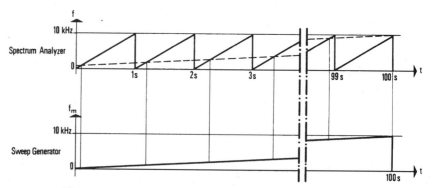

Figure 8.13 Measurement of phase-transfer function $H(j\omega)$. The waveforms explain the timing of the spectrum analyzer and the sweep generator.

modulating frequency f_m should stay quasistationary during one sweep period of the spectrum analyzer. Consequently, the chosen sweep rate of the sweep generator should be much slower than the sweep rate of the spectrum analyzer. In our example the sweep generator scans through the frequency range from 0 to 10 kHz within 100 s, as shown in Fig. 8.13.

The phase-transfer function $H(j\omega)$ recorded with this technique is shown in Fig. 8.14. The spectrum was recorded by means of a Polaroid camera; because the spectrum of 0–10 kHz is scanned in a time interval of 100 s, the shutter of the camera had to remain open for

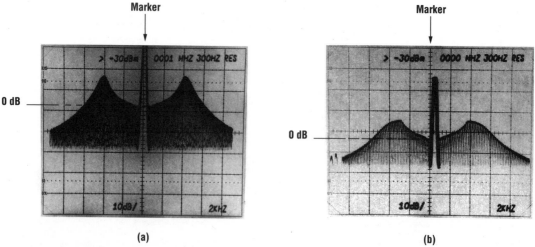

Figure 8.14 Automatic recording of $H(j\omega)$ by a spectrum analyzer. (a) The peak frequency deviation was chosen so that the PD operates in its linear region. (b) The peak frequency deviation is too high, and the transfer function is distorted.

100 s. As an alternative, the spectrum could have been recorded first by a storage oscilloscope and photographed afterward with a normal exposure time.

The phase-transfer function $H(j\omega)$ of the PLL is shown in Fig. 8.14 for two different conditions. Figure 8.14a is the "correct" measurement; the peak frequency deviation was chosen small enough to avoid nonlinear operation of the PD. The frequency scale is symmetrical with respect to the marker, which indicates the origin of the frequency scale. The left half of the spectrum covers the range of negative frequencies from 0 to -10 kHz and can be discarded here.

A low damping factor of $\zeta = 0.08$ was purposely chosen for this measurement. A large resonance peak is therefore observed at the natural frequency of $f_n \approx 4.1$ kHz. Note that this value was measured by a different method in Sec. 8.5.

The 3-dB frequency f_{3dB} is by definition the frequency for which $|H(j\omega)|$ is 3 dB lower than the dc value $|H(j0)|$. From Fig. 8.14a we obtain

$$f_{3dB} \approx 6.8 \text{ kHz}$$

Figure 8.14b shows the same measurement, but the peak frequency deviation $\Delta\omega$ was purposely chosen so high as to cause nonlinear operation of the PD. The measurement is dramatically corrupted by this procedure; the amplitude of the spectrum is heavily compressed, and the curve shows a discontinuity. Nevertheless locations of the natural frequency f_n and the 3-dB frequency f_{3dB} were not shifted to any considerable extent.

The Pull-in Process

Because the pull-in process is a nonlinear phenomenon, it can be calculated to an approximation only. Different authors have derived expressions for pull-in range $\Delta\omega_p$ and pull-in time T_p for one particular linear PLL, i.e., for the LPLL which contains a passive loop filter.[1,2,4] The approximation developed by the author is much more general and can be used to calculate $\Delta\omega_p$ and T_p *for any kind of LPLL and DPLL.* Though the procedure is simpler than those presented by other authors, practical experiments have proved that the new approximations come closer to reality. As we will see, the simplified model derived to calculate $\Delta\omega_p$ and T_p for the LPLL can be adapted to calculate these parameters for the DPLL as well, with the exception of the DPLL using a PFD. As pointed out in Sec. 3.2.3, the model of Fig. 3.15 was used to calculate the pull-in process of this kind of DPLL, so we can restrict ourselves to the DPLLs having an EXOR or a JK-flipflop phase detector. We start with the model for the pull-in process of the LPLL and later extend the method to the DPLL.

A.1 Simplified Model for the Pull-in Range $\Delta\omega_p$ of the LPLL

We assume that a linear PLL system (as shown in Fig. 2.1) is switched on at $t = 0$. The frequency ω_1 of the reference signal is furthermore assumed to be so much higher than the center frequency ω_0 of the PLL that the initial frequency offset $\Delta\omega_0 = \omega_1 - \omega_0$ is greater than the lock range $\Delta\omega_L$. Therefore, the LPLL will not lock immediately, and the VCO will oscillate initially at the center frequency ω_0. As we know from Sec. 2.6.3, the PLL will pull in when the initial frequency offset $\Delta\omega_0$ is smaller than the *pull-in range* $\Delta\omega_p$. We are looking now for a method to compute $\Delta\omega_p$ to an approximation.

In the following computation, the instantaneous (angular) frequency of the VCO is denoted ω_2, with $\omega_2(0) = \omega_0$. As long as the PLL stays unlocked, the output signal of the phase detector is an ac signal given by

$$u_2(t) = K_d \sin\left[(\omega_1 - \omega_2)t + \phi\right] \tag{A.1}$$

where ϕ is the zero phase and is not of concern for the following derivation. To make the calculation simpler, we assume that the initial frequency offset $\Delta\omega_0$ is much larger than the corner frequency $1/\tau_2$ of the loop filter (refer to Fig. 2.3). The gain of the loop filter for these high frequencies is therefore constant and is given by

$$F_H \approx \frac{\tau_2}{\tau_1 + \tau_2} \tag{A.2a}$$

for the passive lag filter (Fig. 2.2a),

$$F_H \approx K_a \frac{\tau_2}{\tau_1} \tag{A.2b}$$

for the active lag filter (Fig. 2.2b), or

$$F_H \approx \frac{\tau_2}{\tau_1} \tag{A.2c}$$

for the active PI filter (Fig. 2.2c).

F_H denotes "gain at high frequencies" in these formulas. Because $\tau_1 \gg \tau_2$ in most cases, all three equations (A.2a) to (A.2c) can be replaced by the approximation

$$F_H \approx K_a \frac{\tau_2}{\tau_1} \tag{A.3}$$

where K_a is the dc gain of the filter and must be set 1 for the passive lag and for the active PI filter. Now the output signal of the loop filter can be written as

$$u_f(t) = F_H u_d(t) = F_H K_d \sin\left[(\omega_1 - \omega_2)t + \phi\right] \tag{A.4}$$

This signal causes a frequency modulation of the VCO output signal (Fig. A.1, bottom curve); its peak frequency deviation is given by $F_H K_0 K_d$. A closer look at this illustration reveals that the difference $\omega_1 - \omega_2$ between reference and output frequencies is no longer a constant but varies with the frequency modulation of the VCO signal. Figure A.1 shows that the difference $\omega_1 - \omega_2$ becomes smaller during

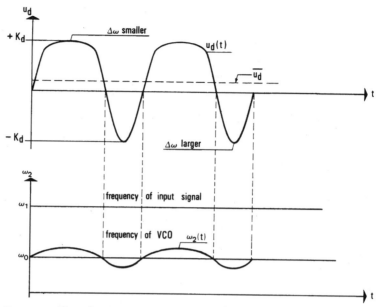

Figure A.1 Plot of $u_d(t)$ and $\omega_2(t)$ against time for the unlocked PLL. It is assumed that the power supply of the PLL has been turned on at $t = 0$.

the time intervals when the frequency of the VCO is increased, and vice versa. The signal $u_d(t)$ therefore becomes inharmonic (see top curve in Fig. A.1), i.e., the duration of the positive half-waves is longer than that of the negative ones. Consequently the average value $\overline{u_d}$ of the phase-detector output signal is *not zero,* but slightly *positive.* This slowly pulls the frequency ω_2 of the VCO in the positive direction.

Under certain conditions to be discussed in the following, this process is *regenerative,* i.e., the VCO is pulled to a frequency close enough to ω_1 that the PLL finally locks. For the calculation of the pull-in process we assume that the frequency ω_2 of the VCO has already been pulled somewhat in the direction of ω_1 (Fig. A.2). The average angular frequency of the VCO is denoted by ω_{20}, and the average frequency offset is $\Delta\omega = \omega_1 - \omega_{20}$. Next the average value $\overline{u_d}$ is calculated as a function of the frequency offset $\Delta\omega$. For an exact solution, the waveform of the signal $u_d(t)$ should be known. This calculation is extremely difficult, however, so we try to get an approximation.

Four distinct points A, B, C, and D of the signal $\omega_2(t)$ are known exactly. Hence, we can derive a simplified expression for its waveforms as well as for the waveform $u_d(t)$, which is proportional to $\omega_2(t)$. At points A and C the instantaneous frequency $\omega_2(t)$ is exactly ω_{20}. At point B, $\omega_2(t)$ is at its positive peak deviation, that is, $\omega_2(t) = \omega_{20} + F_H K_0 K_d$. By analogy the frequency at point D is $\omega_2(t) =$

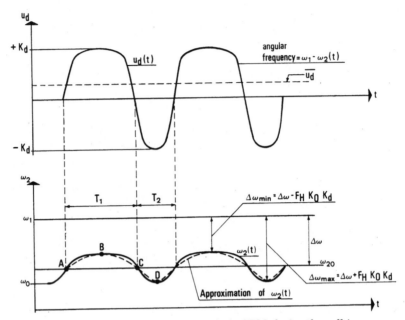

Figure A.2 Plot of $u_d(t)$ and $\omega_2(t)$ for the unlocked PLL during the pull-in process. It is assumed that the frequency of the VCO has already been pulled somewhat toward the reference frequency ω_1.

$\omega_{20} - F_H K_0 K_d$. The function $\omega_2(t)$ is now approximated by two sine half-waves (see the dashed curve in Fig. A.2). This simplification allows us now to calculate the average frequencies $\overline{\omega_{2+}}$ and $\overline{\omega_{2-}}$ during the positive and negative half-waves, respectively. The average value of a half-wave is obtained by multiplying its peak amplitude by $2/\pi$. For $\overline{\omega_{2+}}$ and $\overline{\omega_{2-}}$ we get

$$\overline{\omega_{2+}} = \omega_{20} + \frac{2}{\pi} F_H K_0 K_d \tag{A.5a}$$

$$\overline{\omega_{2-}} = \omega_{20} - \frac{2}{\pi} F_H K_0 K_d \tag{A.5b}$$

The average values of the frequency offset $\Delta\omega(t) = \omega_1 - \omega_{20}(t)$ in the positive and negative half-waves can now be calculated as well.

$$\overline{\Delta\omega_+} = \Delta\omega - \frac{2}{\pi} F_H K_0 K_d \tag{A.6a}$$

$$\overline{\Delta\omega_-} = \Delta\omega + \frac{2}{\pi} F_H K_0 K_d \tag{A.6b}$$

The duration of the half-waves T_1 and T_2 (Fig. A.2) can then be approximated from $\overline{\Delta\omega_{2+}}$ and $\overline{\Delta\omega_{2-}}$:

$$T_1 = \frac{1}{2}\frac{2\pi}{\overline{\Delta\omega_+}} = \frac{\pi}{\Delta\omega - (2/\pi)F_H K_d K_0} \tag{A.7a}$$

$$T_2 = \frac{1}{2}\frac{2\pi}{\overline{\Delta\omega_-}} = \frac{\pi}{\Delta\omega + (2/\pi)F_H K_d K_0} \tag{A.7b}$$

Knowing T_1 and T_2, we can now calculate $\overline{u_d}$. The positive half-wave of $u_d(t)$ contributes an average value of

$$\frac{2}{\pi} K_d \frac{T_1}{T_1 + T_2}$$

to $\overline{u_d}$; the negative half-wave contributes

$$-\frac{2}{\pi} K_d \frac{T_2}{T_1 + T_2}$$

Hence $\overline{u_d}$ is given by

$$\overline{u_d} = \frac{2}{\pi} K_d \frac{T_1 - T_2}{T_1 + T_2} \tag{A.8}$$

To get a simple expression for $\overline{u_d}$, T_1 and T_2 are expanded to a Taylor series and terms of second and higher order are neglected. Then we get for T_1 and T_2:

$$T_1 = \frac{\pi}{\Delta\omega[1 - (2/\pi)F_H K_d K_0/\Delta\omega]} \approx \frac{\pi}{\Delta\omega}\left(1 + \frac{2}{\pi}\frac{F_H K_d K_0}{\Delta\omega}\right) \tag{A.9a}$$

$$T_2 = \frac{\pi}{\Delta\omega[1 + (2/\pi)F_H K_d K_0/\Delta\omega]} \approx \frac{\pi}{\Delta\omega}\left(1 - \frac{2}{\pi}\frac{F_H K_d K_0}{\Delta\omega}\right) \tag{A.9b}$$

With these approximations $\overline{u_d}$ becomes

$$\overline{u_d} = \frac{4F_H K_d^2 K_0}{\pi^2 \Delta\omega} \tag{A.10}$$

This can be written in simplified form as

$$\overline{u_d} = \frac{c_d}{\Delta\omega} \tag{A.10a}$$

with
$$c_d = \frac{4F_H K_d^2 K_0}{\pi^2}$$

Thus the average phase-detector output signal u_d is inversely proportional to $\Delta\omega$. This is valid for large values of $\Delta\omega$ only, because u_d can never become larger than K_d, as is easily seen from Fig. A.2. If the average u_d signal is plotted against $\Delta\omega$ (Fig. A.3), the curve is a hyperbola for large values of $\Delta\omega$ but approaches K_d for small $\Delta\omega$.

Next we derive an expression for the average loop filter output signal u_f as a function of the frequency offset $\Delta\omega$. As defined by Eq. (1.1), the average frequency of ω_{20} of the VCO is given by

$$\omega_{20} = \omega_0 + K_0 \overline{u_f} \tag{A.11}$$

where $\overline{u_f}$ is the average output signal of the loop filter. With the abbreviations $\Delta\omega = \omega_1 - \omega_{20}$ and $\Delta\omega_0 = \omega_1 - \omega_0$ (= initial offset), this can be written as

$$\Delta\omega = \Delta\omega_0 - K_0 \overline{u_f}$$

or

$$\overline{u_f} = \frac{\Delta\omega_0 - \Delta\omega}{K_0} \tag{A.12}$$

Plotting $\overline{u_f}$ against $\Delta\omega$ yields a straight line. This line is also shown in Fig. A.3. From this illustration we can determine whether or not a pull-in process will take place. Depending on the slope of $\overline{u_f}$ against $\Delta\omega$, the line $\overline{u_f}(\Delta\omega)$ can intersect with the curve $\overline{u_d}(\Delta\omega)$ at one point

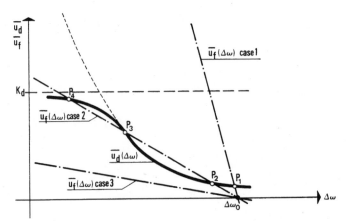

Figure A.3 Plot of the average signals $\overline{u_d}(t)$ and $\overline{u_f}(t)$ against frequency offset $\Delta\omega$ for the unlocked PLL.

(case 1), at three points (case 2), or at no point at all (case 3). Consider case 1 first. The curves $\overline{u}_d(\Delta\omega)$ and $\overline{u}_f(\Delta\omega)$ intersect at point P_1. In this case the frequency of the VCO is pulled up slightly, but the system remains "hung" in point P_1. An analysis of stability shows that point P_1 is a stable point; thus no pull-in process will occur.

In case 2 the curves $\overline{u}_d(\Delta\omega)$ and $\overline{u}_f(\Delta\omega)$ intersect at points P_2, P_3, and P_4. A stability analysis shows that points P_2 and P_4 are stable, but P_3 is unstable. After power-on, the system will thus remain hung at point P_2.

Apparently a pull-in process takes place only when there is no point of intersection, as in case 3. The two curves do not intersect if the equation

$$\overline{u}_f(\Delta\omega) = \overline{u}_d(\Delta\omega) \tag{A.13}$$

does not have a real solution.

Inserting Eqs. (A.10a) and (A.12) into (A.13) yields the quadratic equation

$$\Delta\omega^2 - \Delta\omega_0\,\Delta\omega + K_0 c_d = 0 \tag{A.14}$$

Its solutions are

$$\Delta\omega_{1,2} = \frac{\Delta\omega_0 \pm \sqrt{\Delta\omega_0^2 - 4K_0 c_d}}{2K_0 c_d} \tag{A.15}$$

The roots become complex if the discriminant (the expression under the radical) becomes negative. Putting the discriminant to zero yields the limiting case $\Delta\omega_0 = \Delta\omega_p$; that is, the pull-in range $\Delta\omega_p$ is the initial frequency offset $\Delta\omega_0$ for which the discriminant becomes zero. If we perform this calculation, we obtain

$$\Delta\omega_p = 2\sqrt{K_0 c_d} \tag{A.16}$$

If we insert c_d from Eq. (A.10a) and make use of the substitutions defined in Eqs. (2.15a) to (2.15c), we finally get the expressions

$$\Delta\omega_p = \frac{4}{\pi}\,\sqrt{2\,\zeta\omega_n K_0 K_d - \omega_n^2} \tag{A.17a}$$

for a passive lag filter and a low-gain loop,

$$\Delta\omega_p = \frac{4\sqrt{2}}{\pi}\,\sqrt{\zeta\omega_n K_0 K_d} \tag{A.17b}$$

for a passive lag filter and a high-gain loop,

$$\Delta\omega_p = \frac{4}{\pi}\sqrt{2\,\zeta\omega_n K_0 K_d - \frac{\omega_n^2}{K_a}} \qquad (A.18a)$$

for an active lag filter and a low-gain loop,

$$\Delta\omega_p = \frac{4\sqrt{2}}{\pi}\sqrt{\zeta\omega_n K_0 K_d} \qquad (A.18b)$$

for an active lag filter and a high-gain loop.

Because most loops are high-gain loops, Eq. (A.17b) can be used for most active and passive lag filters. The situation is completely different if the active PI loop filter is used. Since the gain at low frequencies approaches infinity for this type of filter, the pull-in range becomes infinite, too, as explained in Sec. 2.6.3.

We conclude that the model shown in Fig. A.3 enabled us to calculate the pull-in range $\Delta\omega_p$. To get that result, we simply postulated that the two curves for $\overline{u_d}$ and $\overline{u_f}$ shall not intersect. In the following section we will see that a slightly expanded model leads to an approximation for the pull-in time T_p.

A.2   Simplified Model for the Pull-in Time T_p of the LPLL

To get an approximation for the pull-in time, we now try to find a differential equation which tells us how the instantaneous frequency offset $\Delta\omega = \omega_1 - \omega_{20}$ varies with time t. A model for the pull-in process is shown in Fig. A.4. To compute the pull-in process, the transfer functions of the three building blocks must be known for the *unlocked state* of the PLL. The models for the phase detector and for the VCO have already been derived in the preceding section. We found that the average output signal u_d varies inversely proportional to the frequency offset $\Delta\omega = \omega_1 - \omega_{20}$, where ω_{20} is the average instantaneous frequency of the VCO. Equation (A.10a) describes phase-detector performance in the unlocked state. Since the variable $\Delta\omega$ appears in the denominator—and not in the numerator—of Eq. (A.10a), the differential equation yielding $\Delta\omega(t)$ will certainly be nonlinear, which precludes the use of the Laplace transform. The pull-in process must therefore be calculated in the time domain. The operation of the VCO is governed by Eq. (A.12), which is linear. To complete the analysis, we must introduce the differential equation of the loop filter. If we assume for the moment that the passive lag filter is used, its transfer function is given by

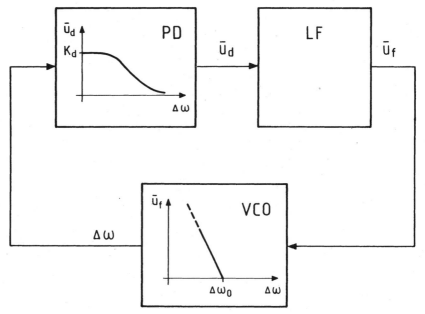

Figure A.4 Mathematical model for the pull-in process of the PLL.

$$F(s) = \frac{1 + s\,\tau_2}{1 + s(\tau_1 + \tau_2)} \qquad (A.19a)$$

When transformed back into time domain, we have

$$\overline{u_f}(t) + (\tau_1 + \tau_2)\frac{d}{dt}\,\overline{u_f}(t) = \overline{u_d}(t) + \tau_2\frac{d}{dt}\,\overline{u_d}(t) \qquad (A.19b)$$

The three equations (A.10a), (A.12), and (A.19b) fully describe the pull-in process. When $\overline{u_d}$ and $\overline{u_f}$ are eliminated, we get the nonlinear differential equation

$$\frac{d}{dt}\,\Delta\omega\left(\frac{\tau_1 + \tau_2}{K_0} - \frac{\tau_2 c_d}{\Delta\omega^2}\right) + \frac{c_d}{\Delta\omega} + \frac{\Delta\omega}{K_0} = \frac{\Delta\omega_0}{K_0} \qquad (A.20)$$

for $\Delta\omega$ vs. t. Of course, such a differential equation can be numerically integrated on a computer, but unfortunately this does not deliver us an explicit expression for the pull-in time. In order to get such a formula, the differential equation should be simplified such that we get an algebraic expression for $\Delta\omega$ vs. time. The solution becomes much simpler if we approximate the transfer function of the loop filter by

$$F(s) = \frac{1}{s\,\tau_1} \tag{A.21a}$$

which is the transfer function of an ideal integrator. The following consideration shows under which circumstances this approximation is acceptable. Figure A.5a shows the step response of a passive lag filter having the transfer function given in Eq. (A.19a) for the case $\tau_1 \gg \tau_2$. In Fig. A.5b, the step response of the ideal integrator [Eq. (A.21a)] is shown. The initial slope of both step responses is the same. It follows that the ideal integrator is a valid approximation to the passive lag filter as long as events are considered whose duration is not much more than the loop filter time constant τ_1. Experience shows that this is not necessarily fulfilled. In cases where T_p turns out to be less than τ_1, the approximation is quite good; i.e., the errors are normally below 10 percent of the predicted value. When T_p shows up to be larger than τ_1, however, the approximation becomes rather crude. Finally, if the

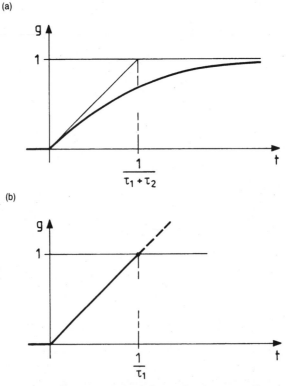

(a)

(b)

Figure A.5 Step response of first-order low-pass filters. (*a*) Step response of a passive lag filter. (*b*) Step response of the ideal integrator.

initial frequency offset $\Delta\omega_0$ comes close to the pull-in range $\Delta\omega_p$, the pull-in time approaches infinity. As a rule of thumb, the formula for T_p, which will be derived below, delivers reasonable estimates (errors < 10 percent) when $\Delta\omega_0$ is less than about $0.8\ \Delta\omega_p$.

Transforming Eq. (A.21a) back into time domain yields

$$\overline{u_d}(t) = \tau_1 \frac{d}{dt}\ \overline{u_f}(t) \tag{A.21b}$$

Inserting Eqs. (A.10a) and (A.21b) into Eq. (A.12) yields the simplified differential equation

$$\frac{d}{dt}\ \Delta\omega\ \frac{\tau_1}{K_0} + \frac{c_d}{\Delta\omega} = 0 \tag{A.22}$$

Although this equation is still nonlinear, the variables $\Delta\omega$ and t can be separated, so we have

$$\frac{\tau_1}{c_d K_0} \int_{\Delta\omega_0}^{\Delta\omega_L} \Delta\omega\, d\,\Delta\omega = -\int_0^{T_p} dt \tag{A.23}$$

The limits of integration on the left side are $\Delta\omega_0$ and $\Delta\omega_L$, which says that the pull-in process ends when the instantaneous frequency offset $\Delta\omega$ falls below the lock range $\Delta\omega_L$. At that time, an ordinary lock-in process follows, as has been shown by the simulations in Secs. 2.10 and 3.6. The limits of integration on the right side are 0 and T_p. In most cases, $\Delta\omega_0$ is much larger than $\Delta\omega_L$, so we finally get

$$T_p \approx \frac{\tau_1\,\Delta\omega_0^2}{c_d K_0} \tag{A.24}$$

Inserting c_d from Eq. (A.10a) and using the definitions of Eqs. (2.15a), we get for the pull-in time

$$T_p \approx \frac{\pi^2}{16}\ \frac{\Delta\omega_0^2}{\zeta\omega_n^3} \tag{A.25a}$$

If the loop filter is an active lag filter, its transfer function can also be replaced by the transfer function of the ideal integrator. When the computation is repeated for the active lag filter, the pull-in time becomes

$$T_p \approx \frac{\pi^2}{16}\ \frac{\Delta\omega_0^2 K_a}{\zeta\omega_n^3} \tag{A.25b}$$

Finally, when the loop filter is an active PI filter, the pull-in time is given by

$$T_p \approx \frac{\pi^2}{16} \frac{\Delta\omega_0^2}{\zeta\omega_n^3} \qquad (A.25c)$$

As we stated in Sec. 3.2.3, the approximations for the pull-in time are valid if the initial frequency offset $\Delta\omega_0$ is markedly below the pull-in range $\Delta\omega_p$, i.e., below about 0.8 $\Delta\omega_p$. If the initial frequency offset comes close to $\Delta\omega_p$, the pull-in time approaches infinity.

A.3   The Pull-in Range $\Delta\omega_p$ of the DPLL

The approach described in Sec. A.1 can be adopted to calculate also the pull-in range of the digital PLL. We assume first that the EXOR gate is used as phase detector. As explained in Sec. 3.2.3 and in Fig. 3.13a, the output signal of the PD in the unlocked state is no longer sinusoidal but is an asymmetrical triangular wave. For a given average frequency offset $\Delta\omega$, we can again calculate the resulting dc component of the phase-detector output signal $\overline{u_d}$. The result is

$$\overline{u_d} = \frac{\pi^2 K_d^2 K_0 F_H}{16 \Delta\omega} \qquad (A.26a)$$

which can be rewritten as

$$\overline{u_d} = \frac{c_d}{\Delta\omega} \qquad (A.26b)$$

with

$$c_d = \frac{\pi^2 K_d^2 K_0 F_H}{16}$$

The characteristics of the loop filter and of the VCO stay the same as in the case of the LPLL, so the pull-in range is obtained simply by inserting the c_d into Eq. (A.16). The result becomes

$$\Delta\omega_p = \frac{\pi}{2} \sqrt{2\zeta\omega_n K_0 K_d - \omega_n^2} \qquad (A.27a)$$

(PD = EXOR, passive lag filter, low-gain loop)

$$\Delta\omega_p = \frac{\pi}{\sqrt{2}} \sqrt{\zeta\omega_n K_0 K_d} \qquad (A.27b)$$

(PD = EXOR, passive lag filter, high-gain loop)

$$\Delta\omega_p = \frac{\pi}{2}\sqrt{2\,\zeta\omega_n K_0 K_d - \frac{\omega_n^2}{K_a}} \qquad (A.28a)$$

(PD = EXOR, active lag filter, low-gain loop)

$$\Delta\omega_p = \frac{\pi}{\sqrt{2}}\sqrt{\zeta\omega_n K_0 K_d} \qquad (A.28b)$$

(PD = EXOR, active lag filter, high-gain loop).

The pull-in range for a DPLL using a *JK*-flipflop phase detector remains to be calculated. As shown in Sec. 3.2.3 and Fig. 3.13b, the phase-detector output signal is an unsymmetrical sawtooth in the unlocked state. Again using the same argumentation as in Sec. A.1, the coefficient c_d can be shown to be

$$c_d = \frac{\pi^2 K_d^2 K_0 F_H}{4}$$

in this case. If we insert c_d into Eq. (A.16) and make use of the substitutions in Eqs. (2.15a) to (2.15c), we get

$$\Delta\omega_p = \pi\sqrt{2\,\zeta\omega_n K_0 K_d - \omega_n^2} \qquad (A.29a)$$

(PD = *JK*-flipflop, passive lag filter, low-gain loop)

$$\Delta\omega_p = \pi\sqrt{2\,\zeta\omega_n K_0 K_d} \qquad (A.29b)$$

(PD = *JK*-flipflop, passive lag filter, high-gain loop)

$$\Delta\omega_p = \pi\sqrt{2\,\zeta\omega_n K_0 K_d - \frac{\omega_n^2}{K_a}} \qquad (A.30a)$$

(PD = *JK*-flipflop, active lag filter, low-gain loop)

$$\Delta\omega_p = \pi\sqrt{2\,\zeta\omega_n K_0 K_d} \qquad (A.30b)$$

(PD = *JK*-flipflop, active lag filter, high-gain loop).

A.4   The Pull-In Time T_p of the DPLL

As we have seen in Sec. A.3, the behavior of LPLLs and DPLLs in the unlocked state is very similar. For all types of phase detectors used, the average output signal can be represented by the same formula

$$\overline{u_d} = \frac{c_d}{\Delta\omega}$$

where only the coefficient c_d depends on the phase-detector type. Therefore, the same differential equation is valid for the pull-in process of DPLLs. The pull-in time T_p is found by inserting the appropriate expression for c_d into Eq. (A.24). Using the substitutions of Eqs. (3.15a) to (3.15c) we finally get

$$T_p \approx \frac{4}{\pi^2} \frac{\Delta\omega_0^2}{\zeta\omega_n^3} \tag{A.31}$$

for the DPLL using the EXOR phase detector and

$$T_p \approx \frac{1}{\pi^2} \frac{\Delta\omega_0^2 K_a}{\zeta\omega_n^3} \tag{A.32}$$

for the DPLL using the JK-flipflop phase detector.

As we stated in Sec. 3.2.3, the approximations for the pull-in time are valid if the initial frequency offset $\Delta\omega_0$ is markedly below the pull-in range $\Delta\omega_p$, i.e., below about $0.8\,\Delta\omega_p$. If the initial frequency offset comes close to $\Delta\omega_p$, the pull-in time approaches infinity.

The Laplace Transform

B.1 Transforms Are the Engineer's Tools

Trying to solve electronic problems without using the Laplace transform is similar to traveling through a foreign country with a globe instead of a map (Fig. B.1). An engineer who tries to find the transient response of an electric network to an impulse function by solving differential equations, for example, certainly is working with inadequate tools (see Fig. B.2). The engineer familiar with the the techniques of the Laplace transform may find a solution very quickly, as shown in Fig. B.3.

A map images a three-dimensional object to a plane. Every spatial point of the three-dimensional object is represented by a unique point in the plane of the map. Things are similar, though different, in the case of the Laplace transform. Here a function in the *time domain* (such as an electric signal) is transformed to another function in the *complex frequency domain*. The trouble with the Laplace transform starts right here: even an electronic hobbyist can imagine what the frequency spectrum of an electric signal is, but what is *complex* frequency?

One need not be a cow to know what milk is, but it is surely easier to understand the term complex frequency if we first consider *real* frequency spectra. The Laplace transform is a more general form of the Fourier transform; in other words, the Fourier transform is a special case of the Laplace transform. In this context it may be easier to start with the special case before proceeding to the more general one.

The Fourier transform is explained best by first looking at a periodic signal such as a square wave (Fig. B.4). The angular frequency of this signal is assumed to be ω_0. This square wave can now be thought to

Figure B.1 Trying to solve electronic problems without using the Laplace transform is as cumbersome as traveling through a foreign country with a globe instead of a map.

be composed of a(n) (infinite) number of sine-wave signals having frequencies ω_0 (fundamental frequency), $2\omega_0$, $3\omega_0$, ... (harmonics). Mathematically, any periodic signal $f(t)$ having a repetition frequency of ω_0 can be written as a sum of its harmonics,

$$f(t) = \sum_{n=-\infty}^{+\infty} F(jn\omega_0) \exp(jn\omega_0 t) \qquad (B.1)$$

The so-called Fourier coefficients $F(jn\omega)$ are calculated from

Figure B.2 Looking at the transient response of electric networks without using the Laplace transform can be tricky . . .

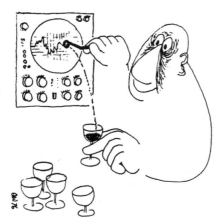

Figure B.3 . . . but the engineer familiar with Laplace techniques may find the solution very quickly.

$$F(jn\omega_0) = \frac{1}{T} \int_{-T/2}^{+T/2} f(t) \exp\left(-jn\omega_0 t\right) dt \qquad \text{(B.2)}$$

where T is the period of the periodic signal $f(t)$, $T = 2\pi/\omega_0$, and $F(jn\omega_0)$ is the amplitude of the harmonic component with frequency $n\omega_0$. (The j operator could be omitted but here we keep this term because any Fourier coefficient F is always a function of $j\omega$ and never of ω alone.) Note that the Fourier coefficients $F(jn\omega_0)$ are generally complex numbers; we can therefore write

$$F(jn\omega_0) = |F(jn\omega_0)| \exp\left(j\phi_n\right) \qquad \text{(B.3)}$$

where $|F(jn\omega_0)|$ is the amplitude and ϕ_n the phase of $F(jn\omega_0)$.

When plotting the Fourier transform of a signal $f(t)$, the amplitude $|F(jn\omega_0)|$ and the phase ϕ_n are normally plotted against ω; these functions are called amplitude and phase spectra, respectively. In the case of periodic functions $f(t)$ the Fourier spectra become discrete, that is, the Fourier coefficients $F(jn\omega_0)$ are defined only at the discrete frequencies $\omega_0, 2\omega_0, 3\omega_0, \ldots$.

Figure B.4 shows the amplitude spectrum $|F(jn\omega_0)|$ of a symmetrical square wave; for a symmetrical waveform it can be shown that the even harmonics disappear. Consequently, the Fourier spectrum of Fig. B.4 only shows lines at $\omega_0, 3\omega_0, 5\omega_0, \ldots$.

In real life we find many signals which are not periodic, such as a single pulse. What about the Fourier transform of such signals? If a signal $f(t)$ is not periodic, we could state that its period approaches infinity, which means its fundamental frequency approaches zero. If the fundamental frequency approaches zero, the spectral lines in Eq.

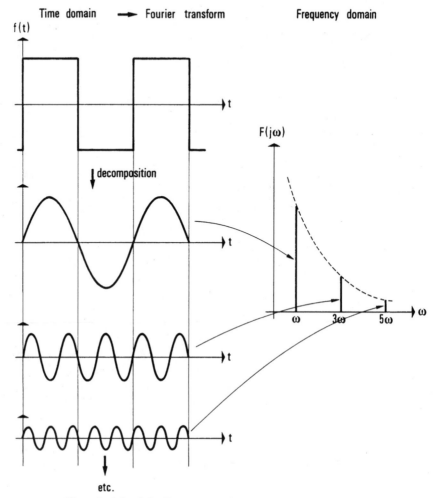

Figure B.4 The principle of the Fourier transform.

(B.1) come closer and closer, and finally the sum is replaced by an integral,

$$f(t) = \frac{1}{2\pi} \int_{-\infty}^{+\infty} F(j\omega)e^{j\omega t} \, dt \tag{B.4}$$

For an aperiodic signal $f(t)$ the Fourier spectrum $F(j\omega)$ becomes continuous; $F(j\omega)$ is called the *Fourier transform* of the signal $f(t)$. The Fourier transform is calculated in the same way as the Fourier coefficients $F(jn\omega)$ in Eq. (B.2). If T approaches ∞ in this equation, we get

$$F(j\omega) = \int_{-\infty}^{\infty} f(t)e^{-j\omega t}\, dt \tag{B.5}$$

This is the definition of the *continuous* Fourier transform.

Note that the Fourier transform $F(j\omega)$ for any given signal $f(t)$ can be calculated by evaluating the integral in Eq. (B.5). If the Fourier transform of a signal $f(t)$ is given first, the corresponding signal $f(t)$ in the time domain can be obtained by applying Eq. (B.4). This equation is also called the *inverse* Fourier transform. The Fourier transforms $F(j\omega)$ for the most important signal waveforms are tabulated in many reference books.[36]

The usefulness of the Fourier transform is limited by a serious drawback: if we try to evaluate the Fourier integral in Eq. (B.5), we find that the integral does *not converge* for most signals $f(t)$. You will find it impossible to find a solution of the Fourier integral by conventional methods, even for such a simple signal as $f(t) = \sin \omega t$. The Laplace transform offers a way to circumvent this problem.

B.2 A Laplace Transform Is the Key to Success

Imagine we would like to know the Fourier transform of an extremely simple signal, $f(t) = \sin \omega_0 t$ [with $f(t) = 0$ for $t < 0$]. Using the definition of the Fourier transform, Eq. (B.5), we get

$$F(j\omega) = \int_{0}^{\infty} f(t)e^{-j\omega t}\, dt = \int_{0}^{\infty} \sin \omega_0 t\, e^{-j\omega t}\, dt$$

Using an integral table we find for $F(j\omega)$

$$F(j\omega) = \frac{-j\omega e^{-j\omega t}\sin \omega_0 t - \omega_0 e^{-j\omega t}\cos \omega_0 t}{\omega_0^2 - \omega^2}\Bigg|_{0}^{\infty}$$

Introducing the limits of integration (here 0 and ∞) yields

$$F(j\omega) = \frac{-j\omega e^{-j\infty}\sin \infty - \omega_0 e^{-j\infty}\cos \infty + \omega_0}{\omega_0^2 - \omega^2}$$

But what is $\sin \infty$, and what is $\cos \infty$? Both functions are periodic and are within a range of -1 to $+1$; hence $\sin \infty$ and $\cos \infty$ are not defined.

The evaluation of the Fourier integral would be simpler if the function $f(t)$ would approach zero for very large values of t. The solution of the Fourier integral becomes possible if $f(t)$ is multiplied by a damping function $e^{-\sigma t}$, where σ is a positive real number:

$$F'(j\omega, \sigma) = \int_0^\infty [f(t)e^{-\sigma t}]e^{-j\omega t} \, dt \qquad (\text{B.6})$$

The notation $F'(j\omega, \sigma)$ is chosen to emphasize that now F' is also dependent on σ. The prime is furthermore chosen to differentiate F' from the Fourier integral in Eq. (B.5). The product $[f(t)e^{-\sigma t}]$ approaches zero for large t. This is true even when $f(t)$ is an exponential function $f(t) = e^{at}$ with positive a. If σ is chosen larger than a, the product approaches zero for $t \rightarrow \infty$. The modified Fourier integral in Eq. (B.6) is now easily evaluated. Let us perform the calculation for the previous example $f(t) = \sin \omega_0 t$. We then have

$$F'(j\omega, \sigma) = \int_0^\infty \sin \omega_0 t e^{-\sigma t} e^{-j\omega t} \, dt$$

Again using an integral table, we get

$$F'(j\omega, \sigma) = \frac{-(\sigma + j\omega)e^{-(\sigma+j\omega)t} \sin \omega_0 t - \omega_0 e^{-(\sigma+j\omega)t} \cos \omega_0 t}{\omega_0^2 + (\sigma + j\omega)^2} \bigg|_0^\infty$$

If the limits of integration are now inserted, the term

$$e^{-(\sigma+j\omega)\infty}$$

becomes zero for positive σ. Then $F'(j\omega)$ becomes

$$F'(j\omega, \sigma) = \frac{\omega_0}{\omega_0^2 + (\sigma + j\omega)^2}$$

Of course $F'(j\omega)$ contains the damping factor σ. The Fourier integral $F(j\omega)$ is now obtained simply by letting $\sigma \rightarrow 0$:

$$F(j\omega) = \lim_{\sigma \rightarrow 0} F'(j\omega, \sigma) = \frac{\omega_0}{\omega_0^2 - \omega^2}$$

If σ is set equal to zero in Eq. (B.6), this equation is transformed into Eq. (B.5). *Thus the Fourier transform of any function $f(t)$ is obtained by first introducing a damping function $e^{-\sigma t}$ evaluating the integral [Eq. (B.6)] for $\sigma > 0$ and finally letting $\sigma \rightarrow 0$.*

Let us now see what the Laplace transform really is. Equation (B.6) can also be written in a different form;

$$F'(j\omega, \sigma) = \int_0^\infty f(t)[e^{-\sigma t}e^{-j\omega t}] \, dt$$

In contrast to Eq. (B.6), we now combine the damping function $e^{-\sigma t}$ with the exponential function $e^{-j\omega t}$. This can be written as

$$F'(j\omega,\,\sigma) = \int_0^\infty f(t)e^{-(\sigma+j\omega)t}\,dt$$

We now define

$$s = \sigma + j\omega$$

and call the new variable s complex frequency. Because the two variables σ and ω appear only in the form $\sigma + j\omega$, $F'(j\omega,\,\sigma)$ can be written as $F(\sigma + j\omega) = F(s)$. We then have

$$F(s) = \int_0^\infty f(t)e^{-st}\,dt \tag{B.7}$$

This is the definition of the Laplace transform.

In the case of the Fourier transform, $F(j\omega)$ was a *complex* function of the *real* variable ω. To plot $F(j\omega)$ against ω we had to plot amplitude $|F(j\omega)|$ and phase $\phi(j\omega)$ as a function of ω. (This plot is commonly called a Bode diagram.) In the case of the Laplace transform, however, $F(s)$ is a *complex* function of the *complex* variable s. Plotting $F(s)$ is much more difficult than plotting $F(j\omega)$. To plot $F(s)$, a relief of the amplitude $|F(s)|$ and of the phase $\phi(s)$ could be constructed; a relief of $|F(s)|$ is shown in Fig. B.5.

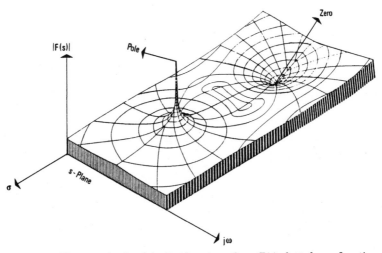

Figure B.5 The magnitude of the Laplace transform $F(s)$ plotted as a function of complex frequency s yields a relief. This illustration shows a pole and a zero of the transfer function $F(s)$.

The construction of such a relief is a cumbersome procedure. As we shall see later, it is sufficient to know the locations of some singular points of $F(s)$ only, namely, the *poles* and *zeros* of $F(s)$. A pole is a point in the s plane where $F(s)$ becomes infinity, and a zero is a point in the s plane where $F(s)$ is zero. The so-called pole-zero plot (Fig. B.6) shows the locations in the s plane where $F(s)$ has poles (●) or zeros (○). [Note that not every function $F(s)$ necessarily has poles or zeros.]

We see immediately that the Fourier transform [Eq. (B.5)] is a special case of the Laplace transform [Eq. (B.7)]. The Fourier transform $F(j\omega)$ is simply obtained by finding the Laplace transform $F(s)$ and then putting $\sigma = 0$

$$F(j\omega) = \lim_{\sigma \to 0} F(s), \qquad s = \sigma + j\omega \tag{B.8}$$

In other words, $F(j\omega)$ is equal to the function $F(s)$ on the imaginary axis of the s plane.

Equation (B.7) shows us how to calculate the Laplace transform from a given function $f(t)$ in the time domain. As was the case with the Fourier transform, it is also possible to calculate $f(t)$ if $F(s)$ is given first. This transformation is called *inverse* Laplace transform and is defined by

$$f(t) = \frac{1}{2\pi j} \int_{c-j\infty}^{c+j\infty} F(s)e^{-st}\, ds \tag{B.9}$$

where c is a real constant. As is seen from Eq. (B.9) the path of integration is a line parallel to the imaginary axis.[1,3,18]

In most cases, computation of the Laplace transform of the inverse Laplace transform by evaluating Eqs. (B.7) and (B.9) is no longer nec-

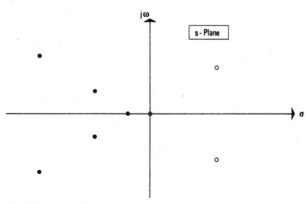

Figure B.6 Plot of the poles and zeroes of a given transfer function $F(s)$.

essary since transformation tables are available (see Table B.1). To conclude this section, let us introduce some convenient abbreviations:

$$F(s) = \mathcal{L}\{f(t)\} \qquad (B.10a)$$

$$f(t) = \mathcal{L}^{-1}\{F(s)\} \qquad (B.10b)$$

Equation (B.10a) says that $F(s)$ is the Laplace transform of the signal $f(t)$; Eq. (B.10b) states that $f(t)$ is the inverse Laplace transform of $F(s)$.

B.3 A Numerical Example of the Laplace Transform

To see how the Laplace transform $F(s)$ of a signal $f(t)$ could be obtained by evaluating Eq. (B.7), let us calculate an example.

Assume $f(t)$ is the unit step function

$$f(t) = u(t)$$

as plotted in Fig. B.7. For the Laplace transform $F(s)$ we have, according to Eq. (B.7),

$$F(s) = \int_0^\infty u(t)e^{-st}\, dt = \int_0^\infty e^{-st}\, dt = \left.\frac{e^{-st}}{-s}\right|_0^\infty$$

$$= -\frac{1}{s}(e^{-\infty} - e^0) = -\frac{1}{s}(0 - 1) = \frac{1}{s}$$

that is,

$$F(s) = \frac{1}{s}$$

$F(s)$ has one single pole at $s = 0$, that is, at the origin of the s plane. Evaluating the Laplace integral in Eq. (B.7) for the most common test signals $f(t)$ yields Table B.1.

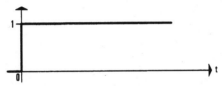

Figure B.7 Plot of the unit step function against time.

TABLE B.1 Laplace Transform $f(t) = 0$ for $t < 0$

	Signal $f(t)$	Laplace transform $F(s)$
General rules:		
Differentiation	$\dfrac{df(t)}{dt}$	$sF(s) - f(0)$
Integration	$\displaystyle\int_0^t f(t)\,dt$	$\dfrac{F(s)}{s} + \dfrac{f^{(-1)}(0)}{s}$
Time delay	$f(t - \tau)$	$F(s)e^{-s\tau}$
Multiplication by constant	$kf(t)$	$kF(s)$
Convolution	$f_1(t) \cdot f_2(t)$ $f_1(t) * f_2(t)$	$F_1(s) * F_2(s)$ $F_1(s) \cdot F_2(s)$
Functions:	0	0
Unit step	$u(t)$	$\dfrac{1}{s}$
Delta function	$\delta(t)$	1
Ramp function	t	$\dfrac{1}{s^2}$
Parabola	$\dfrac{t^2}{2}$	$\dfrac{1}{s^3}$
Polynomial in t	$\dfrac{t^{n-1}}{(n-1)!}$	$\dfrac{1}{s^n}$, $n > 0$ (can be fractional)
Exponential functions	$e^{\alpha t}$	$\dfrac{1}{s - \alpha}$
	$\dfrac{1}{\alpha}(e^{\alpha t} - 1)$	$\dfrac{1}{s(s - \alpha)}$
	$\dfrac{1}{\alpha}(1 - e^{-\alpha t})$	$\dfrac{1}{s(s + \alpha)}$
	$\dfrac{t^n - 1}{(n-1)!e^{\alpha t}}$	$\dfrac{1}{(s - \alpha)^n}$, $n > 0$ (can be fractional)
Trigonometric functions	$\dfrac{1}{\alpha}\sin \alpha t$	$\dfrac{1}{s^2 + \alpha^2}$
	$\cos \alpha t$	$\dfrac{s}{s^2 + \alpha^2}$
	$\dfrac{1}{\alpha^2}(1 - \cos \alpha t)$	$\dfrac{1}{s(s^2 + \alpha^2)}$

TABLE B.1   Laplace Transform $f(t) = 0$ for $t < 0$ (Continued)

	Signal $f(t)$	Laplace transform $F(s)$
General rules:		
Hyperbolic functions	$\dfrac{1}{\alpha}\sinh \alpha t$	$\dfrac{1}{s^2 - \alpha^2}$
	$\cosh \alpha t$	$\dfrac{s}{s^2 - \alpha^2}$
	$\dfrac{1}{\alpha^2}(\cosh \alpha t - 1)$	$\dfrac{1}{s(s^2 - \alpha^2)}$
Second-order systems (aperiodic)	$\dfrac{e^{\beta t} - e^{\alpha t}}{\beta - \alpha}$	$\dfrac{1}{(s - \alpha)(s - \beta)}$
	$\dfrac{\beta e^{\beta t} - \alpha e^{\alpha t}}{\beta - \alpha}$	$\dfrac{s}{(s - \alpha)(s - \beta)}$
	$\dfrac{\beta e^{\alpha t} - \alpha e^{\beta t}}{\alpha\beta(\alpha - \beta)} + \dfrac{1}{\alpha\beta}$	$\dfrac{1}{s(s - \alpha)(s - \beta)}$
Second-order systems (periodic)	$\dfrac{e^{-\zeta\omega_n t}\sin\sqrt{1 - \zeta^2}\,\omega_n t}{\sqrt{1 - \zeta^2}\,\omega_n}$	$\dfrac{1}{s^2 + 2s\,\zeta\omega_n + \omega_n^2}$
	$\left[\cos\sqrt{1 - \zeta^2}\,\omega_n t - \dfrac{\zeta}{\sqrt{1 - \zeta^2}}\sin\sqrt{1 - \sigma^2}\,\omega_n t\right]e^{-\zeta\omega_n t}$	$\dfrac{s}{s^2 + 2s\,\zeta\omega_n + \omega_n^2}$
	$\omega_n e^{-\zeta\omega_n t}\left[\dfrac{2\zeta^2 - 1}{\sqrt{1 - \zeta^2}}\sin\sqrt{1 - \zeta^2}\,\omega_n t \right.$ $\left. - 2\zeta\cos\sqrt{1 - \zeta^2}\,\omega_n t\right]$	$\dfrac{s^2}{s^2 + 2s\,\zeta\omega_n + \omega_n^2}$
	$\dfrac{1}{\omega_n^2}\left[1 + \left(\cos\sqrt{1 - \zeta^2}\,\omega_n t \right.\right.$ $\left.\left. + \dfrac{\zeta}{1 - \zeta^2}\sin\sqrt{1 - \zeta^2}\,\omega_n t\right)e^{-\zeta\omega_n t}\right]$	$\dfrac{1}{s(s^2 + 2s\,\zeta\omega_n + \omega_n^2)}$
Third-order systems	$\dfrac{e^{\alpha t} - [1 + (\alpha - \beta)t]e^{\beta t}}{(\alpha - \beta)^2}$	$\dfrac{1}{(s - \alpha)(s - \beta)^2}$
	$\dfrac{\alpha e^{\alpha t} - [\alpha + \beta(\alpha - \beta)t]e^{\beta t}}{(\alpha - \beta)^2}$	$\dfrac{s}{(s - \alpha)(s - \beta)^2}$
	$\dfrac{\alpha^2 e^{\alpha t} - [2\alpha - \beta + \beta(\alpha - \beta)t]\beta e^{\beta t}}{(\alpha - \beta)^2}$	$\dfrac{s^2}{(s - \alpha)(s - \beta)^2}$
	$\dfrac{(\beta - \gamma)e^{\alpha t} + (\gamma - \alpha)e^{\beta t} + (\alpha - \beta)e^{\gamma t}}{(\alpha - \beta)(\beta - \gamma)(\gamma - \alpha)}$	$\dfrac{1}{(s - \alpha)(s - \beta)(s - \gamma)}$
Various functions	$\dfrac{\alpha\sin\beta t - \beta\sin\alpha t}{\alpha\beta(\alpha^2 - \beta^2)}$	$\dfrac{1}{(s^2 + \alpha^2)(s^2 + \beta^2)}$
	$\dfrac{\cos\beta t - \cos\alpha t}{\alpha^2 - \beta^2}$	$\dfrac{s}{(s^2 + \alpha^2)(s^2 + \beta^2)}$
	$\dfrac{1}{\sqrt{\pi t}}$	$\dfrac{1}{\sqrt{s}}$

TABLE B.1 Laplace Transform $f(t) = 0$ for $t < 0$ (Continued)

	Signal $f(t)$	Laplace transform $F(s)$
General rules:	$2\sqrt{\dfrac{t}{\pi}}$	$\dfrac{1}{s\sqrt{s}}$
	$\dfrac{n!}{(2n)!} \cdot \dfrac{4^n}{\sqrt{\pi}}\, t^{n-1/2}$	$\dfrac{1}{s^n\sqrt{s}}$
	$\dfrac{1}{\sqrt{\pi t}}\, e^{\alpha t}$	$\dfrac{1}{\sqrt{s - \alpha}}$
Error functions (erf)	$\dfrac{2}{\sqrt{\alpha \pi}} \displaystyle\int_0^{\sqrt{\alpha t}} e^{-\zeta^2}\, d\zeta$	$\dfrac{1}{s\sqrt{s + \alpha}}$
	$\dfrac{2e^{-\alpha t}}{\sqrt{\pi(\beta - \alpha)}} \displaystyle\int_0^{\sqrt{(\beta - \alpha)t}} e^{-\zeta^2}\, d\zeta$	$\dfrac{1}{(s + \alpha)\sqrt{s + \beta}}$
	$\dfrac{e^{-\alpha t}}{\sqrt{\pi t}} + 2\sqrt{\dfrac{\alpha}{\pi}} \displaystyle\int_0^{\sqrt{\alpha t}} e^{-\zeta^2}\, d\zeta$	$\dfrac{\sqrt{s + \alpha}}{s}$
Bessel function of zero order	$I_0(\alpha t)$	$\dfrac{1}{\sqrt{s^2 + \alpha^2}}$
Modified Bessel function of zero order	$J_0(\alpha t)$	$\dfrac{1}{\sqrt{s^2 - \alpha^2}}$

B.4 Some Basic Properties of the Laplace Transform

For practical applications it is useful to be familiar with some basic properties of the Laplace transform. The most important ones are discussed below.

B.4.1 Addition theorem

The Laplace transform as defined by Eq. (B.7) is linear with respect to $f(t)$. If two signals $f_1(t)$ and $f_2(t)$ are given, the Laplace transform of the sum of $f_1 + f_2$ is therefore equal to the sum of the individual Laplace transforms $F_1(s)$ and $F_2(s)$;

$$\mathcal{L}\{f_1(t) + f_2(t)\} = \mathcal{L}\{f_1(t)\} + \mathcal{L}\{f_2(t)\} = F_1(s) + F_2(s) \qquad \text{(B.11)}$$

B.4.2 Multiplication by a constant factor k

For the same reason we have

$$\mathcal{L}\{k(f(t))\} = k\mathcal{L}\{f(t)\} = kF(s) \qquad \text{(B.12)}$$

If the signal $f(t)$ is multiplied by a constant factor k, the Laplace transform $F(s)$ is simply multiplied by the same factor.

B.4.3 Multiplication of signals

If two signals $f_1(t)$ and $f_2(t)$ are multiplied together in the time domain, the Laplace transform of $f_1(t) \cdot f_2(t)$ is *not* given by multiplying the individual Laplace transform $F_1(s) = \mathcal{L}\{f_1(t)\}$ and $F_2(s) = \mathcal{L}\{f_2(t)\}$. Thus

$$\mathcal{L}\{f_1(t) \cdot f_2(t)\} \neq F_1(s) \cdot F_2(s)$$

The operation in the complex frequency domain which corresponds to the multiplication in the time domain is much more complicated; it is called *complex convolution*.[3,13,14,37] We will not consider this operation in detail here, but will define it for completeness:

$$\mathcal{L}\{f_1(t) \cdot f_2(t)\} = \frac{1}{2\pi j} \oint F_1(\chi) \cdot F_2(s - \chi)\, d\chi \qquad (\text{B.13})$$

where

$$F_1(s) = \mathcal{L}\{f_1(t)\}$$

$$F_2(s) = \mathcal{L}\{f_2(t)\}$$

and χ is an auxiliary complex variable. The integral on the right-hand side of Eq. (B.13) is called a *complex convolution integral*. The contour of integration is a closed loop and must be chosen on the basis of mathematical considerations[3,37] that will not be discussed here. For the complex convolution integral the simplified form

$$\frac{1}{2\pi j} \oint F_1(\chi) \cdot F_2(s - \chi)\, d\chi = F_1(s) * F_2(s)$$

is often used.

Now we determine which operation in the time domain corresponds to a multiplication in the complex frequency domain. The result is given without a mathematical derivation:

$$\mathcal{L}^{-1}\{F_1(s) \cdot F_2(s)\} = \int_0^t f_1(\tau) f_2(t - \tau)\, dt \qquad (\text{B.14})$$

The integral on the right-hand side of Eq. (B.14) is called *convolution* and is often written as

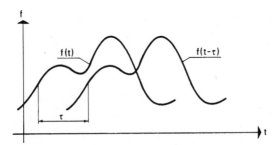

Figure B.8 Plot of a function $f(t)$ displaced by a time delay τ.

$$\int_0^t f_1(\tau) f_2(t - \tau)\, dt = f_1(t) * f_2(t)$$

B.4.4 Delay in the time domain

A signal $f(t)$ and its Laplace transform $F(s)$ are given. The signal is now delayed by the time interval τ (Fig. B.8). What is the Laplace transform of the delayed signal $f(t - \tau)$? The derivation[3,37] yields

$$\mathcal{L}\{f(t - \tau)\} = F(s)e^{-s\tau} \qquad (\text{B.15})$$

We shall calculate the Laplace transform of a rectangular pulse (Fig. B.9) as a numerical example. The pulse is first decomposed into two unit step functions, one starting at $t = 0$ and the other being delayed by the time interval τ. The Laplace transform of the first unit function is given by $1/s$, that of the second by $-(1/s) \cdot e^{-s\tau}$. Hence the Laplace transform of the pulse becomes

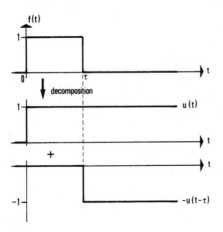

Figure B.9 If the Laplace transform of a single square-wave pulse (top trace) has to be calculated, the pulse is first decomposed into two unit step functions (center and bottom traces).

$$F(s) = \frac{1 - e^{-s\tau}}{s}$$

It is interesting to see that the Laplace transform describes a function with one single expression, whereas three separate equations would be required to describe the pulse in the time domain,

$$f(t) = \begin{cases} 0 & t < 0 \\ 1 & 0 \leq t \leq \tau \\ 0 & t > \tau \end{cases}$$

B.4.5 Differentiation and integration in the time domain

A signal $f(t)$ and its Laplace transform $F(s)$ are given. In many cases we are interested to know the Laplace transform of the derivative df/dt or of the integral $\int_0^t f(t)\, dt$. The Laplace transform of the derivative is[3,37]

$$\mathcal{L}\left\{\frac{df(t)}{dt}\right\} = sF(s) - f(0) \tag{B.16}$$

where $f(0)$ is the value of the signal $f(t)$ at $t = 0$. A differentiation in the time domain corresponds to a multiplication by s in the complex frequency domain. The second term in Eq. (B.16) can be dropped if $f(0)$ is zero.

A similar rule applies to integration. The Laplace transform of the integral of a function $f(t)$ is[3]

$$\mathcal{L}\left\{\int_0^t f(t)\, dt\right\} = \frac{F(s)}{s} - \frac{f^{(-1)}(0)}{s} \tag{B.17}$$

The term $f^{(-1)}(0)$ is by definition the integral of $f(t)$ over the time interval between $-\infty$ and 0:

$$f^{(-1)}(0) = \int_{-\infty}^0 f(t)\, dt$$

$f^{(-1)}$ is equal to the shaded area in Fig. B.10. If $f(t)$ is zero for $t < 0$, the second term in Eq. (B.17) can be dropped. Integration in the time domain corresponds to a division by s in the complex frequency domain.

Let us now apply the rule of integration to an example (Fig. B.11). This figure shows a unit step function $u(t)$ and its first and second integrals. The first integral is a ramp function given by $f(t) = t$ in the time domain, and the second integral is a parabola given by $f(t) =$

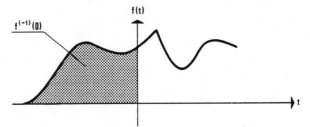

Figure B.10 The term $f^{(-1)}(0)$ is given by the shaded area under the curve $f(t)$.

$t^2/2$. Because $f(t) = 0$ for $t < 0$ in the case of the unit step function, $f^{(-1)}$ is zero. The Laplace transform of the unit step has been shown to be $F(s) = 1/s$ (see Sec. B.3 or Table B.1). The Laplace transform of the ramp function (second row in Fig. B.11) is therefore obtained by a division by s,

$$F(s) = \frac{1}{s^2}$$

Finally the Laplace transform of the parabola (third row in Fig. B.11) is given by

$$F(s) = \frac{1}{s^3}$$

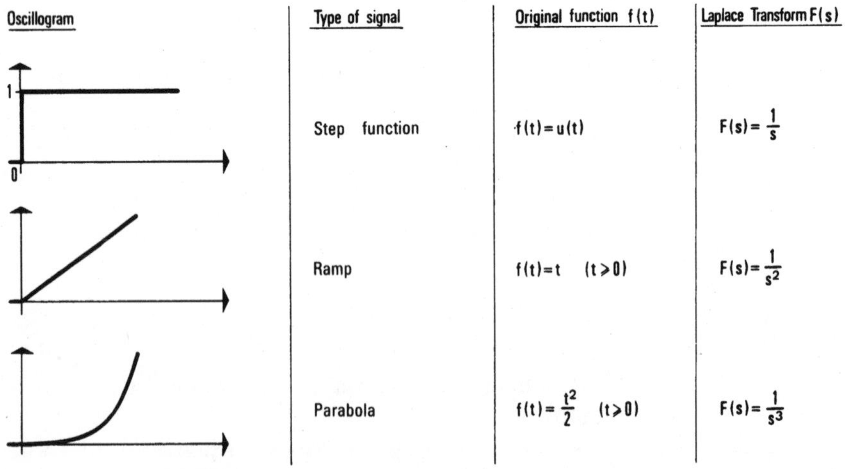

Oscillogram	Type of signal	Original function f(t)	Laplace Transform F(s)
	Step function	$f(t) = u(t)$	$F(s) = \frac{1}{s}$
	Ramp	$f(t) = t \quad (t > 0)$	$F(s) = \frac{1}{s^2}$
	Parabola	$f(t) = \frac{t^2}{2} \quad (t > 0)$	$F(s) = \frac{1}{s^3}$

Figure B.11 Example of applying the rule of integration. The Laplace transforms of a unit step, a ramp, and a parabola are determined.

Let us now work out an example of the differentiation rule [refer to Eq. (B.16)]. The unit step function (see, for example, Fig. B.7) has been used throughout this text. Its Laplace transform has been shown to be $F(s) = 1/s$. But what is the first derivative of the step function, and what is its Laplace transform?

The first derivative of the unit step is called the *delta function* $\delta(t)$ (or impulse function). For $t < 0$ the unit step function is 0. Hence its derivative is also 0 in this range. For $t > 0$ the unit step function is 1. Its derivative must therefore also be 0. At $t = 0$ the unit step function shows a transient from 0 to 1. Consequently its derivative, the delta function, must be infinite at $t = 0$. The delta function $\delta(t)$ is shown in Fig. B.12.

The amplitude of the delta function is ∞; the pulse width is 0. The area under the delta function is obtained by multiplying the amplitude by the pulse width. The result must be unity because the area of $\delta(t)$ must be equal to the amplitude of the unit step function. An impulse of infinite amplitude and zero pulse width is hard to imagine, of course. But there is another way to understand better what a delta function really is (Fig. B.13).

The first row of this figure shows a flattened unit step function. Its amplitude is still 1, but it takes a time of one unit (such as 1 second) to reach this amplitude. Consequently the derivative of this function is an impulse having an amplitude of 1 and a pulse width of 1. The area under the pulse is 1. Assume now that our unit step becomes a little steeper (second row in Fig. B.13); therefore it rises from 0 to 1 within 0.5 unit of time. The derivative of this function is an impulse having an amplitude of 2 and a pulse width of 0.5. The area under the pulse is still 1, of course. If the unit step becomes even steeper, it may rise from 0 to 1 in as little as 0.1 unit of time. Its derivative then will be a pulse having an amplitude of 10 and a pulse width of 0.1. Of course the area of the pulse still is 1. This process can be continued as long as desired, until we finally end up with a pulse of infinite amplitude, a duration of zero, and the area under the peak is still 1.

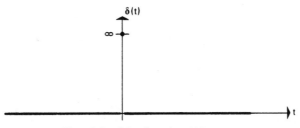

Figure B.12 Plot of the delta function $\delta(t)$.

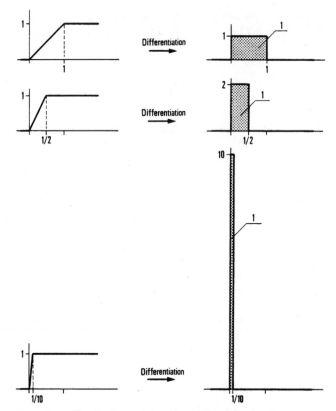

Figure B.13 Practical approximation of the delta function.

What is the Laplace transform of the delta function now? The answer is very simple, because we only have to multiply the Laplace transform of the unit step function by s. The Laplace transform of the unit step has been shown to be

$$F(s) = \frac{1}{s} \qquad \text{unit step}$$

Hence the Laplace transform of the delta function is

$$F(s) = 1$$

B.4.6   The initial- and final-value theorems

In many cases where the Laplace transform $F(s)$ of a signal $f(t)$ is given, we are interested in knowing only the *initial value* $f(0)$ or the *final value* $f(\infty)$ of $f(t)$. The initial and final values $f(0)$ and $f(\infty)$, respectively, can be obtained immediately from the Laplace transform $F(s)$ without performing the inverse Laplace transform. The initial- and final-value theorems are given here without proof.[3] The initial- value theorem reads:

$$f(0) = \lim_{s \to \infty} sF(s) \tag{B.18}$$

The final-value theorem reads:

$$f(\infty) = \lim_{s \to 0} sF(s) \tag{B.19}$$

A numerical example is given to illustrate these theorems. The Laplace transform of a (hitherto) unknown signal $f(t)$ is given by

$$F(s) = \frac{1}{s(1 + sT)}$$

where T is a time constant. What are the values of $f(0)$ and $f(\infty)$? Using the initial-value theorem [Eq. (B.18)], we get

$$f(0) = \lim_{s \to \infty} sF(s) = \lim_{s \to \infty} \frac{\cancel{s}}{\cancel{s}(1 + sT)} = 0$$

On the other hand, the final value, according to Eq. (B.19), is

$$f(\infty) = \lim_{s \to 0} sF(s) = \lim_{s \to 0} \frac{\cancel{s}}{(\cancel{s}(1 + sT)} = 1$$

B.5   Using the Table of Laplace Transforms

Table B.1 lists the Laplace transforms of the most commonly used signals $f(t)$. The table also includes some of the most important the- orems of the Laplace transform. When using the table we should be aware that the Laplace transform $F(s)$ was obtained by integrating the Laplace integral of Eq. (B.7) over the time interval $0 \le t \le \infty$. The values of the signal $f(t)$ at negative t therefore did not contribute to $F(s)$. It is equivalent to state that $f(t)$ is effectively 0 for negative t.

This fact has an effect on the inverse Laplace transform. Performing the inverse Laplace transform for a given function $F(s)$ yields signal values $f(t)$ for positive t only. If the Laplace transform of a signal $f(t)$ is given by, say,

$$F(s) = \frac{1}{s + a}$$

the table gives the corresponding signal $f(t)$ as

$$f(t) = e^{-at}$$

This holds true for positive t only. For negative t, $f(t) = 0$ by definition.

B.6 Applying the Laplace Transform to Electric Networks

The Laplace transform is the most effective tool for analyzing the transient response of electric networks. All linear electric devices, from electric motors to operational amplifiers, are modeled by a configuration of passive elements (resistor R, inductor L, and capacitor C) and active elements (voltage and current sources, controlled voltage and current sources). The transient response of such electric networks is analyzed in the time domain by writing the differential equations for the branch currents and voltages. Voltages and currents in the R, L, and C elements are related by Ohm's law as follows: For resistors,

$$u(t) = Ri(t) \tag{B.20a}$$

for inductors,

$$u(t) = L\frac{di}{dt} \tag{B.20b}$$

and for capacitors,

$$u(t) = \frac{1}{C}\int i\,dt \tag{B.20c}$$

When analyzing a network in the complex frequency domain [using the rules of differentiation, Eq. (B.16), and of integration, Eq. (B.17)], we obtain, for resistors,

$$U(s) = RI(s) \tag{B.21a}$$

for inductors,

$$U(s) = L[sI(s) + i(0)] \tag{B.21b}$$

and for capacitors,

$$U(s) = \frac{1}{C}\left[\frac{I(s)}{s} + \frac{i^{(-1)}(0)}{s}\right] \tag{B.21c}$$

If the initial current $i(0)$ in an inductor L is zero, the second term in Eq. (B.21b) is also zero. In this case we have

$$U(s) = sLI(s)$$

and the quotient

$$\frac{U(s)}{I(s)} = sL \tag{B.22a}$$

can be defined as the *impedance* of the inductor. If the Laplace transform is replaced by the Fourier transform, s is replaced by $j\omega$, and this expression becomes

$$\frac{U(j\omega)}{I(j\omega)} = j\omega L$$

which is the familiar ac impedance of the inductor known from the theory of alternating currents.

If the initial charge $i^{(-1)}/C$ in a capacitor C is zero, the second term in Eq. (B.21c) is also zero. In this case we have

$$U(s) = \frac{1}{sC}\, I(s)$$

and the quotient

$$\frac{U(s)}{I(s)} = \frac{1}{sC} \tag{B.22b}$$

is defined as the *impedance* of the capacitor. If the Laplace transform is again replaced by the Fourier transform, s is replaced by $j\omega$, and Eq. (B.22b) becomes

$$\frac{U(j\omega)}{I(j\omega)} = \frac{1}{j\omega C}$$

which is the familiar ac impedance of the capacitor known from the theory of alternating currents.

Let us now apply the Laplace transform to the analysis of a simple electric network.

Numerical Example: Transient Response of the Passive RC "Differentiator" We want to find the transient response $u_2(t)$ of the differentiator in Fig. B.14a on a single square-wave pulse $u_1(t)$. First we introduce the variables in the time domain on the right-hand side of Fig. B.14a. These variables are transformed into the complex frequency domain in Fig. B.14b. It is assumed that there is no initial charge on the capacitor.

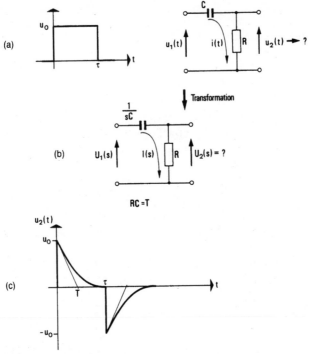

Figure B.14 Calculating the transient response of the passive RC differentiator using the Laplace transform. (a) Defining the variables in the time domain. (b) Introducing variables in the complex-frequency domain. (c) Plot of the output signal u_2 against time.

Solution We can now write the node and mesh equations of the network directly in the complex frequency domain. Making use of Eqs. (B.21a) and (B.22b), we have

$$U_1(s) = RI(s) + \frac{1}{sC} I(s)$$

$$U_2(s) = RI(s)$$

After the elimination of $I(s)$ we obtain

$$U_2(s) = \frac{sRC}{1 + sRC} U_1(s)$$

This can be written as

$$U_2(s) = F(s)U_1(s) \qquad\qquad (B.23)$$

where $F(s)$ is the transfer function of the network. If $F(s)$ is known, the transient response of the network in any input signal $u_1(t)$ may be calculated.

In our example, $u_1(t)$ is a single square-wave pulse whose Laplace transform was previously determined in Sec. B.4.4. For $U_1(s)$ we can therefore write

$$U_1(s) = u_0 \frac{1 - e^{-s\tau}}{s}$$

where τ is the duration of the pulse and u_0 is its amplitude. Then $U_2(s)$ becomes

$$U_2(s) = u_0 \frac{1 - e^{-s\tau}}{s + 1/T} \qquad (B.24a)$$

where $T = RC$.

To transform this expression back into the time domain, we decompose it as follows:

$$U_2(s) = u_0 \frac{1}{s + 1/T} - u_0 \frac{e^{-s\tau}}{s + 1/T} \qquad (B.24b)$$

From the table of Laplace transforms (Table B.1), the signal corresponding to the first term of Eq. (B.24b) is

$$u_{21}(t) = u_0 e^{-t/T} \qquad t \geq 0 \qquad (B.25a)$$

The second term in Eq. (B.24b) corresponds to a decaying exponential function delayed by τ,

$$u_{22}(t) = -u_0 e^{(t-\tau)/T} \qquad (B.25b)$$

Because the signal $u_{21}(t)$ is zero for $t < 0$, the *delayed* signal $u_{22}(t)$ is defined only for $t \geq \tau$, but is zero for $t < \tau$. Therefore, for the combined output signal $u_2(t)$ of the differentiator we obtain

$$u_e(t) = \begin{cases} 0 & t < 0 \\ u_0 e^{-t/T} & 0 \leq t \leq \tau \\ u_0[e^{-t/T} - e^{-(t-\tau)/T}] & t > \tau \end{cases}$$

The waveform of $u_2(t)$ is plotted in Fig. B.14c.

B.7   Closing the Gap between the Time Domain and the Complex-Frequency Domain

As demonstrated in the previous section, the Laplace transforms of input and output signals of any linear electric network are related by the transfer function $F(s)$,

$$U_2(s) = U_1(s)F(s) \qquad\qquad \text{(B.23)}$$

Equation (B.23) enables us to determine the transient response of the network for any input signal $u_1(t)$.

Assume now that the network is excited by a delta function, $u_1(t) = \delta(t)$. The transient response of the network on a delta function $\delta(t)$ will be hereafter denoted by $u_2(t) = h(t)$. As shown in Sec. B.4.5, the Laplace transform of the delta function is

$$\mathscr{L}\{\delta(t)\} = 1$$

Introducing this expression into Eq. (B.23) we obtain

$$U_2(s) = F(s)$$

From this expression we learn that the transfer function $F(s)$ of an electric network is the Laplace transform of the transient response $h(t)$ on a delta function

$$F(s) = \mathscr{L}\{h(t)\} \qquad\qquad \text{(B.26)}$$

A practical example will clarify the correspondence given by Eq. (B.26).

Numerical Example A passive RC low-pass filter (commonly called an RC integrator), as shown in Fig. B.15, is excited by a delta function

$$u_1(t) = \delta(t)$$

What is the transient response $u_2(t)$ on this input signal?

Solution Applying Ohm's law to the resistor and capacitor [Eqs. (B.21a) and (B.22b)], we obtain for the transfer function

$$F(s) = \frac{U_2(s)}{U_1(s)} = \frac{1}{1 + sT}$$

where $T = RC$. According to Eq. (B.26), the transient response of $u_2(t)$ of the integrator on a delta function applied to its input is

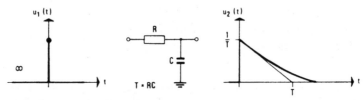

Figure B.15 Calculating the transient response of the passive RC integrator using the Laplace transform.

$$u_2(t) = \mathcal{L}^{-1}\{F(s)\} = \mathcal{L}^{-1}\left\{\frac{1}{1+sT}\right\}$$

Using the table of Laplace transforms (Table B.1) we obtain

$$u_2(t) = h(t) = \frac{1}{T}e^{-t/T}$$

This is the impulse response plotted on the right-hand side of Fig. B.15.

Some readers will have observed that the impulse response given by Eq. (B.15) has the dimension of s^{-1} instead of the expected V. This stems from the definition of the delta function which is the time derivative of a unit step and has the dimension s^{-1} in fact. To get the correct physical unit, we would have to multiply the delta function by a factor $U_0 \cdot T_0$, where $U_0 = 1$ V and $T_0 = 1$ s. For simplicity, this factor is omitted here and in following calculations.

B.8 Networks with Nonzero Stored Energy at $t = 0$

We now apply the Laplace transform to electric networks having either an inductor in which a nonzero current $i(0)$ flows at $t = 0$ or a capacitor on which there is a nonzero voltage $u(0)$ at $t = 0$. In both cases nonzero initial energy is stored in the network at $t = 0$. Let us again calculate the transient response of the RC integrator in Fig. B.15 on a delta function, assuming now that the initial voltage across the capacitor is $u(0) \neq 0$.

Two equations in the time domain can be written for the RC integrator:

$$u_1(t) = i(t)R + u_2(t)$$

$$u_2(t) = \frac{1}{C}\int i\, dt$$

Applying the Laplace transform to these equations [refer to Eqs. (B.21)], we obtain

$$U_1(s) = I(s)R + U_2(s)$$

$$U_2(s) = \frac{1}{C}\left[\frac{I(s)}{s} + \frac{i^{(-1)}(0)}{s}\right]$$

We can eliminate $I(s)$ from the first of these equations. We then get

$$U_2(s) = \frac{1}{C}\left[\frac{U_1(s) - U_2(s)}{Rs} + \frac{i^{(-1)}(0)}{s}\right]$$

After some manipulation, we have

$$U_2(s) = U_1(s)\frac{1}{1 + sT} + \frac{i^{(-1)}(0)}{C}\frac{T}{1 + sT}$$

where $T = RC$ and $i^{(-1)}(0)$ is the integral of current which flowed into the capacitor in the time interval $-\infty < t < 0$, that is, $i^{(-1)}(0)$ is the initial *charge* stored in the capacitor at $t = 0$. Hence $i^{(-1)}(0)/C$ is simply the initial voltage $u(0)$ across the capacitor. Consequently we obtain

$$U_2(s) = U_1(s)\frac{1}{1 + sT} + u(0)\frac{T}{1 + sT}$$

Because $u_1(t)$ has been assumed to be a delta function, $U_1(s) = 1$, and we have

$$U_2(s) = \frac{1}{1 + sT} + u(0)\frac{T}{1 + sT}$$

Transforming this equation back into the time domain (see Table B.1), we obtain for $u_2(t)$

$$u_2(t) = \frac{1}{T}e^{-t/T} + u(0)e^{-t/T}$$

Note that the first term is identical with the response of the RC integrator obtained for zero initial voltage across the capacitor. The second term is due to the initial charge stored in the capacitor.

B.9   Analyzing Dynamic Performance by the Pole-Zero Plot

The transfer function of any linear network built from lumped elements such as resistors, inductors, capacitors, and amplifiers is given by a polynomial fraction in s,

$$F(s) = \frac{a_n s^n + a_{n-1}s^{n-1} + \cdots + a_1 s + a_0}{b_m s^m + b_{m-1}s^{m-1} + \cdots + b_1 s + b_2} \qquad m \geq n \quad (B.27a)$$

This transfer function can also be written in the factored form:

$$F(s) = \frac{(s - \alpha_1)(s - \alpha_2)\cdots(s - \alpha_n)}{(s - \beta_1)(s - \beta_2)\cdots(s - \beta_m)} \qquad (B.27b)$$

Here the α_i values are the zeros and the β_i values the poles of the transfer function.

The α_i and β_i can be either real or complex. For example, a real value of β_i corresponds to a pole located on the real axis (σ axis) in the complex s plane (refer to Fig. B.16). If a complex pole is located at $\beta_i = A + jB$, another conjugate complex pole will exist at $\beta_i^* = A - jB$, where the asterisk denotes the conjugate value of β_i. Hence complex poles always exist as pairs of conjugate complex poles.

We will now see that the transient response of a network is very easily found if the locations of the poles and zeros of the network transfer function $F(s)$ are known. To transform Eq. (B.27b) back into the time domain, it is most convenient to decompose this expression into partial fractions;

$$F(s) = \underbrace{\frac{R_1}{s - \beta_1} + \frac{R_2}{s - \beta_2} + \cdots}_{\substack{\text{partial fractions generated} \\ \text{by single real poles}}}$$

$$+ \underbrace{\frac{R_1}{s - (A_i + jB_i)} + \frac{R_i^*}{s - (A_i - jB_i)} + \cdots}_{\substack{\text{2 partial fractions generated by} \\ \text{one pair of conjugate complex poles}}}$$

(B.27c)

The terms $R_1, R_2, \ldots$ are constants and are called *residues*.[3,37] In Eq. (B.27c) we separated two groups of partial fractions, those emanating from the single real poles, and those emanating from the pairs of conjugate complex poles.

Let us first look at the transient response $f(t)$ due to the real poles of $F(s)$. The inverse Laplace transform of the term

$$F(s) = \frac{R_i}{s - \alpha_i}$$

is

$$f(t) = R_i \exp{(\alpha_i t)}$$

Hence the contribution of the single real poles to the signal $f(t)$ is

$$f(t)_{\text{single poles}} = R_1 \exp{(\alpha_1 t)} + R_2 \exp{(\alpha_2 t)} + \cdots \qquad \text{(B.28)}$$

Note that the residues of the partial fractions due to single real poles are always real numbers. [If they are not, the signal $f(t)$ would become complex, which is physically impossible.]

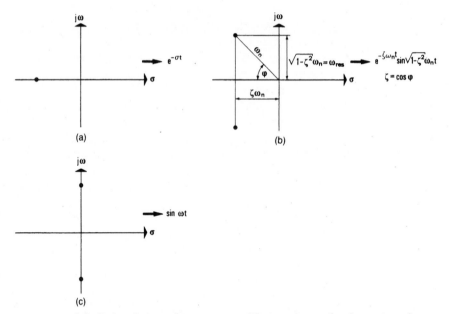

Figure B.16 Calculating the transient response of first- and second-order systems from the pole positions. (*a*) First-order system, pole on the negative σ axis. (*b*) Second-order system, complex-conjugate pole pair in the negative half-plane (damped oscillation). (*c*) Second-order system, poles on the imaginary axis (undamped oscillation).

The next portion of the transient response $f(t)$ is contributed by the complex pole pairs. For simplicity we isolate one complex pole pair:

$$F(s)_{\text{pole pair}} = \frac{R_i}{s - (A_i + jB_i)} + \frac{R_i^*}{s - (A_i - jB_i)} \qquad \text{(B.29a)}$$

Here the residues R_i and R_i^* must not necessarily be real, but can be complex. Moreover, if R_i is a complex number, R_i^* is its conjugate value; i.e., in the most general case we have

$$R_i = a + jb$$

$$R_i^* = a - jb$$

where a and b are real constants.

This is easily proved by combining the two terms in Eq. (B.29a) over a common denominator.

$$F(s)_{\text{pole pair}} = \frac{s(R_i + R_i^*) - A(R_i + R_i^*) + jB(R_i - R_i^*)}{s^2 - 2As + (A^2 + B^2)} \qquad \text{(B.29b)}$$

All individual coefficients in the numerator of Eq. (B.29b) must be real, that is,

$$R_i + R_i^* \longrightarrow \text{real}$$

$$j(R_i - R_i^*) \longrightarrow \text{real or } R_i - R_i^* \longrightarrow \text{imaginary}$$

These conditions are met only if R_i and R_i^* form a conjugate complex pair of numbers, as stated above. Equation (B.29a) can therefore be rewritten as

$$F(s)_{\text{pole pair}} = \frac{a + jb}{s - (A + jB)} + \frac{a - jb}{s - (A - jB)} \qquad \text{(B.29c)}$$

Referring to Table B.1, we recall that the inverse Laplace transform of $1/(s - \alpha)$ is $f(t) = e^{\alpha t}$. This also applies for complex α. Applying this to Eq. (B.29c), we obtain

$$f(t)_{\text{pole pair}} = a[e^{At}e^{jBt} + e^{At}e^{-jBt}] + jb[e^{At}e^{jbt} - e^{At}e^{-jBt}] \qquad \text{(B.29d)}$$

Using the well-known Euler theorem,

$$\sin x = \frac{e^{jx} - e^{-jx}}{2j}$$

$$\cos x = \frac{e^{jx} + e^{-jx}}{2}$$

we can write

$$f(t)_{\text{pole pair}} = 2ae^{At} \cos Bt - 2be^{At} \sin Bt \qquad \text{(B.29e)}$$

This can be brought into the more general form

$$f(t)_{\text{pole pair}} = Ce^{At} \cos (Bt + \phi) \qquad \text{(B.29f)}$$

where
$$C = 2\sqrt{a^2 + b^2}$$

$$\phi = -\tan^{-1} \frac{b}{a}$$

The results are summarized in Table B.2.

These results allow a simple interpretation of the term complex frequency, introduced earlier, which will be discussed in the next section.

TABLE B.2 Laplace Transforms of Generalized First- and Second-Order Systems

$F(s)$	$f(t)$
$\dfrac{R_i}{s - \alpha_i}$	$\longrightarrow R_i e^{\alpha_i t}$
$\dfrac{R_i}{s - (A + jB)} + \dfrac{R_i^*}{2 - (A - jB)}$	$\longrightarrow Ce^{At} \cos (Bt + \phi)$

B.10 A Simple Physical Interpretation of "Complex Frequency"

Refer again to the pole-zero plot in Fig. B.16. A single pole located on the negative σ axis ($\alpha < 0$) gives rise to the decaying exponential function (see Table B.2)

$$f(t) \sim e^{\alpha t} \qquad \alpha < 0$$

This is an exponential function having a real exponent; hence α is called a *real frequency*.

A complex pole pair located on the imaginary axis [$A = 0$ in Eq. (B.29f)] gives rise to an undamped oscillation:

$$f(t) \sim \cos Bt$$

This is an exponential function having a purely imaginary exponent; hence B is called an *imaginary frequency*.

A pole pair located in the left half of the s plane, with $A < 0$ and $B \neq 0$, gives rise to a damped oscillation:

$$f(t) = e^{At} \cos (Bt + \phi)$$

This is an exponential function having a complex exponent; hence we speak of an oscillation with a *complex frequency*.

It has proven useful to write partial fractions generated by complex conjugate pole pairs in the so-called normalized form. When doing so, we introduce the substitution [refer to Eq. (B.29c)]:

$$F(s)_{\text{pole pair}} = \frac{2a + jb}{s - (A + jB)} + \frac{a - jb}{s - (A - jB)}$$

$$\longrightarrow \frac{a + jb}{s - (-\zeta\omega_n + j\sqrt{1 - \zeta^2}\omega_n)}$$

$$+ \frac{a - jb}{s - (-\zeta\omega_n - j\sqrt{1 - \zeta^2}\omega_n)} \qquad (B.29g)$$

Comparing the coefficients on both sides of the arrow, we obtain the equalities

$$A \equiv -\zeta\omega_n$$

$$B \equiv \sqrt{1 - \zeta^2}\,\omega_n \qquad\qquad\text{(B.30)}$$

where ζ and ω_n are the damping factor and the natural frequency, respectively.

If a transfer function $F(s)$ has a conjugate complex pole pair located at $A \pm jB$ in the complex s plane, this pole pair will generate a transient response $f(t)$ of the form

$$f(t) = \exp\left(-\zeta\omega_n t\right)\cos\left(\sqrt{1 - \zeta^2}\,\omega_n t + \phi\right) \qquad\qquad\text{(B.31)}$$

The time constant of the decaying exponential function in Eq. (B.31) is given by $1/(\zeta\omega_n)$; the frequency of the damped oscillation is given by $\omega_\text{res} = \sqrt{1 - \zeta^2}\,\omega_n$. If the complex pole pair is plotted in the complex s plane, the distance of the poles from the origin is seen to be exactly ω_n, as shown in Fig. B.16b.

Digital Filter Basics ·

In the age of all-digital PLLs and software PLLs, increasing use of digital filters is made. This appendix is a short overview on digital filter design.

C.1 The Transfer Function $H(z)$ of Digital Filters

Analog filters are mostly described by their frequency response $H(j\omega)$ or by their transfer function $H(s)$. $H(s)$ is defined to be

$$H(s) = \frac{O(s)}{I(s)} \qquad (C.1)$$

where $O(s)$ is the Laplace transform of the output signal $o(t)$ and $I(s)$ is the Laplace transform of the input signal $i(t)$. If the input signal is a delta function

$$i(t) = \delta(t)$$

the response of the filter is called *impulse response* $h(t)$. Because $I(s) = 1$ in this case (refer also to Appendix B), we have

$$O(s) = I(s)H(s) = H(s) \qquad (C.2)$$

i.e., the transfer function $H(s)$ of the analog filter is the Laplace transform of the impulse response $h(t)$. This simple property is extensively used to build digital filters. Often digital filters are designed to have an impulse response similar to that of an analog filter. This is explained by Fig. C.1. Figure C.1a shows the impulse response $h(t)$ of an analog filter; a low-pass filter has been chosen in this example. Of

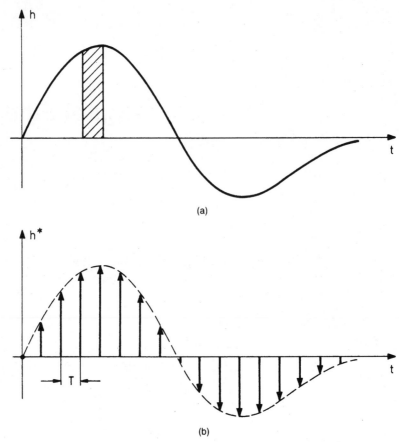

(a)

(b)

Figure C.1 Impulse responses of analog and digital filters. (a) Impulse response $h(t)$ of an analog filter. (b) Impulse response $h*(t)$ of a digital filter. $h*(t)$ is a sampled version of $h(t)$.

course, $h(t)$ is a continuous function of time. For most analog filters, the impulse response $h(t)$ is a damped oscillation or a decaying exponential function; in any case the duration of the impulse response is infinite, because $h(t)$ asymptotically approaches zero (or a constant value) for $t \rightarrow \infty$. With a digital filter, both input and output signals are sampled signals. Its impulse response $h*(t)$ must therefore be a sampled signal, too. It shows up that the transfer function $H*(s)$ of a digital filter comes close to the transfer function $H(s)$ of a corresponding analog filter (with some distinctions, as will be shown later), if the impulse response $h*(t)$ of the digital filter is a sampled version of the impulse response $h(t)$ of the analog filter; this is illustrated by Fig. C.1b. We will see later that the sampling frequency $f_s = 1/T$ cannot be arbitrarily chosen but must satisfy the so-called *sampling theorem*.

The sampled impulse response shown in Fig. C.1*b* has infinite duration as well. (It will be discussed later that the duration of the impulse response can be made finite by truncation.) For the sampled impulse response in Fig. C.1*b* we can write

$$h^*(t) = T \sum_{n=0}^{\infty} h(nT)\delta(t - nT) \tag{C.3}$$

where $h(nT)$ is the amplitude of the nth sample of $h^*(t)$. We observe that each sample of the impulse response has been multiplied by the sampling interval T in Eq. (C.3). This is necessary because the "area" under the delta function at sampling instant $t = nT$ should be identical with the area under the continuous impulse response (shaded area in Fig. C.1*a*) in the interval $nT \le t < (n + 1)T$. If the factor T were omitted in Eq. (C.3), the dimension of the impulse response $h^*(t)$ would be wrong. To simplify the expressions, we drop the sampling interval T in the following formulas; i.e., we write $h(n)$ for $h(nT)$, etc. Because the Laplace transform of a time-shifted delta function $\delta(t - \tau)$ is given by $e^{-s\tau}$, the transfer function $H^*(s)$ of the digital filter becomes

$$H^*(s) = T \sum_{n=0}^{\infty} h(n)e^{-snT} \tag{C.4}$$

We note that the complex frequency s appears only in the form of e^{-snT}. To simplify the notation, we introduce the substitution

$$z = e^{sT} \tag{C.5}$$

This is the definition of the *z-transform*. Using Eq. (C.5), the transfer function of the digital filter can be rewritten as

$$H(z) = H^*(s) \big|_{z=e^{sT}} = T \sum_{n=0}^{\infty} h(n)z^{-n} \tag{C.6}$$

The star symbol in $H(z)$ is dropped for simplicity, and $H(z)$ is called *z-transfer function* of the digital filter. We observe that the z-transform is nothing more than an alternative notation of the Laplace transform. By the z-transform in Eq. (C.5), the complex s-plane is projected onto the complex z-plane. When working with the z-transform, we are performing operations in the *z-domain*. The variable z is called *z-operator*. z has a very simple physical interpretation: since $z^{-1} = e^{-sT}$, multiplication with z^{-1} in the z-domain corresponds to a time shift by one sampling interval in the time domain. Assume that $x(t)$ is a sampled signal which only exists at the sampling instants $t = 0, T, 2T, \ldots, nT$, and that $X(z)$ is the corresponding z-transform. Then $z^{-1}X(z)$ is the z-transform of the time-shifted signal $x(t - T)$, $z^{-2}X(z)$ is the z-transform of the time-shifted signal $x(t - 2T)$, etc.

As we mentioned above, the impulse response of the most general digital filter is an infinite series of delta functions. Digital filters having an infinite impulse response are called *IIR filters* (infinite impulse response filters). There is another class of digital filters where the impulse response is truncated when the amplitudes of the delta functions have fallen below a given threshold. This class of filters is called *FIR filters* (finite impulse response filters). Though FIR filters look simpler at a first glance, the corresponding theory is more complex, so we start the discussion with the IIR filter.

C.2   IIR Filters

Most IIR filters are designed to perform like a known analog filter. Things are complicated by the fact that there are different approaches to transforming an analog filter into a digital. Only two of the various design procedures have survived, however: (1) the impulse-invariant z-transform and (2) the bilinear z-transform. The second method is by far more important, but it is easier to start with the first.

C.2.1   The impulse-invariant z-transform

We suppose that an IIR filter is to be built whose impulse response is a sampled version of the impulse response of a known analog filter, as shown in Fig. C.1. The transfer function of the analog filter is assumed to be $H(s)$. The impulse response $h^*(t)$ of the IIR filter is therefore given by Eq. (C.3). $h^*(t)$ is obtained by multiplying the impulse response $h(t)$ of the analog filter by the sampling function $w(t)$, which is an infinite series of delta functions, all having the amplitude 1. This is illustrated by Fig. C.2. Figure C.2a shows the continuous function $h(t)$ again, Fig. C.2b the sampling function $w(t)$, and Fig. C.2c the result of the multiplication. Obviously, the transfer function $H^*(s)$ of the IIR filter is obtained by transforming the impulse response $h^*(t) = h(t)w(t)T$ into the s-domain. As we see from Appendix B (Sec. B.4.3), a multiplication of two signals corresponds to the complex convolution of their Laplace transforms in the s-domain, i.e., we have

$$H^*(s) = \frac{T}{2\pi j} H(s)*W(s) = \frac{T}{2\pi j} \oint_c H(\chi)W(s - \chi) \, d\chi \qquad (C.7)$$

In this integral, χ is an auxiliary variable for complex frequency and C is the contour along which the integration has to be performed. First

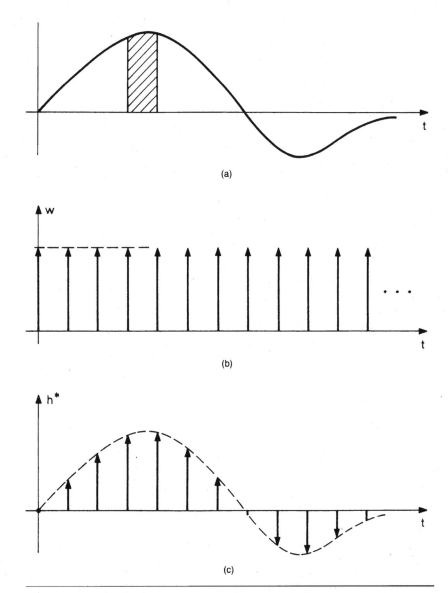

Figure C.2 The impulse response of a digital filter can be obtained by multiplying the continuous impulse response $h(t)$ of an analog filter with a sampling function $w(t)$. (a) Continuous impulse response $h(t)$ of an analog filter. (b) Sampling function $w(t)$. (c) Result of the multiplication $h^*(t) = h(t)w(t)\, T$.

we must know the Laplace transform $W(s)$ of the sampling function $w(t)$. Applying the Laplace transform to $w(t)$ yields

$$W(s) = 1 + e^{-sT} + e^{-2sT} + \cdots \qquad (C.8a)$$

This is clearly a geometric series, and using the formula for the sum of geometric series leads to

$$W(s) = \frac{1}{1 - e^{-sT}} \qquad (C.8b)$$

When Eq. (C.8b) is inserted into Eq. (C.7), $H^*(s)$ becomes

$$H^*(s) = \frac{T}{2\pi j} \oint_c \frac{H(\chi)}{1 - e^{-(s-\chi)T}} \, d\chi \qquad (C.9)$$

To determine the contour C for integration, we plot the poles of $H(\chi)$ and $W(s - \chi)$ in the χ-plane (Fig. C.3). The locations of the poles of $H(\chi)$ in the χ-plane are identical with the locations of the poles of $H(s)$ in the s-plane. For a stable system, they must be in the left half of the χ-plane. The poles of $W(s - \chi)$ can be shown to be at locations $\chi_k = s + kj\omega_s$, where $\omega_s = 2\pi/T$ is the angular sampling frequency and k is an integer in the range $-\infty < k < \infty$.[37] s is an arbitrary displacement here; in the example of Fig. C.3 s is assumed to be real and positive. The contour C of integration must be chosen such that the integrand in Eq. (C.9) is *analytic* everywhere. According to function theory this is true when the contour is a closed curve as shown by Fig. C.3. The vertical branch of C must be located right from the poles of $H(\chi)$ and left from the poles of $W(s - \chi)$. We note that only the poles of $H(\chi)$ are enclosed by the integration path C. The integral can be solved by applying Cauchy's residue theorem to Eq. (C.9),[37] and we get

$$H^*(s) = T \sum_{\substack{\text{Poles} \\ \text{of } H(\chi)}} \frac{\text{Res } H(\chi)}{1 - e^{-(s-\chi)T}} \qquad (C.10a)$$

In Eq. (C.10a), Res $H(\chi)$ denotes the residue of the function $H(\chi)$ at $\chi = \chi_0$, where χ_0 is the location of a pole of $H(\chi)$. Since $H^*(s)$ is the transfer function of a sampled system, the s operator appears only in the form of e^{-nsT}, so we can again substitute e^{sT} by z and obtain

$$H(z) = H^*(s) \big|_{z=e^{sT}} = T \sum_{\substack{\text{Poles} \\ \text{of } H(\chi)}} \frac{\text{Res } H(\chi)}{1 - z^{-1}e^{\chi T}} \qquad (C.10b)$$

Here again, the star symbol in $H(z)$ has been omitted. Equation

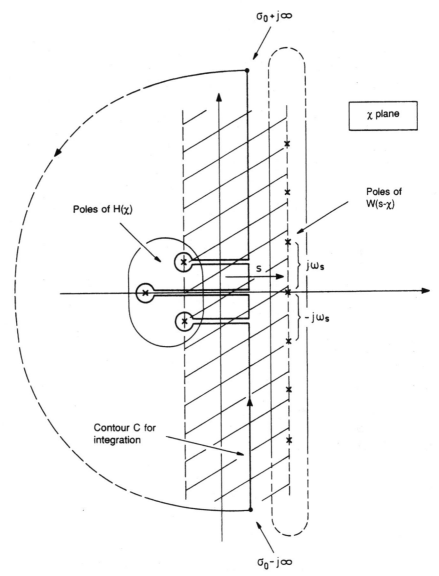

Figure C.3 Locations of the poles of $H(\chi)$ and $W(s - \chi)$ in the χ-plane. The closed curve is the integration path for the convolution integral. For details, refer to text.

(C.10b) is the definition of the impulse-variant z-transform. It enables us to calculate the z-transfer function $H(z)$ of an IIR filter from the transfer function $H(s)$ of a corresponding analog filter. This computation has been done for the most important transfer functions $H(s)$. The result is shown in Table C.1. From the definition of $H(z)$ in Eq. (C.10b)

TABLE C.1 Table of the Impulse-Invariant z-Transform for Filters of Order 0–2

Filter type	$H(s)$	$H(z)$
Zero-order term	1	1
Integrator	$\dfrac{1}{sT_i}$	$\dfrac{T}{T_i}\dfrac{1}{1-z^{-1}}$
First-order lag	$\dfrac{1}{1+s/\omega_0}$	$\omega_0 T\,\dfrac{1}{1-z^{-1}\exp(-\omega_0 T)}$
Second-order lag	$\dfrac{1}{1+2\zeta(s/\omega_0)+(s^2/\omega_0^2)}$	$\dfrac{z^{-1}(\omega_0 T/\sqrt{1-\zeta^2})\exp(-\zeta\omega_0 T)\sin(\omega_0\sqrt{1-\zeta^2}\,T)}{1-z^{-1}\,2\exp(-\zeta\omega_0 T)\cos(\omega_0\sqrt{1-\zeta^2}\,T)+z^{-2}\exp(-2\zeta\omega_0 T)}$
Second-order lag, linear term in numerator	$\dfrac{s/\omega_1}{1+2\zeta(s/\omega_0)+(s^2/\omega_0^2)}$	$\dfrac{\omega_0^2 T}{\omega_1}\dfrac{1-z^{-1}\exp(-\zeta\omega_0 T)[\cos(\omega_0\sqrt{1-\zeta^2}\,T)+(\zeta/\sqrt{1-\zeta^2})\sin(\omega_0\sqrt{1-\zeta^2}\,T)]}{1-z^{-1}\,2\exp(-\zeta\omega_0 T)\cos(\omega_0\sqrt{1-\zeta^2}\,T)+z^{-2}\exp(-2\zeta\omega_0 T)}$
Second-order lag, constant linear term in numerator	$\dfrac{1+s/\omega_1}{1+2\zeta(s/\omega_0)+(s^2/\omega_0^2)}$	$\dfrac{\omega_0^2 T}{\omega_1}\dfrac{1-z^{-1}\exp(-\zeta\omega_0 T)\{\cos(\omega_0\sqrt{1-\zeta^2}\,T)+[(\omega_1/\omega_0-\zeta)/\sqrt{1-\zeta^2}]\sin(\omega_0\sqrt{1-\zeta^2}\,T)\}}{1-z^{-1}\,2\exp(-\zeta\omega_0 T)\cos(\omega_0\sqrt{1-\zeta^2}\,T)+z^{-2}\exp(-2\zeta\omega_0 T)}$

it follows that the order of $H(z)$ is identical with the order of $H(s)$; i.e., the impulse-invariant IIR filter has the same number of poles as the corresponding analog filter.

It follows from Eq. (C.10b) that the z-transfer function can always be represented as the ratio of two polynomials in z, whose order is finite, i.e., we have

$$H(z) = \frac{O(z)}{I(z)} = \frac{b_0 + b_1 z^{-1} + \cdots + b_M z^{-M}}{1 + a_1 z^{-1} + \cdots + a_N z^{-N}} \qquad (C.10c)$$

The a_i and b_i ($i = 0, 1, 2, \ldots$) are called *filter coefficients*. There is no restriction on the size of the orders N and M of the polynomials in Eq. (C.10c). M can be smaller, equal to, or larger than N. Taking all terms containing $O(z)$ to the left side of the equation gives

$$O(z) + a_1 z^{-1} O(z) + \cdots = b_0 I(z) + b_1 z^{-1} I(z) + \cdots \qquad (C.10d)$$

If this is transformed back into time domain and the sampling interval T is dropped everywhere, we get

$$o(n) = b_0 i(n) + b_1 i(n - 1) + \cdots - a_1 o(n - 1) - a_2 o(n - 2) - \cdots$$
$$(C.10e)$$

which is a recursion for the output signal at sampling instant $t = nT$. The IIR filter is therefore also referred to as *recursive filter*. Equation (C.10e) says that $o(n)$ is calculated from a weighted sum of one or several input signals $i(n)$, $i(n - 1)$, $\ldots$ and from one or more terms of the output signals which have been calculated in previous sampling instants.

To see why the impulse-invariant z-transform is an inconvenient tool for the design of digital filters, we consider the frequency response of the impulse-invariant IIR filter. Let $H(j\omega)$ be the frequency response of the analog and $H^*(j\omega)$ be the frequency response of the IIR filter. Then, using Eq. (C.7) and setting $s = j\omega$ and $d\chi = j\,d\eta$, we have

$$H^*(j\omega) = \frac{T}{2\pi} \oint_c H(j\eta) W[j(\omega - \eta)]\,d\eta \qquad (C.11)$$

where η is an auxiliary variable for angular frequency. Equation (C.11) says that the frequency response of the IIR filter is obtained by convolving $H(j\omega)$, i.e., with the spectrum of the sampling function $w(t)$. The sampling function $w(t)$ is shown once more in Fig. C.4a, its spectrum in Fig. C.4b. Let us investigate the result of the convolution by the example of Fig. C.5. Figure C.5a shows the frequency response

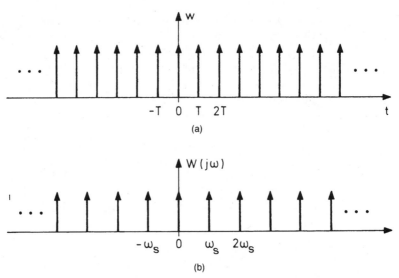

Figure C.4 Sampling function $w(t)$ and its spectrum $W(j\omega)$. (a) Sampling function $w(t)$. (b) Spectrum $W(j\omega)$.

$H(j\omega)$ of a low-pass filter. Figure C.5b presents the spectrum $W(j\omega)$ of the sampling function $w(t)$. The gain of a low-pass filter is defined to be 1 at $\omega = 0$ and 0 at very high frequencies. The gain of the filter rolls off at frequencies higher than the 3-dB corner frequency. In the given example the corner frequency has been chosen such that the gain of the filter is near 0 for frequencies higher than $\omega_n = \omega_s/2$; ω_n is called *Nyquist frequency*. The result of the convolution is shown in Fig. C.5c. Because the impulse response of the IIR filter is a sampled function, its frequency response becomes *periodic*. The period on the frequency axis is equal to the angular sampling frequency ω_s. The solid curve in Fig. C.5c is the so-called *main lobe;* the dashed curve represents the *side lobes* of the frequency response. Because the gain of the analog filter is nearly zero at frequencies above the Nyquist frequency, the side lobes do not markedly overlap with the main lobe of the frequency response $H^*(j\omega)$. We are going to consider now an example, where the gain $H(j\omega)$ has a value much greater than 0 at the Nyquist frequency (Fig. C.6a). Figure C.6b once more shows the spectrum $W(j\omega)$ of the sampling function $w(t)$, and Fig. C.6c is the result of the convolution $H(j\omega)*W(j\omega)$. Now the main and side lobes strongly overlap, and the transfer function of the digital filter (in the frequency range $-\omega_s/2 \le \omega < \omega_s/2$) markedly deviates from the transfer function of the analog filter. This effect is called *aliasing*. The sampling theorem (also called Shannon or Nyquist theorem) dictates that the sampling frequency must be chosen at least twice a critical fre-

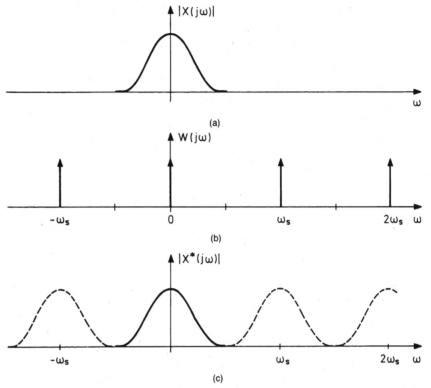

Figure C.5 Result of the convolution of $X(j\omega)$ and $W(j\omega)$ for the case of negligible aliasing. (a) Frequency response $X(j\omega)$ of an analog filter. (b) Spectrum $W(j\omega)$ of the sampling function $w(t)$. (c) Result of the convolution $X(j\omega)*W(j\omega)$.

quency ω_c, where the gain $H(j\omega_c)$ is so low that it can be neglected. A numerical example will demonstrate the implications of the sampling theorem.

Numerical Example   Assume that we want to design a digital two-pole low-pass filter having a 3-dB corner frequency of 1 kHz. A corresponding analog low-pass filter would roll off its gain above 1 kHz with 40 dB/decade, so the filter would have an attenuation of 40 dB at 10 kHz, 80 dB at 100 kHz, etc. If we claim that the IIR filter actually reaches an attenuation of 80 dB, its Nyquist frequency must be at least 100 kHz. Thus the sampling frequency must be *at least 200 kHz,* i.e., 200 times as high as the 3-dB corner frequency.

The requirement for very high sampling frequencies is the main drawback of the impulse-invariant IIR filter. Fortunately, the bilinear z-transform enables us to realize IIR filters, where the main and side lobes of the transfer function do not overlap and can therefore be realized with a much lower sampling frequency.

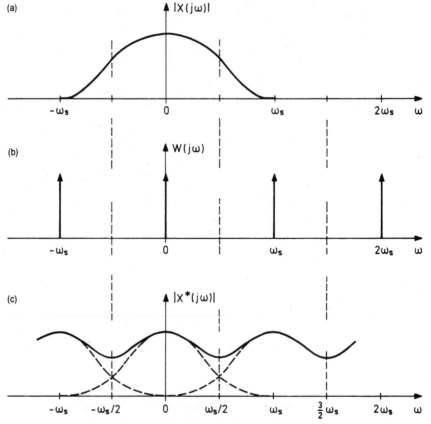

Figure C.6 Similar to Fig. C.5, but with severe aliasing. The gain $X(j\omega)$ of the analog filter is too high at the Nyquist frequency $\omega_n = \omega_s/2$. (*a*) Frequency response $X(j\omega)$ of an analog filter. (*b*) Spectrum $W(j\omega)$ of the sampling function $w(t)$. (*c*) Result of the convolution $X(j\omega)*W(j\omega)$.

C.2.2 The bilinear *z*-transform

The bilinear *z*-transform also converts the transfer function $H(s)$ of a given analog filter into the transfer function $H(z)$ of a digital filter but uses a different mathematical procedure. Figure C.7*a* shows the frequency response of an analog filter; again a low-pass filter has been chosen. Its 3-dB (angular) corner frequency is denoted ω_0. We now introduce a new angular frequency variable, the so-called pseudo-frequency ω_ψ. We define that the pseudo-frequency ω_ψ and the frequency ω are related by a tangent transform

$$\omega = \frac{2}{T} \tan \frac{\omega_\psi T}{2} \tag{C.12}$$

(a)

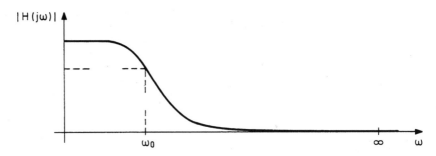

(b)

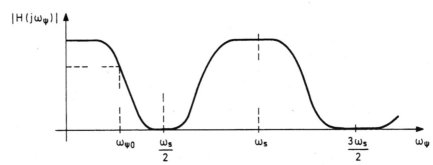

Figure C.7 Principle of the bilinear z-transform. (*a*) Frequency response $H(j\omega)$ of an analog filter. (*b*) The frequency axis of (*a*) is projected onto the pseudo-frequency axis ω_ψ. Transforming the frequency response of the filter in (*a*) into the pseudo-frequency domain results in the periodic frequency response $H^*(j\omega)$.

i.e., the frequency range $-\infty < \omega < \infty$ is projected onto the pseudo-frequency interval $-\omega_s/2 \le \omega_\psi \le \omega_s/2$, which is called the *Nyquist interval*. Because the tangent transform is not unambiguous, the frequency range $-\infty < \omega < \infty$ is also projected onto the other Nyquist intervals, i.e., onto the pseudo-frequency intervals $\omega_s/2 \le \omega_\psi \le 3\omega_s/2$, $3\omega_s/2 \le \omega_\psi \le 5\omega_s/2$, etc. If we transform the frequency response of the analog filter into the pseudo-frequency domain, we get the periodic transfer function $H^*(j\omega_\psi)$ as plotted in Fig. C.7*b*. Between $H(j\omega)$ and $H^*(j\omega_\psi)$ there is the relation

$$H(j\omega_\psi) = H(j\omega)\big|_{\omega=(2/T)\tan(\omega_\psi T/2)} \qquad (C.13)$$

which says that the gain of the new filter at pseudo-frequency ω_ψ is identical with the gain of the analog filter at frequency ω, where ω and ω_ψ are related by Eq. (C.12). Because the frequency response of the new filter is periodic, it is a *digital filter*. The 3-dB corner frequency of the digital filter is denoted $\omega_{\psi 0}$ in Fig. C.7*b*, which is called the *pseudo corner frequency*. By the tangent transform, the pseudo corner frequency becomes smaller than the original corner frequency of the analog filter. Normally it is required that the digital filter should have the same corner frequency as the original analog filter; i.e., the pseudo corner frequency should be ω_0. This is easily accomplished by *prewarping* the frequency response of the analog filter. If the prewarped corner frequency of the analog filter is denoted ω_{0p}, we have

$$\omega_{0p} = \frac{2}{T} \tan \frac{\omega_0 T}{2} \tag{C.14}$$

It is easily seen then that the pseudo corner frequency becomes ω_0.

To realize the digital filter, we must know its z-transfer function $H(z)$. To get $H(z)$, we first introduce complex frequency variables. We assume that the transfer function of the analog filter is $H(s)$; its frequency response $H(j\omega)$ is obtained simply by setting $s = j\omega$. In the same way, the transfer function of the digital filter is given by $H^*(s_\psi)$; its frequency response $H^*(j\omega_\psi)$ is also obtained by setting $s_\psi = j\omega_\psi$. To transform the complex frequency domain into the complex pseudo-frequency domain, we must use now the transformation

$$s = \frac{2}{T} \tanh \frac{s_\psi T}{2} \tag{C.15}$$

When we set $s = j\omega$ and $s_\psi = j\omega_\psi$ in Eq. (C.15), Eq. (C.12) is obtained again. Using Eq. (C.15), the transfer function $H^*(s_\psi)$ of the digital filter is calculated from

$$H^*(s_\psi) = H(s)\big|_{s = 2/T) \tanh(s_\psi T/2)} \tag{C.16}$$

Using Euler's relations, the tanh function can be written in the form of exponential functions:

$$\tanh \frac{s_\psi T}{2} = \frac{1 - e^{-s_\psi T}}{1 + e^{-s_\psi T}} \tag{C.17}$$

Equation (C.16) can therefore be rewritten as

$$H^*(s_\psi) = H(s)\big|_{s = (2/T)(1 - e^{-s_\psi T})/(1 + e^{-s_\psi T})} \tag{C.18a}$$

Because the complex pseudo-frequency appears exclusively in the form $\exp(s_\psi T)$, this expression can be replaced by z:

$$z = e^{s_\psi T} \tag{C.18b}$$

i.e., we apply the z-transform to the pseudo-frequency domain. Using the substitution of Eq. (C.18b) we finally get

$$H(z) = H(s)|_{s=(2/T)(1-z^{-1})/(1+z^{-1})} \tag{C.19}$$

The star symbol in $H(z)$ has been omitted for simplicity. Equation (C.19) defines the *bilinear z-transform*. It enables us to calculate the z-transfer function $H(z)$ of a digital filter directly from a given transfer function $H(s)$ of an analog filter just by substituting s by

$$s = \frac{2}{T} \frac{1 - z^{-1}}{1 + z^{-1}}$$

Table C.2 shows the bilinear z-transforms for filters of order 0 to 2. As with the impulse-invariant z-transform, the order of the bilinear z-transform $H(z)$ is identical with the order of the corresponding transfer function $H(s)$ of the analog filter. When comparing Table C.2 with Table C.1 (impulse-invariant z-transform), we observe that the expressions are quite similar but that the numerator polynomials are different. When the z-transfer function $H(z)$ is transformed back into time domain, a recursion for the output signal $o(n)$ is obtained which has the same form as the impulse-invariant IIR filter [Eq. (C.10e)].

To see the real benefit of the bilinear z-transform, let us have a look again at Fig. C.7. The filter in this example is a low-pass filter. The gain $H(j\omega)$ of the analog filter (Fig. C.7a) approaches 0 only for $\omega \to \infty$. Because of the tangent transformation, however, the gain of the digital filter becomes *exactly* 0 at the Nyquist pseudo-frequency $\omega_c = \omega_s/2$, as shown in Fig. C.7b. Consequently, the main and side lobes of the pseudo-frequency response $H^*(j\omega_\psi)$ *do not overlap,* which means that this digital filter does *not exhibit aliasing*. This enables us to choose the sampling frequency much lower than with the impulse-invariant IIR filter. The lack of aliasing is specially important for the realization of high-pass filters and differentiators. For both of these types, the gain at $\omega = 0$ should be zero. When realized by the impulse-invariant z-transform, however, the gain of the digital filter becomes nonzero at $\omega_\psi = 0$ owing to aliasing. This is the reason why practically all IIR filters are designed by the bilinear z-transform.

Numerical Example We are going to design a two-pole Butterworth low-pass digital filter. The 3-dB corner frequency of the IIR filter is required to be $f_0 = 200$ Hz. The transfer function of the corresponding analog filter is given by

$$H(s) = \frac{1}{1 + 2\zeta(s/\omega_0) + (s/\omega_0)^2} \tag{C.20a}$$

where $\zeta = 0.707$.[3] The sampling frequency is chosen $f_s = 1000$ Hz, which is not

TABLE C.2 Table of the Bilinear z-Transform for Filters of Order 0–2

Filter type	$H(s)$	$H(z)$
Zero-order term	1	1
Integrator	$\dfrac{1}{sT_i}$	$\dfrac{T}{2T_i}\dfrac{1+z^{-1}}{1-z^{-1}}$
First-order lag	$\dfrac{1}{1+s/\omega_0}$	$\dfrac{1+z^{-1}}{1+2/(\omega_0 T)+z^{-1}[1-2/(\omega_0 T)]}$
Second-order lag	$\dfrac{1}{1+s2\zeta/\omega_0+(s/\omega_0)^2}$	$\dfrac{1+2z^{-1}+z^{-2}}{1+4\zeta/(\omega_0 T)+4/(\omega_0 T)^2+z^{-1}[2-8/(\omega_0 T)^2]+z^{-2}[1+4\zeta/(\omega_0 T)+4/(\omega_0 T)^2]}$
Second-order lag, linear term in numerator	$\dfrac{s/\omega_1}{1+s2\zeta/\omega_0+(s/\omega_0)^2}$	$\dfrac{2}{\omega_1 T}\cdot\dfrac{1-z^{-2}}{1+4\zeta/(\omega_0 T)+4/(\omega_0 T)^2+z^{-1}[2-8/(\omega_0 T)^2]+z^{-2}[1+4\zeta/(\omega_0 T)+4/(\omega_0 T)^2]}$
Second-order lag, constant + linear term in numerator	$\dfrac{1+s/\omega_1}{1+s2\zeta/\omega_0+(s/\omega_0)^2}$	$\dfrac{(1+2/\omega_1 T)+2z^{-1}+z^{-2}(1-2/\omega_1 T)}{1+4\zeta/(\omega_0 T)+4/(\omega_0 T)^2+z^{-1}[2-8/(\omega_0 T)^2]+z^{-2}[1+4\zeta/(\omega_0 T)+4/(\omega_0 T)^2]}$

considerably higher than the corner frequency. First the 3-dB corner frequency of the corresponding analog filter must be prewarped; thus we have

$$\omega_0 = 2\pi \cdot 200 = 1256 \ s^{-1}$$

$$\omega_{0p} = \frac{2}{T} \tan \frac{\omega_0 T}{2} = 1453 \ s^{-1}$$

The bilinear z-transform of the digital filter is found in Table C.2 and reads

$$H(z) = \frac{1 + z^{-1}}{1 + 2/(\omega_{0p}T) + z^{-1}[1 - 2/(\omega_{0p}T)]} \qquad \text{(C.20b)}$$

Entering numerical values yields and normalizing the constant term in the denominator to 1, we get

$$H(z) = \frac{0.4208 + 0.4208z^{-1}}{1 - 0.1584z^{-1}}$$

Transforming back into time domain, we get the recursion for the IIR filter

$$o(n) = 0.1584 \ o(n - 1) + 0.4208 \ i(n) + 0.4208 \ i(n - 1)$$

In the most general case, the filter to be realized can have a large number of poles and zeros. The analog filter from which the design starts then will have a transfer function of the form

$$H(s) = \frac{(1 + s/\omega_{01}) \cdot (1 + 2s\zeta_3/\omega_{03} + (s/\omega_{03})^2) \cdot (\cdots)}{(1 + s/\omega_{02}) \cdot (1 + 2s\zeta_2/\omega_{04} + (s/\omega_{04})^2) \cdot (\cdots)}$$

Elliptic filters for example, have transfer functions of this kind.[13] Their transfer function may include a real-axis zero (first term in the numerator); if this zero does not exist, the corresponding term does not show up. The second term in the numerator represents a complex-conjugate zero pair. There may be further complex-conjugate zero pairs, as indicated by the empty bracket in the numerator. In a similar way, the transfer function may have a real-axis pole (first term in denominator) and one or several complex-conjugate pole pairs. The corresponding corner frequencies of this filter would be ω_{01}, ω_{03}, $\ldots$, ω_{02}, ω_{04}, $\ldots$, etc. When an IIR filter having this transfer function has to be designed, all corner frequencies in the numerator and in the denominator have to be prewarped using Eq. (C.14).

A user who wants to design an IIR filter today probably will no longer use z-transform tables, but rather one of the numerous software tools which are available now.[25,35]

C.3 FIR Filters

Like IIR filters, FIR filters are often designed on the base of an existing analog filter. As we will see later, this must not necessarily be the case. Let us start again by plotting the impulse response $h(t)$ of an existing analog filter as shown in Fig. C.1a. The digital filter is assumed to have an impulse response as shown in Fig. C.1b, but we now set all samples $h(n) = 0$ for $n \geq N$. N is a positive integer and is also referred to as *length of the FIR filter*. The impulse response of the digital filter now reads

$$h^*(t) = T \sum_{n=0}^{N-1} h(n)\delta(t - nT) \tag{C.21}$$

The z-transfer function of the FIR filter then becomes [Eq. (C.6)]

$$H(z) = T \sum_{n=0}^{N-1} h(n)z^{-n} \tag{C.22}$$

Transforming back into time domain, the output signal of the FIR filter at sampling instant $t = nT$ is obtained from

$$o(n) = T \sum_{k=0}^{N-1} i(n)h(n - k) \tag{C.23}$$

This is not a recursion, since $o(n)$ does not depend on previously calculated samples $o(n - 1)$, $o(n - 2)$, etc. For this reason, FIR filters are often referred to as *nonrecursive filters*. The signal flow diagram of an FIR filter is shown in Fig. C.8; in this drawing, T has been set 1 for

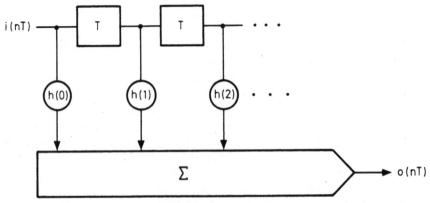

Figure C.8 Signal flow diagram of the FIR filter. The blocks marked with T are delay blocks.

simplicity. Because the FIR filter looks like a tapped delay line, the filter length N is also called *number of taps*. The filter coefficients $h(0)$, $h(1)$, . . . , $h(N − 1)$ are nothing more than the values of the impulse response $h^*(t)$ at sampling times $t = 0, T, 2T, ...$ For the FIR filter, the coefficients $h(n)$, $n = 0 . . . N − 1$, can be chosen arbitrarily, so the impulse response may have any desired shape. This degree of freedom was not attainable with the IIR filter. We conclude therefore that FIR filters can be built which do not have an analog counterpart.

Let us first consider two kinds of impulse response, which are of particular interest. Figure C.9a shows a finite impulse response, which is an even function of time t; in Fig. C.9b another impulse response is plotted, which is an odd function of time. From the theory of Fourier transform it follows that the spectrum of a signal, which is an even function of time, is purely real.[12,13,14] The spectrum of a signal, however, which is an odd function of time, is purely imaginary. Because the frequency response $H(j\omega)$ of the FIR filter is by definition the Fourier transform of its impulse response, the frequency response

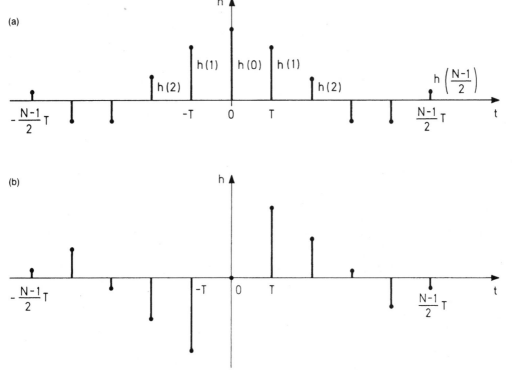

Figure C.9 Two special cases of impulse response of an FIR filter. (*a*) The impulse response is an even function of time. (*b*) The impulse response is an odd function of time.

of an FIR filter having an even impulse response would be purely real. This means that the FIR filter provides frequency-dependent gain with zero phase shift. Such a filter is called a *zero-phase filter;* unfortunately it is not realizable in real-time operation. Moreover, the frequency response of an FIR filter, which has an odd impulse response, would be purely imaginary. Its phase shift would be 90° independent of the frequency, and the amplitude response could be any desired function of frequency. This kind of FIR filter can also be considered as a zero-phase filter; sometimes they are referred to as *zero-phase filters of the second kind.* If zero-phase filters were realizable, they could be used to build *ideal filters.* To give an example, an ideal low-pass filter would have a real frequency response $H(j\omega)$, which is exactly 1 within the passband and 0 at all other frequencies. The impulse response $h(t)$ of the ideal low-pass filter would be a sine function

$$h(t) = \frac{\sin \omega_0 t}{\pi t} \tag{C.24}$$

where ω_0 is the angular corner frequency of the ideal low-pass filter. The frequency response of the ideal low-pass filter is shown in Fig. C.10a, the impulse response in Fig. C.10b. Such a filter is not realizable, of course, because the impulse response starts at $t = -\infty$. The filter becomes realizable to an approximation after two important modifications. First, a *window function* $w(t)$—as shown in Fig. C.10c —is used to cut out a finite portion of the impulse response; i.e., the finite impulse response h^*—as shown in Fig. C.10d—is obtained by multiplying the continuous impulse response h with the window function:

$$h^*(t) = h(t)w(t)$$

The impulse response shown in Fig. C.10d is finite. If the filter were realizable, its frequency response could be made to approximate very closely that of the ideal low-pass filter by choosing a large number of samples N. Let $Z_{zph}(j\omega)$ be the frequency response of this (unrealizable) filter. To get a realizable filter, a second modification must be made. The impulse response of Fig. C.10d is delayed such that it starts at $t = 0$. If the required time delay is τ, the frequency response of the modified FIR filter is given by

$$H_{lph}(j\omega) = H_{zph}(j\omega)e^{-j\omega\tau} \tag{C.25}$$

The phase of term $H_{zph}(j\omega)$ is zero by definition. The phase of the second term in Eq. (C.25) varies linearly with frequency, i.e., we have

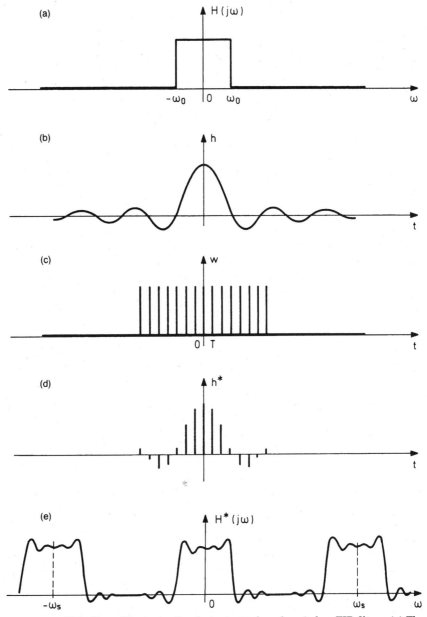

Figure C.10 This figure illustrates the design procedure for window FIR filters. (a) The design starts by defining the frequency response $H(j\omega)$ of an "ideal" analog filter. (b) The impulse response $h(t)$ of the "ideal" analog filter is calculated. (c) To get a finite and sampled impulse response, a window function $w(t)$ is defined which will be used to cut out a finite portion of the impulse response $h(t)$. (d) Result of the multiplication of $h(t)$ and $w(t)$. This is the impulse response of the FIR filter.

$$\phi(\omega) = -\omega\tau \qquad\qquad (C.26)$$

Consequently, the obtained FIR filter has linear phase. Linear-phase filters are the most important applications of the FIR filter. Zero-phase filters can be realized when the signal to be filtered is first recorded and filtered thereafter. As the filter algorithm of Eq. (C.23) demonstrates, the filtered output sample at time $t = nT$ must then be calculated from a number of samples, which have occurred prior to t, and from a number of samples, which will occur only later. This becomes possible now because they are already recorded. We can conclude this introduction to FIR filters by stating that the FIR filter is able to approximate an ideal filter to any desired precision with the exception that the filter exhibits a time delay. Two important design techniques for FIR filters are in common use today: (1) the *window* technique and (2) the *Parks-McClellan algorithm,* also called *Remez algorithm.* When the window technique is used, the filter is effectively designed in the time domain; i.e., its impulse response is derived from the impulse response of the fictive ideal filter. When using the Parks-McClellan algorithm, however, the filter is designed directly in the frequency domain. The term *frequency sampling* was formerly used for this approach.

C.3.1   Window-FIR filters

When using the window method, the FIR filter is designed in the time domain. The design procedure is illustrated by Fig. C.10. It starts by defining the frequency response of an "ideal" analog filter, which means that we set a goal which should be approximated within a given error specification. The "ideal" frequency response is plotted in Fig. C.10a. Next, the impulse response $h(t)$ of the ideal filter is calculated (Fig. C.10b). To get a finite impulse response, a portion of $h(t)$ is cut out using the window function $w(t)$ shown in Fig. C.10c. In this example the simplest window is used: the *rectangular window*. The finite impulse response $h^*(t)$ of the FIR filter is given by the multiplication of $h(t)$ and $w(t)$ and is plotted in Fig. C.10d. To get a filter which is capable of working in real time, the impulse response is delayed by half its duration, so the first nonzero sample of $h^*(t)$ starts at $t = 0$. Two parameters of the window have still to be determined: (1) the sampling frequency $f_s = 1/T$ and (2) the filter length N (number of taps). The sampling frequency must be chosen large enough to avoid aliasing; it must be larger than twice the corner frequency f_0 of the ideal filter. The effect of filter length is discussed in the following. To see its impact, we must calculate the frequency response $H^*(j\omega)$ of the

FIR filter. As the impulse response of the FIR filter is obtained by the multiplication of $h(t)$ with $w(t)$, its frequency response is obtained by convolving the spectra $H(j\omega)$ and $W(j\omega)$. The spectrum of the rectangular window can be shown to be[12,13,14]

$$W(j\omega) = \frac{\text{sinc }(\omega NT/2)}{\text{sinc }(\omega T/2)} \qquad \text{(C.27)}$$

hence is periodic on the frequency scale with the period $\omega_s = 2\pi/T$. Figure C.11 shows a portion of the spectrum $W(j\omega)$. It consists of a main lobe whose width is

$$\Delta\omega_W = \frac{2\omega_s}{N} \qquad \text{(C.28)}$$

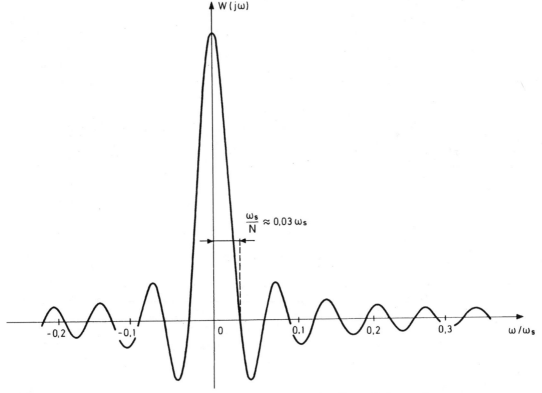

Figure C.11 Spectrum $W(j\omega)$ of the rectangular window $w(t)$ in Fig. C.10c. Only a portion of $W(j\omega)$ is shown here. The function is periodic in ω.

hence is inversely proportional to filter length N. The amplitudes of the side lobes decay slowly with increasing frequency, i.e., proportional to $1/f$. For an ideal low-pass filter, the result of the convolution $H(j\omega)*W(j\omega)$ is shown in Fig. C.12. In the example shown, the sampling frequency has been chosen 4 times the corner frequency of the filter. The filter length has arbitrarily been chosen $N = 31$. Figure C.12a shows the frequency response $H^*(j\omega)$ with logarithmic amplitude scale, Fig. C.12b with linear amplitude scale. (For calculation of $H^*(j\omega)$ it was assumed that the impulse response $h^*(t)$ is an even function of time, i.e., that the FIR filter is a zero-phase filter. When the FIR filter is required to operate in real time, its amplitude response would simply be the absolute value of the curve shown in Fig. C.12a). Two effects of the convolution are easily recognized in Fig. C.12a: (1) the steep transition from passband to stopband is widened by an amount which equals the width of the main lobe of $W(j\omega)$ and is given by Eq. (C.28). The transition width can be made arbitrarily narrow by increasing the filter length N, but we should keep in mind that increasing N means also increasing the number of computations to be made in every sampling interval. (2) Owing to convolution, ripple is superimposed to the frequency response of the FIR filter. The ripple amplitude is not constant, moreover. Starting with $f = 0$, the ripple in the passband increases with frequency and reaches a maximum near the corner frequency. In the stopband, the ripple amplitude decays only slowly with frequency. The impact of ripple is better recognized from the semilogarithmic representation in Fig. C.12a. Each ripple maximum on the linear amplitude scale corresponds to a damping minimum on the logarithmic scale. The first damping minimum (i.e., the first peak right from the transition in Fig. C.12a) is a poor -21 dB. At the Nyquist frequency, the damping minima are slightly larger but do not exceed the very modest value of -35 dB. Unfortunately, increasing the filter length does not improve the size of the damping minima. To get improved filter performance in the stopband, we must utilize alternate window functions. Windows must be found, therefore, whose spectrum decays faster with increasing frequency. A great many different windows are known, e.g., the Hanning (also called von Hann), Hamming, Bartlett, Blackman, Kaiser, and flat-top windows.[12,13,14,19] Each of these windows has its particular pros and cons. To save space, we just demonstrate the application of the simplest of these, the Hanning window (Fig. C.13). To compare it with the known rectangular window, Fig. C.13a shows once more the function of the rectangular window, which is now denoted w_R. Figure C.13b shows again its spectrum $W_R(j\omega)$. In Fig. C.13c, the Hanning window is plotted. Its envelope is simply a cosine function; for this reason the Hanning window is often referred to as *raised cosine window*. Figure C.13d shows the spectrum $W_{\mathrm{HAN}}(j\omega)$ of the Hanning window. When compar-

(a)

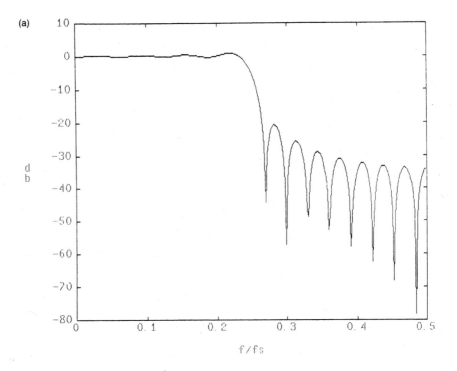

(b)

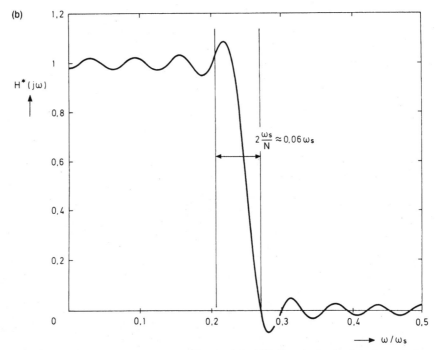

Figure C.12 The frequency response $\dot{H}^*(j\omega)$ of the FIR filter is obtained by the complex convolution $H(j\omega)*W(j\omega)$. (a) Frequency response $H^*(j\omega)$ on a logarithmic scale. (b) Linear plot of $H^*(j\omega)$.

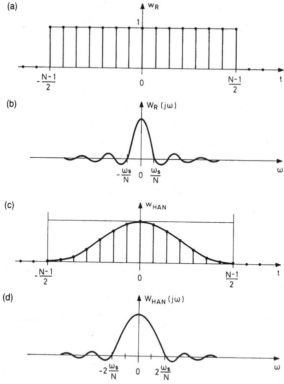

Figure C.13 Comparison of different window functions and their spectra. (*a*) The rectangular window $w_R(t)$. (*b*) The spectrum $W_R(j\omega)$ of the rectangular window. (*c*) The Hanning window $w_{HAN}(t)$. (*d*) The spectrum $W_{HAN}(j\omega)$ of the Hanning window.

ing it with the spectrum of the rectangular window, we note that the main lobe of the Hanning window is twice that of the rectangular. This is clearly a drawback, as it widens the transition region (refer to Fig. C.12) by a factor of 2. The advantage of the Hanning window is in the fact, however, that the side lobes (refer to Fig. C.13*d*) decay much faster with increasing frequency, i.e., with $1/f^3$. When the same FIR low-pass filter as above is realized using the Hanning window with filter length $N = 31$, the frequency response shown in Fig. C.14 is obtained. It is easily recognized that the transition has become wider, but the first damping minimum is now at -44 dB (instead of -21 dB for the rectangular window). Still better performance in the stopband is obtained by making use of more sophisticated windows; the Kaiser window is considered the "best" available window because it provides largest damping minima with shortest filter lengths.[12] A common drawback of all window FIR filters is the fact, however, that the ripple

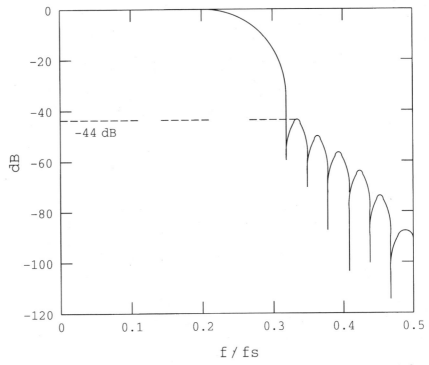

Figure C.14 Frequency response of a low-pass FIR filter, which uses a Hanning window.

in the passband and in the stopband is not constant over frequency. The FIR filters designed with the Parks-McClellan algorithm show up constant ripple in both stopband and passband, hence are often called *equiripple filters*. Moreover, the ripple amplitudes can be specified independently for the pass- and stopbands. The Parks-McClellan approach is explained in the next section.

C.3.2 Designing FIR filters with the Parks-McClellan algorithm

The Parks-McClellan algorithm enables us to design FIR filters which have constant ripple in the passband and in the stopband. The method furthermore allows design of multiband filters, differentiators, and Hilbert transformers. A multiband filter is a filter having more than one passband and stopband. We could think, for example, of a filter passing the frequencies in the range of 0 to 1 kHz, rejecting frequencies in the band from 1 to 4 kHz, passing frequencies from 4 to 6 kHz, etc. When used to design a differentiator, the frequency response of a bandlimited ideal differentiator[12] can be approximated to any desired degree of precision, of course, to the account of filter length. The same holds true for the Hilbert transformer,[13] which is a filter having a gain of 1 and a constant phase shift of 90° over the whole frequency range. To save space, we concentrate on FIR filters with only one passband, such as the low-pass filter. Figure C.15 illustrates the design procedure. It starts again by plotting the ideal frequency response which has to be approximated (Fig. C.15a). Next, transition width and ripple amplitude are specified by the boundary plot of Fig. C.15b. The FIR filter is required to have constant ripple ε_1 in the passband and constant ripple ε_2 in the stopband. The actual frequency must lie entirely

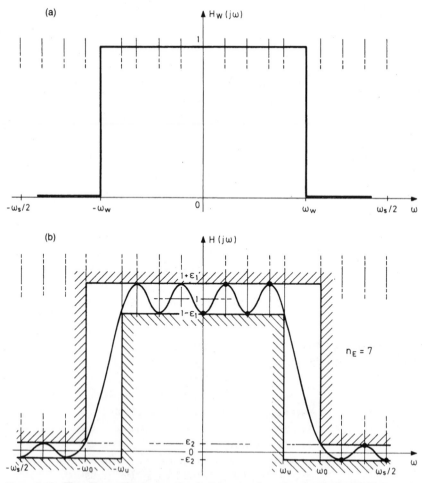

Figure C.15 This figure illustrates the design procedure for the Parks-McClellan algorithm. (a) The frequency response $H(j\omega)$ of an ideal analog filter is specified. (b) A boundary plot specifies the deviations of the actual filter from the ideal. In this plot, transition width and maximum ripple in the passband and stopband are defined.

within the boundaries, as indicated in the drawing. The (unweighted) error function of the FIR filter is defined to be the difference between ideal and actual frequency response. Consequently, the (unweighted) error function has maxima and minima of $\pm\varepsilon_1$ in the passband and of $\pm\varepsilon_2$ in the stopband. Normally, it is desired that ε_2 be much smaller than ε_1. If we wish the FIR filter to have minimum damping of 60 dB in the stopband, for example, ε_2 becomes 0.001. It would be exaggerated, however, to claim that the ripple in the passband could perhaps be as large as 0.1, which is 100 times as much. We therefore introduce a weighting function $W(j\omega)$, which is shown in Fig. C.15c. In our example, the allowed ripple in the stopband is half the ripple in the passband; hence the weighting function is chosen 1 in the stopband and 0.5 in the passband. Finally, Fig. C.15d shows the weighted error function $F_w(j\omega)$ of the FIR filter. The weighted error function is obtained by multiplying the unweighted error function by $W(j\omega)$. Obviously, the weighted error function has the same ripple in passband and stopband. The famous *Chebyshev polynomials* are known to behave like the weighted error curve shown in Fig. C.15d.

The Parks-McClellan algorithm is an iterative procedure which tailors a function in the frequency domain such that it comes closer and

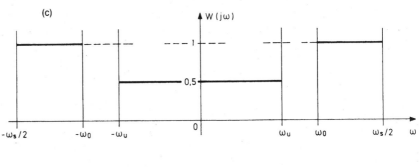

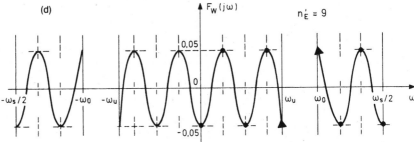

Figure C.15 (*Continued*) (*c*) Because different amounts of ripple are tolerated in passband and stopband, a weighting function $W(j\omega)$ is introduced. (*d*) The weighted error curve $F_w(j\omega)$ is given by the product of absolute error (from Fig. C.15*b*) and weighting function $W(j\omega)$ (from Fig. C.15*c*).

closer to the desired weighted error function $F_w(j\omega)$. It can be shown that for a filter length N the number of extrema (i.e., minima plus maxima) of the weighted error function in the frequency range 0 to f_n (Nyquist frequency) is around $N/2$. (The exact figure depends somewhat on the type of FIR filter to be realized; we will not go into details here.[13]) The Parks-McClellan algorithm starts with an "initial guess" of the locations of the extrema of $F_w(j\omega)$, see the filled dots in Fig. C.15d. Because it is very difficult to predict these locations accurately, the Parks-McClellan algorithm determines in the first pass a polynomial which passes through the filled dots; to simplify the mathematics, a *Lagrange polynomial* is used.[13] (When N points of a curve are specified, it is relatively simple to determine a polynomial of degree $N - 1$ which exactly goes through these points; this kind of polynomial is called Lagrange polynomial and is often used for the interpolation of functions.) If the initial guess would have been perfect, the extrema of the fitted Lagrange polynomial would be exactly there, where they were supposed to be. Probably the true extrema of the Lagrange polynomial will be at different locations, however. A maxima/minima search is started now which determines the locations where the extrema really are. This set of new values is used now as an improved initial guess. In the next run, the Parks-McClellan algorithm fits another Lagrange polynomial through the new points. It performs so many passes, until the extrema found in the Kth pass do not markedly deviate from the extrema found in the $K - 1$th pass. Having found a sufficiently good fit to the weighted error function $F_w(j\omega)$, the Parks-McClellan algorithm now calculates the unweighted error function and finally the actual frequency response $H^*(j\omega)$ of the FIR filter. Actually, a sampled version of $H^*(j\omega)$ is known at this time. Using the IFFT (inverse FFT), the actual impulse response $h^*(t)$ is calculated. This yields the required FIR filter coefficients $h(0)$, $h(1)$, . . . , $h(N - 1)$, where N is the filter length. Figure C.16 is an example of an FIR low-pass filter of length $N = 253$ which has been designed by the DISPRO program.[35]

The Parks-McClellan algorithm is an extremely computation-intensive procedure. Only computer programs are capable of performing such an approach in reasonable time. Fortunately many software tools are available which enable the design of any kind of digital filters, i.e., IIR filters, FIR filters with the window method and with the Parks-McClellan algorithm.[25,35] When using such a software tool, the designer will enter first the error specifications of the FIR filter. Knowing ripple amplitudes and transition width(s), most programs estimate the filter length N required to meet the design goals. The filter length depends to a large extent on the required transition width. The narrower the transition, the longer the filter will be.

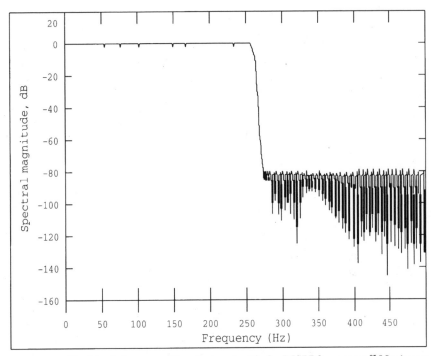

Figure C.16 Equiripple low-pass filter designed with the DISPRO program.[35] Maximum ripple in the passband is 0.5 dB, minimum stopband attenuation is 80 dB, and transition width is 10 Hz. To meet these specifications, a filter length $N = 253$ is required.

Another problem also deserves attention: the number of bits which are necessary to represent the filter coefficients in the hardware platform, which is used to perform the filtering function. The mentioned computer programs calculate the filter coefficients with the precision of floating-point numbers, i.e., usually to a precision of 32, 64, or even 80 bits. (Note that these numbers include the bits used for mantissa and exponent.) In many cases the user will try to implement the filter algorithm with fixed-point arithmetic, which may use a 16-bit (or even an 8-bit) format. The actual filter coefficients have to be rounded therefore. Because of rounding, the frequency response can be corrupted, however. The filter performance in the stopband is most likely to be deteriorated. To check numerical resolution, most filter design tools offer an option to round the filter coefficients and to recalculate the frequency response with the truncated coefficients.

Reference

1. Gardner, Floyd M.: *Phaselock Techniques,* 2d ed., John Wiley and Sons, New York, 1979.
2. Richman, D.: "Color Carrier Reference Phase Synchronization Accuracy in NTSC Color Television," *Proc. IRE,* vol. 42, January 1954, pp. 106–133.
3. Izawa, K.: *Introduction to Automatic Control,* Elsevier, New York, 1963.
4. Viterbi, Andrew J.: *Principles of Coherent Communication,* McGraw-Hill, New York, 1966.
5. Frazier, J. P., and J. Page: "Acquisition and Tracking Behaviour of Phase-Locked Loops," *IRE Trans. Space Electr. Telem.,* vol. SET-8, September 1962, pp. 210–227.
6. Sanneman, R. W., and J. R. Rowbotham: "Unlock Characteristic of the Optimum Type II Phase-Locked Loop," *IEEE Trans. Aerosp. Navig. Electron.,* vol. ANE-11, March 1964, pp. 15–24.
7. *Phase-Locked Loop Data Book,* 2d ed., Motorola Semiconductor Products Inc., Phoenix, AZ, August 1973.
8. Lindsey, William C., and Chak Ming Chie: "A Survey of Digital Phase-Locked Loops," *Proc. IEEE,* vol. 69, April 1981.
9. Troha, Donald G., and James D. Gallia: "Digital Phase-Locked Loop Design Using SN54/74LS297," *Application Note AN 3216,* Texas Instruments Inc., Dallas, TX.
10. Rohde, Ulrich L.: "Low-Noise Frequency Synthesizers Using Fractional *N* Phase-Locked Loops," *r. f. design,* January/February 1981.
11. Rohde, Ulrich L.: *Digital PLL Frequency Synthesizers, Theory and Design,* Prentice-Hall, Englewood Cliffs, NJ, 1983.
12. Hamming, R. W.: *Digital Filters,* 2d ed., Prentice-Hall, Englewood Cliffs, NJ, 1983.
13. Rabiner, L. R., and B. Gold: *Theory and Application of Digital Signal Processing,* Prentice-Hall, Englewood Cliffs, NJ, 1975.
14. Oppenheim, A. V., and R. W. Schafer: *Discrete-Time Signal Processing,* Prentice-Hall, Englewood Cliffs, NJ, 1989
15. Volgers, R.: "Phase-Locked Loop Circuits: 74HC/HCT4046A & 74HC/HCT7046A," Philips Components, 1989. (Available in the United States from Signetics Corporation, 811 East Arques Ave., Sunnyvale, CA 94088-3409.)
16. Rosink, W. B.: "All-Digital Phase-Locked Loops Using the 74HC/HCT297," Philips Components, 1989. (Available in the United States from Signetics Corporation, 811 East Arques Ave., Sunnyvale, CA 94088-3409.)
17. Mullins, Mark: "How to Measure Signal Jitter," *Electronic Products,* July 1992, pp. 41–44.
18. Higgins, Richard J.: *Digital Signal Processing in VLSI,* Prentice-Hall, Englewood Cliffs, NJ, 1990.
19. Kuc, Roman: *Introduction to Digital Signal Processing,* McGraw-Hill, New York, 1988.
20. Sklar, Bernard: *Digital Communications, Fundamentals and Applications,* Prentice-Hall, Englewood Cliffs, NJ, 1988.
21. Tzafestas, Spyros G.: *Walsh Functions in Signal and Systems Analysis and Design,* Van Nostrand, New York, 1985.

References

22. De Bellescize, H.: "La réception synchrone," *L'onde électrique,* vol. 11, May 1932, pp. 225–240.
23. Best, Roland E.: *Phase-Locked Loops, Theory, Design and Applications,* 1st ed., McGraw-Hill, New York, 1984, appendix D.
24. Selle, D.: "Theoretische Grundlagen, Dimensionierung und charakteristische Kenngrössen von PLL-Schaltungen (Phasenregelkreisen)," internal report, available from Prof. Dr. Dieter Selle, Fachhochschule Braunschweig-Wolfenbüttel, Institut für Nachrichtentechnik, D - 3302 Cremlingen (Germany).
25. MATLAB, The MathWorks Inc., 21 Eliot Street, South Natick, MA 01760.
26. Control Kit, Software package executing under MATLAB (ref. 25), available from Rapid Data Ltd., Crescent Road, Worthing, West Sussex, BN11 5RW, UK.
27. Bently, W. E., and S. G. Varsos: "Squeeze More Data onto Mag Tape by Use of Delay-Modulation Encoding and Decoding," *Electron. Des.,* October 11, 1975.
28. McNamara, John E.: *Technical Aspects of Data Communication,* Copyright 1977, 1978 by Digital Equipment Corporation, Bedford, MA 01730.
29. Larimore, W. E.: "Synthesis of Digital Phase-Locked Loops," in *EASCON Rec.,* October 1968, pp. 14–20.
30. Langston, J. Leland: "μC Chip Implements High-Speed Modems Digitally," *Electron. Des.,* June 24, 1982.
31. HiJaak Version 2, Graphics Conversion and Screen Capture Program, Inset Systems, 71 Commerce Drive, Brookfield, CT 06804.
32. 8253 Programmable Interval Timer (Data Sheet), Intel Literature Sales, P. O. Box 58130, Santa Clara, CA 95052-8130.
33. 9513 System Timing Controller (Data Sheet), Advanced Micro Devices, Inc., 901 Thompson Place, Sunnyvale, CA 94088.
34. MCS-51 Microprocessor Family, Intel Literature Sales, P. O. Box 58130, Santa Clara, CA 95052-8130.
35. DISPRO V2.0 (Digital Filter Design Program), Signix Corporation, 19 Pelham Island Rd., Wayland, MA 01178.
36. Brigham, E. O.: *The Fast Fourier Transform,* Prentice-Hall, Englewood Cliffs, NJ, 1974.
37. Tou, Julius T.: *Digital and Sampled-Data Control Systems,* McGraw-Hill, New York, 1959.
38. *Turbo Pascal Version 6.0, Programmer's Guide,* Copyright 1983, 1990 by Borland International, Inc., 1800 Green Hills Road, Scotts Valley, CA 95067-0001.
39. Stofka, Marian: "Digital-Only PLL Exhibits No Overshoot," *EDN,* May 26, 1982.
40. Nyquist, H.: "Certain Topics of Telegraph Transmission Theory", Trans. Am. Inst. Electr. Eng., vol. 47, Apr. 1928, pp. 617–644.
41. David Grieve, Hewlett-Packard Ltd, Queensferry Microwave Division, South Queensferry, Lothian UK EH30 9TG, private communication. David Grieve analyzed impulse responses of raised cosine and root raised cosine filters by a MathCad worksheet. For correspondence contact his home page at http://wkweb5.cableinet.co.uk/dwg/.
42. Data sheet HSP50210 (Digital Costas Loop), Harris Semiconductor, 1025 W. NASA Boulevard, Melbourne FL 32919.
43. Data sheet HSP50307 (Burst QPSK Modulator), Harris Semiconductor, 1025 W. NASA Boulevard, Melbourne FL 32919.
44. Data sheet HSP50110 (Digital Quadrature Tuner), Harris Semiconductor, 1025 W. NASA Boulevard, Melbourne FL 32919.
45. Application Note AN9661.1, Implementing Polyphase Filtering with the HSP50110 (DQT), HSP50210 (DCL) and the HSP43168 (DFF), January 1999, Harris Semiconductor, 1025 W. NASA Boulevard, Melbourne FL 32919.
46. Functional Block Diagram of a Demodulator Using the HSP50110/HSP50210 SATCOM Modem Chip Set, Harris Semiconductor, 1025 W. NASA Boulevard, Melbourne FL 32919. Can be downloaded from http://www.semi.harris.com.
47. David Grieve: DVB-C, Article was downloaded from David Grieve's home page http://wkweb5.cableinet.co.uk.

Index

---

*Boldface numbers indicate where topics are discussed in full.

ABOUT THE AUTHOR

Roland E. Best is the founder of Best Engineering and a world-renowned authority on phase-locked loops, circuit design, and microprocessor applications. Dr. Best has also worked for Sandoz A.G. and the IBM Research Laboratory in Zurich, where his company is now headquartered.

Program Installation

Insert the CD into the CD-ROM drive. If you have a plug-and-play CD-ROM, installation starts automatically. Follow the instructions on screen. If the CD-ROM is not "plug-and-play," start the setup program by double-clicking D:\SETUP.EXE in Windows Explorer or in My Computer. (Users of Windows 3.1 or 3.11 can use File Manager to start the setup program.)

SOFTWARE AND INFORMATION LICENSE

The software and information on this diskette (collectively referred to as the "Product") are the property of The McGraw-Hill Companies, Inc. ("McGraw-Hill") and are protected by both United States copyright law and international copyright treaty provision. You must treat this Product just like a book, except that you may copy it into a computer to be used and you may make archival copies of the Products for the sole purpose of backing up our software and protecting your investment from loss.

By saying "Just like a book," McGraw-Hill means, for example, that the Product may be used by any number of people and may be freely moved from one computer location to another, so long as there is no possibility of the Product (or any part of the Product) being used at one location or on one computer while it is being used at another. Just as a book cannot be read by two different people in two different places at the same time, neither can the Product be used by two different people in two different places at the same time (unless, of course, McGraw-Hill's rights are being violated).

McGraw-Hill reserves the right to alter or modify the contents of the Product at any time.

This agreement is effective until terminated. The Agreement will terminate automatically without notice if you fail to comply with any provisions of this Agreement. In the event of termination by reason of your breach, you will destroy or erase all copies of the Product installed on any computer system or made for backup purposes and shall expunge the Product from your data storage facilities.

LIMITED WARRANTY

McGraw-Hill warrants the physical diskette(s) enclosed herein to be free of defects in materials and workmanship for a period of sixty days from the purchase date. If McGraw-Hill receives written notification within the warranty period of defects in materials or workmanship, and such notification is determined by McGraw-Hill to be correct, McGraw-Hill will replace the defective diskette(s). Send request to:

Customer Service
McGraw-Hill
Gahanna Industrial Park
860 Taylor Station Road
Blacklick, OH 43004-9615

The entire and exclusive liability and remedy for breach of this Limited Warranty shall be limited to replacement of defective diskette(s) and shall not include or extend to any claim for or right to cover any other damages, including but not limited to, loss of profit, data, or use of the software, or special, incidental, or consequential damages or other similar claims, even if McGraw-Hill has been specifically advised as to the possibility of such damages. In no event will McGraw-Hill's liability for any damages to you or any other person ever exceed the lower of suggested list price or actual price paid for the license to use the Product, regardless of any form of the claim.

THE McGRAW-HILL COMPANIES, INC. SPECIFICALLY DISCLAIMS ALL OTHER WARRANTIES, EXPRESS OR IMPLIED, INCLUDING BUT NOT LIMITED TO, ANY IMPLIED WARRANTY OF MERCHANTABILITY OR FITNESS FOR A PARTICULAR PURPOSE. Specifically, McGraw-Hill makes no representation or warranty that the Product is fit for any particular purpose and any implied warranty of merchantability is limited to the sixty day duration of the Limited Warranty covering the physical diskette(s) only (and not the software of in-formation) and is otherwise expressly and specifically disclaimed.

This Limited Warranty gives you specific legal rights; you may have others which may vary from state to state. Some states do not allow the exclusion of incidental or consequential damages, or the limitation on how long an implied warranty lasts, so some of the above may not apply to you.

This Agreement constitutes the entire agreement between the parties relating to use of the Product. The terms of any purchase order shall have no effect on the terms of this Agreement. Failure of McGraw-Hill to insist at any time on strict compliance with this Agreement shall not constitute a waiver of any rights under this Agreement. This Agreement shall be construed and governed in accordance with the laws of New York. If any provision of this Agreement is held to be contrary to law, that provision will be enforced to the maximum extent permissible and the remaining provisions will remain in force and effect.